Cambridge IGCSE® & O Level
Essential Chemistry

Third Edition

Roger Norris
Editor: Lawrie Ryan
Onn May Ling, Mei Chew
Bhavna Narayanan

OXFORD
UNIVERSITY PRESS

Great Clarendon Street, Oxford, OX2 6DP, United Kingdom

Oxford University Press is a department of the University of Oxford. It furthers the University's objective of excellence in research, scholarship, and education by publishing worldwide. Oxford is a registered trade mark of Oxford University Press in the UK and in certain other countries

© Oxford University Press 2020

The moral rights of the authors have been asserted

First published in 2020

All rights reserved. No part of this publication may be reproduced, stored in a retrieval system, or transmitted, in any form or by any means, without the prior permission in writing of Oxford University Press, or as expressly permitted by law, by licence or under terms agreed with the appropriate reprographics rights organization. Enquiries concerning reproduction outside the scope of the above should be sent to the Rights Department, Oxford University Press, at the address above.

You must not circulate this work in any other form and you must impose this same condition on any acquirer

British Library Cataloguing in Publication Data
Data available

978-1-38-200612-5 (standard)

10 9 8 7

978-1-38-200611-8 (enhanced)

10 9 8 7 6 5 4 3 2

The manufacturing process conforms to the environmental regulations of the country of origin.

Printed and bound in Great Britain by Bell and Bain Ltd, Glasgow

Acknowledgements

The publisher and authors would like to thank the following for permission to use photographs and other copyright material:

Cover: Jirik V/Shutterstock **Photos: p2:** Claude Nuridsany & Marie Perennou/Science Photo Library; **p4:** Jorge Salcedo/Shutterstock; **p9:** Martyn F. Chillmaid/Science Photo Library; **p10** t&b: sciencephotos/Alamy Stock Photo; **p14:** Martyn F. Chillmaid/Science Photo Library; **p18:** itanistock/Alamy Stock Photo; **p20:** Martyn Chillmaid; **p22:** Roger Hutchings/Alamy Stock Photo; **p29:** Catherine Pouedras/Science Photo Library; **p33:** Charles D. Winters/Science Photo Library; **p34:** Martyn F. Chillmaid/Science Photo Library; **p38:** Charles D. Winters/Science Photo Library; **p42:** kasza/123RF; **p46:** everything possible/Shutterstock; **p48:** KDR In-Focus Productions/Shutterstock; **p52:** Science Photo Library; **p58:** Martyn F. Chillmaid/Science Photo Library; **p63:** ART Collection/Alamy Stock Photo; **p66:** DigiStu/E+/Getty Images; **p68:** Matthew Ashmore/Alamy Stock Photo; **p73:** Kirza/Istock/Getty Images; **p74:** sciencephotos/Alamy Stock Photo; **p76:** Martyn Chillmaid; **p80:** Ehrman Photographic/Shutterstock; **p82:** Pitchal frederic/Sygma/Getty Images; **p86:** Sheila Terry/Science Photo Library; **p88:** Simon Fraser/Northumbria Circuits/Science Photo Library; **p90:** Hugh Sitton/The Image Bank Unreleased/Getty Images; **p92:** Howard Davies/Alamy Stock Photo; **p96:** Can Bakcioglu; **p99:** Martyn F. Chillmaid/Science Photo Library; **p100:** Shutterstock; **p102:** Angelo Giampiccolo/Shutterstock; **p104:** Danita Delimont/Alamy Stock Photo; **p108:** Mark Davidson/Alamy Stock Photo; **p110:** John E. Straub, Boston University; **p112:** Aleksandar Varbenov/123RF; **p115:** Astrid & Hanns-Frieder Michler/Science Photo Library; **p116:** Sue Prideaux/Science Photo Library; **p119:** bjones27/E+/Getty Images; **p122:** Martyn F. Chillmaid/Science Photo Library; **p126:** Nostalgia for Infinity/Shutterstock; **p128:** Ysbrand Cosijn/Shutterstock; **p130:** 1990 Chip Clark/Fundamental Photographs; **p135:** Andrew Lambert Photography/Science Photo Library; **p136:** Dmitri Ma/Shutterstock; **p138:** mihail39/123RF; **p141:** Martyn Chillmaid; **p143:** Andrew Lambert Photography/Science Photo Library; **p144:** Wayne HUTCHINSON/Alamy Stock Photo; **p148:** Martyn F. Chillmaid/Science Photo Library; **p150:** helfei/Shutterstock; **p152:** Kekyalyaynen/Shutterstock; **p154:** Martyn Chillmaid; **p156:** Andrew Lambert Photography/Science Photo Library; **p158: A**ndrew Lambert Photography/Science Photo Library; **p163:** ChemistryGod/Alamy Stock Photo; **p164c: A**ndrew Lambert Photography/Science Photo Library; **p164br:** Charles D. Winters/Science photo library; **p164bl:** E.R. Degginger/Alamy Stock Photo; **p166cl:** By Ian Miles-Flashpoint Pictures/Alamy Stock Photo; **p166cr:** Lester V. Bergman/Corbis/Getty Images; **p166b:** Andrew Lambert Photography/Science Photo Library; **p168:** Shutterstock; **p170:** tratong/Shutterstock; **p175:** Martyn Chillmaid; **p176:** Jerry Mason/Science Photo Library; **p178:** kaspargallery/iStock/Getty Images; **p181:** Peticolas/Megna/Fundamental Photos/Science Photo Library; **p184:** Oleksiy Mark/Shutterstock; **p186:** Zoonar GmbH/Alamy Stock Photo; **p187:** Ian Cartwright/LGPL/Alamy Stock Photo; **p189:** pixinoo/Shutterstock; **p190:** mozcann/iStock/Getty Images; **p191:** wernimages/Shutterstock; **p194:** Claver Carroll/Photolibrary/Getty Images; **p196:** Nigel Cattlin/Alamy Stock Photo; **p198:** TonyV3112/Shutterstock; **p200:** John Shaw/NHPA/Photoshot; **p204c:** Georgette Douwma/Science Photo Library; **p204bl:** eClick/Shutterstock; **p204br:** L.F/Shutterstock; **p205:** EvijaF/Shutterstock; **p206:** 1000 Words/Shutterstock; **p208:** Brisbane/Shutterstock; **p212t:** Boaz Rottem/Alamy Stock Photo; **p212b:** Richard Whitcombe/Shutterstock; **p215:** blickwinkel/Alamy Stock **Photo; p218:** Martyn Chillmaid; **p220:** CHAIWATPHOTOS/Shutterstock; **p224:** Anan Kaewkhammul/Shutterstock; **p226:** Gerard Koudenburg/Shutterstock; **p230:** Martyn Chillmaid; **p233:** Kaj R. Svensson/Science Photo Library; **p234:** Martyn Chillmaid; **p236:** bioraven/Shutterstock; **p238:** satit_srihin/Shutterstock; **p240:** christophe_cerisier/iStock/Getty Images; **p244t:** Martyn Chillmaid; **p244b:** Rich Carey/Shutterstock; **p246:** Jeffrey Coolidge/Corbis/Getty Images; **p248: A**nthony Malone/Shutterstock; **p250:** gnomeandi/Shutterstock.

Artwork by: Aptara, GreenGate Publishing Services and OUP.

Every effort has been made to contact copyright holders of material reproduced in this book. Any omissions will be rectified in subsequent printings if notice is given to the publisher.

This Student Book refers to the Cambridge O Level Chemistry (5070) Syllabus and Cambridge IGCSE Chemistry (0620) Syllabus published by Cambridge Assessment International Education.

This work has been developed independently from and is not endorsed by or otherwise connected with Cambridge Assessment International Education.

Contents

Introduction		v
Syllabus matching grid		vi

Unit 1 Particle theory — 2
- 1.1 Solids, liquids and gases — 2
- 1.2 Using the kinetic particle theory — 4
- 1.3 Heating and cooling curves — 6
- 1.4 Solvents, solutes and solutions — 8
- 1.5 Diffusion — 10
- Summary and Practice questions — 12

Unit 2 Separating substances — 14
- 2.1 Apparatus for measuring — 14
- 2.2 Paper chromatography — 16
- 2.3 Is that chemical pure? — 18
- 2.4 Separation and purification — 20
- 2.5 More about separation and purification — 22
- Summary and Practice questions — 24

Unit 3 Atoms, elements and compounds — 26
- 3.1 Inside the atom — 26
- 3.2 Isotopes — 28
- 3.3 Electronic structure and the Periodic Table — 30
- 3.4 Elements, compounds and mixtures — 32
- 3.5 Metals and non-metals — 34
- Summary and Practice questions — 36

Unit 4 Structure and bonding — 38
- 4.1 Ionic bonding — 38
- 4.2 Covalent bonding (1): simple molecules — 40
- 4.3 Covalent bonding (2): more complex molecules — 42
- 4.4 Ionic or simple molecular? — 44
- 4.5 Giant covalent structures — 46
- 4.6 Metallic bonding — 48
- Summary and Practice questions — 50

Unit 5 Formulae and equations — 52
- 5.1 Chemical formulae — 52
- 5.2 Working out the formula — 54
- 5.3 Chemical equations — 56
- 5.4 More about equations — 58
- Summary and Practice questions — 60

Unit 6 Chemical calculations — 62
- 6.1 Masses and molecules — 62
- 6.2 Chemical calculations and the mole — 64
- 6.3 More chemical calculations — 66
- 6.4 Amount of product — 68
- 6.5 Gas volume calculations — 70
- 6.6 Yield and purity — 72
- 6.7 Empirical formula and molecular formula — 74
- 6.8 Titrations — 76
- Summary and Practice questions — 78

Unit 7 Electricity and chemistry — 80
- 7.1 Conductors and electrolysis — 80
- 7.2 The products of electrolysis — 82
- 7.3 Electrolysis of aqueous solutions — 84
- 7.4 Explaining electrolysis — 86
- 7.5 Purifying copper — 88
- 7.6 Electroplating — 90
- 7.7 Extracting aluminium — 92
- Summary and Practice questions — 94

Unit 8 Chemical energetics — 96
- 8.1 Energy transfer in chemical reactions — 96
- 8.2 Reaction pathway diagrams — 98
- 8.3 Bond energy calculations — 100
- 8.4 Fuels and energy production — 102
- 8.5 Fuel cells — 104
- Summary and Practice questions — 106

Unit 9 Rates of reaction — 108
- 9.1 Investigating rate of reaction — 108
- 9.2 Evaluating experiments — 110
- 9.3 Interpreting data — 112
- 9.4 Surfaces and reaction rate — 114
- 9.5 Concentration and rate of reaction — 116
- 9.6 Temperature and rate of reaction — 118
- Summary and Practice questions — 120

Unit 10 Chemical reactions — 122
- 10.1 Reversible reactions — 122
- 10.2 Shifting the equilibrium — 124
- 10.3 Redox reactions — 126
- 10.4 More about redox reactions — 128
- 10.5 Oxidising agents and reducing agents — 130
- Summary and Practice questions — 132

Unit 11 Acids and bases — 134
- 11.1 How acidic? — 134
- 11.2 Properties of acids — 136
- 11.3 Bases — 138

Contents

11.4	More about acids and bases	140
11.5	Acid–base titrations and indicators	142
11.6	Oxides	144
	Summary and Practice questions	146

Unit 12 Making and identifying salts — 148

12.1	Making salts from metals, bases or carbonates	148
12.2	Making salts by titration	150
12.3	Making salts by precipitation	152
12.4	What's that gas?	154
12.5	Testing for cations	156
12.6	Testing for anions	158
	Summary and Practice questions	160

Unit 13 The Periodic Table — 162

13.1	The Periodic Table	162
13.2	Group I metals	164
13.3	Group VII elements	166
13.4	Noble gases and periodic trends	168
13.5	Transition elements	170
	Summary and Practice questions	172

Unit 14 Metals and reactivity — 174

14.1	The metal reactivity series	174
14.2	Metal oxides and their reduction	176
14.3	From metal compounds to metals	178
14.4	More about metal reactivity	180
	Summary and Practice questions	182

Unit 15 More about metals — 184

15.1	Extracting iron	184
15.2	The rusting of iron	186
15.3	Alloys	188
15.4	Uses of metals	190
	Summary and Practice questions	192

Unit 16 Compounds of nitrogen and sulfur — 194

16.1	Making ammonia	194
16.2	Fertilisers	196
16.3	Sulfuric acid	198
16.4	Acid rain	200
	Summary and Practice questions	202

Unit 17 Environmental chemistry — 204

17.1	Air pollution	204
17.2	Carbon dioxide in the atmosphere	206
17.3	Global warming and climate change	208
17.4	Reducing environmental pollution	210
17.5	Clean water	212
17.6	Water pollution	214
	Summary and Practice questions	216

Unit 18 Organic chemistry and petrochemicals — 218

18.1	A variety of organic compounds	218
18.2	Formulae of organic compounds	220
18.3	Structural formulae and isomerism	222
18.4	Fuels	224
18.5	Petroleum fractionation	226
	Summary and Practice questions	228

Unit 19 Some homologous series of organic compounds — 230

19.1	Alkanes	230
19.2	Cracking alkanes	232
19.3	Alkenes	234
19.4	Alcohols	236
19.5	Fermentation	238
19.6	Carboxylic acids and esters	240
	Summary and Practice questions	242

Unit 20 Polymers — 244

20.1	What are polymers?	244
20.2	More about polymer structure	246
20.3	Polyamides and polyesters	248
20.4	Amino acids and proteins	250
	Summary and Practice questions	252

Alternative to practical section	254
Exam skills practice	262
Glossary	297
Index	301

Introduction

This book is designed specifically for Cambridge IGCSE® Chemistry 0620. Experienced teachers have been involved in all aspects of the book, including detailed planning to ensure that the content gives the best match possible to the syllabus.

Using this book will ensure that you are well prepared for studies beyond the IGCSE level in pure sciences, in applied sciences or in science-dependent vocational courses. The features of the book outlined below are designed to make learning as interesting and effective as possible:

LEARNING OUTCOMES
- These are at the start of each spread and will tell you what you should be able to do at the end of the spread.
- Some outcomes will be needed only if you are taking a supplement paper and these are clearly labelled, as is any content in the spread that goes beyond the syllabus.

EXAM TIP
Experienced teachers give you suggestions on how to avoid common errors or give useful advice on how to tackle questions.

KEY POINTS
These summarise the most important things to learn from the spread.

PRACTICAL
These show the opportunities for practical work. The results are included to help you if you do not actually tackle the experiment or are studying at home.

SUMMARY QUESTIONS
These questions are at the end of each spread and allow you to test your understanding of the work covered in the spread.

At the end of each unit there is a double page of examination-style questions written by the author.

At the end of the book you will also find:

'Alternative to practical' section – this provides guidance if you are doing this examination paper instead of coursework or the practical examination.

Assessment structure

Paper 1: Multiple Choice (Core)

Paper 2: Multiple Choice (Supplement)

Paper 3: Theory (Core)

Paper 4: Theory (Supplement)

Paper 5: Practical Test

Paper 6: Alternative to Practical

Extra resources, including **answers**, and a Revision Checklist:

www.oxfordsecondary.com/essential-igcse-science

Contents

Syllabus matching grid

Topic number in this book	Topic title	IGCSE Syllabus section
1.1	Solids, liquids and gases	1.1
1.2	Using the kinetic particle theory	1.1
1.3	Heating and cooling curves	1.1
1.4	Solvents, solutes and solutions	1.1
1.5	Diffusion	1.2
2.1	Apparatus for measuring	12.1
2.2	Paper chromatography	12.3
2.3	Is that chemical pure?	12.3/12.4
2.4	Separation and purification	12.4
2.5	More about separation and purification	12.4
3.1	Inside the atom	2.2
3.2	Isotopes	2.3
3.3	Electronic structure and the Periodic Table	2.2
3.4	Elements, compounds and mixtures	2.1
3.5	Metals and non-metals	9.1
4.1	Ionic bonding	2.4
4.2	Covalent bonding (1): simple molecules	2.5
4.3	Covalent bonding (2): more complex molecules	2.5
4.4	Ionic or simple molecular?	2.4/2.5
4.5	Giant covalent structures	2.6
4.6	Metallic bonding	2.7
5.1	Chemical formulae	3.1
5.2	Working out the formula	3.1
5.3	Chemical equations	3.1
5.4	More about equations	3.1
6.1	Masses and molecules	3.2
6.2	Chemical calculations and the mole	3.2/3/3
6.3	More chemical calculations	3.3
6.4	Amount of product	3.3

Contents

Topic number in this book	Topic title	IGCSE Syllabus section
6.5	Gas volume calculations	3.3
6.6	Yield and purity	3.3
6.7	Empirical formula and molecular formula	3.3
6.8	Titrations	3.3/12.2
7.1	Conductors and electrolysis	4.1
7.2	The products of electrolysis	4.1
7.3	Electrolysis of aqueous solutions	4.1
7.4	Explaining electrolysis	4.1
7.5	Purifying copper	4.1
7.6	Electroplating	4.1
7.7	Extracting aluminium	4.1
8.1	Energy transfer in chemical reactions	5.1/6.1
8.2	Reaction pathway diagrams	5.1
8.3	Bond energy calculations	5.1
8.4	Fuels and energy production	10.3
8.5	Fuel cells	4.2
9.1	Investigating rate of reaction	6.2
9.2	Evaluating experiments	6.2
9.3	Interpreting data	6.2
9.4	Surfaces and reaction rate	6.2
9.5	Concentration and rate of reaction	6.2
9.6	Temperature and rate of reaction	6.2
10.1	Reversible reactions	6.3
10.2	Shifting the equilibrium	6.3
10.3	Redox reactions	6.4
10.4	More about redox reactions	6.4
10.5	Oxidising agents and reducing agents	6.4
11.1	How acidic?	7.1
11.2	Properties of acids	7.1

Contents

Topic number in this book	Topic title	IGCSE Syllabus section
11.3	Bases	7.1
11.4	More about acids and bases	7.1
11.5	Acid–base titrations and indicators	7.1
11.6	Oxides	7.2
12.1	Making salts from metals, bases or carbonates	7.3
12.2	Making salts by titration	7.3
12.3	Making salts by precipitation	7.3
12.4	What's that gas?	12.5
12.5	Testing for cations	12.5
12.6	Testing for anions	12.5
13.1	The Periodic Table	8.1
13.2	Group I metals	8.2
13.3	Group VII elements	8.3
13.4	Noble gases and periodic trends	8.5
13.5	Transition elements	8.4
14.1	The metal reactivity series	9.1/9.4
14.2	Metal oxides and their reduction	9.4/9.6
14.3	From metal compounds to metals	9.6
14.4	More about metal reactivity	9.4
15.1	Extracting iron	9.6
15.2	The rusting of iron	9.5
15.3	Alloys	9.3
15.4	Uses of metals	9.2
16.1	Making ammonia	6.3

Contents

Topic number in this book	Topic title	IGCSE Syllabus section
16.2	Fertilisers	10.2
16.3	Sulfuric acid	6.3
16.4	Acid rain	10.3
17.1	Air pollution	10.3
17.2	Carbon dioxide in the atmosphere	10.3
17.3	Global warming and climate change	10.3
17.4	Reducing environmental pollution	10.3
17.5	Clean water	10.3
17.6	Water pollution	10.3
18.1	A variety of organic compounds	11.1
18.2	Formulae of organic compounds	11.1/11.2
18.3	Structural formulae and isomerism	11.1/11.2
18.4	Fuels	11.3
18.5	Petroleum fractionation	11.3
19.1	Alkanes	11.4
19.2	Cracking alkanes	11.5
193	Alkenes	11.5
19.4	Alcohols	11.6
19.5	Fermentation	11.6
19.6	Carboxylic acids and esters	11.7
20.1	What are polymers?	11.8
20.2	More about polymer structure	11.8
20.3	Polyamides and polyesters	11.8
20.4	Amino acids and proteins	11.8

1.1 Solids, liquids and gases

LEARNING OUTCOMES
- State the general properties of solids, liquids and gases
- Describe the structures of solids, liquids and gases in terms of particle separation, arrangement and motion
- State the effect of temperature and pressure on the volume of gases using the kinetic particle theory

The three states of matter

Substances can be solids, liquids or gases. These are the three **states of matter**. Most substances can exist in all three states. For example, water can exist as ice, liquid water or steam.

All matter is made up of particles. Three types of particles make up most matter – atoms, molecules and ions.

An **atom** is the smallest particle that cannot be broken down by chemical means.

A **molecule** is an uncharged particle made of two or more atoms joined together.

An **ion** is an atom or group of atoms that carries a positive or negative electrical charge.

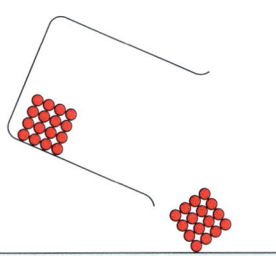
A solid has a definite shape and volume, but cannot flow.

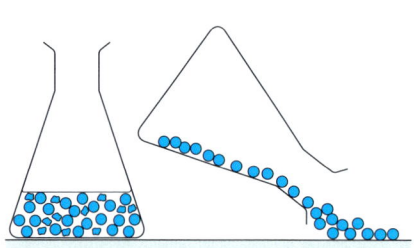
A liquid has a definite volume but takes the shape of its container. It can flow.

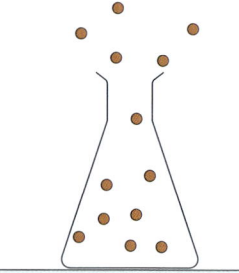
A gas has no definite volume. It can spread everywhere throughout its container.

Figure 1.1.1 The three states of matter: solid, liquid and gas.

Figure 1.1.2 When solid iodine is heated it forms a liquid at 114°C, then soon turns to gas at 184°C.

EXAM TIP
Remember that in liquids the particles slip and slide over each other. They are not completely free to move anywhere.

We can explain the properties of solids, liquids and gases by looking at their particle arrangement, motion and separation (how close they are together).

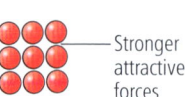 Stronger attractive forces
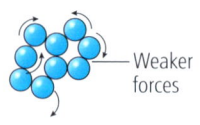 Weaker forces
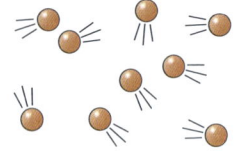

SOLID: arrangement: fixed pattern; motion: only vibrate; separation: close together

LIQUID: arrangement: random – no fixed pattern; motion: slide past each other; separation: close together

GAS: arrangement: random; motion: move every where rapidly; separation: far apart

Figure 1.1.3 The general properties of solids, liquids and gases.

Unit 1: Particle theory

The kinetic particle theory

In liquids and gases, the particles are constantly moving and changing directions as they hit other particles. The idea that particles are constantly in motion is called the **kinetic particle theory**.

The kinetic particle theory states that:

- particles in gases, liquids and solid behave as hard spheres
- particles in gases and liquids move randomly (in any direction)
- particles in gases do not attract each other.

Compressing gases

We can picture a gas as a collection of randomly moving particles which collide with each other and with the walls of their container. When we increase the pressure, the particles get closer to each other. So the volume of the gas decreases.

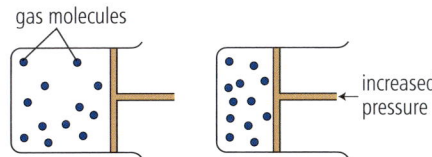

Figure 1.1.4 When the volume of the container is decreased, the gas molecules are squashed closer together and hit the walls of the container more often.

Heating gases

When the volume of gas is not fixed, for example in a gas syringe, the volume of gas increases as the temperature increases. At higher temperatures, the particles have more **kinetic energy**. So they move faster and hit the walls of the syringe with greater force. So the plunger in the syringe is pushed outwards, and the volume of the gas increases.

SUMMARY QUESTIONS

1. Which of these phrases refers to:
   a gases   b liquids   c solids?
   - The particles are close together.
   - The particles are randomly arranged.
   - The particles only vibrate.

2. Describe how a solid differs from a gas in its general properties.

3. State how the motion and average distance between the gas particles changes in a closed container when **a** the temperature increases and **b** the pressure increases.

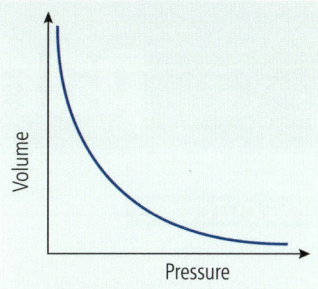

Figure 1.1.5 As the pressure of a gas increases, its volume decreases (at constant temperature).

EXAM TIP

Note that increasing the temperature increases the average speed of the gas particles, but increasing the pressure at constant temperature has no effect on the speed of the gas particles.

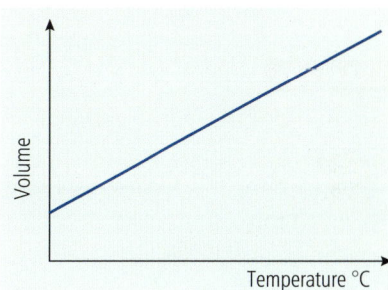

Figure 1.1.6 As the temperature of a gas increases, its volume increases (at constant pressure).

KEY POINTS

- Solids, liquids and gases can be distinguished by their shapes and how easily they flow or spread out.
- The properties of solids, liquids and gases can be explained in terms of the separation, arrangement and motion of their particles.
- The closer the particles in a gas, the higher the pressure and the lower the volume.
- Increasing the temperature of a gas increases the speed at which its particles move.

1.2 Using the kinetic particle theory

LEARNING OUTCOMES
- Describe the changes of state in terms of melting, boiling, evaporation, freezing and condensation
- **S** Explain the effects of pressure and temperature on the volume of a gas using the kinetic particle theory

Energy changes are looked at in more detail in Topic 8.1

EXAM TIP

Remember that most of the particles in liquids are touching one another. It is a common error to think that all the particles are separated.

Figure 1.2.1 Water changes to steam at its boiling point, 100°C.

EXAM TIP

Remember the difference between boiling and evaporation. Boiling takes place at the boiling point of the liquid. Evaporation takes place at temperatures below its boiling point.

Changes of state

Melting

When we heat a solid, energy is transferred to the solid. Its particles gain energy and their vibrations are stronger. The **forces of attraction** between the particles are weakened and the solid **melts**. The solid turns to a liquid. The **melting point** is the temperature at which a solid turns to a liquid.

Boiling and evaporation

Heating a liquid to a higher temperature weakens the forces of attraction between particles further. When a high enough temperature is reached, the attractive forces keeping the liquid's particles grouped together are broken. The particles can form bubbles of gas that escape from the surface of the liquid. So as the bubbling liquid turns into a gas, we say that the liquid **boils**. The **boiling point** is the temperature at which a liquid turns to a gas.

At temperatures below the boiling point of a liquid, some of its particles have enough energy to escape from the surface. They form a vapour. This process is called **evaporation**. Energy needs to be taken in to melt, boil or evaporate a substance.

Condensing and freezing

Cooling a gas makes it condense into a liquid. Further cooling results in the liquid **freezing (solidifying)**. Energy is released (given out or transferred) to the surroundings when a substance condenses or freezes. The surroundings include the air as well as the container in which the solid, liquid or gas is placed.

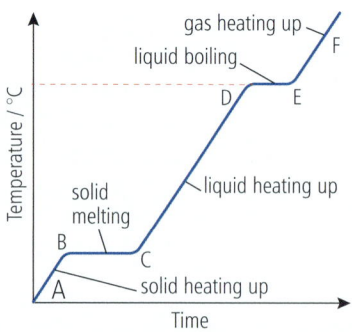

Figure 1.2.2 The changes of state. Energy must be absorbed to melt and boil a substance. Energy is released on condensing and freezing.

Unit 1: Particle theory

Supplement

Gases and the kinetic particle theory

Pressure and volume

The gas particles exert a force on the walls of their container, causing pressure. When we decrease the volume of a fixed mass of gas, the molecules get closer together and hit the walls of the container more frequently. This causes an increase in gas pressure (see Figure 1.2.3). The higher the pressure, the closer the particles are to each other.

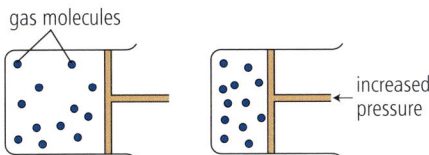

Figure 1.2.3 When the volume of the container is decreased, the gas molecules are squashed closer together and hit the walls of the container more frequently.

Pressure and temperature

A closed container has a fixed volume. If we heat a gas in a closed container, as the temperature increases, the gas particles move faster and hit the walls of the container with increased force. We say that the molecules have greater kinetic energy (energy associated with movement) at a higher temperature. Since the volume of a closed container does not change, the pressure increases when the temperature increases.

Temperature and volume

If the volume of the gas is not fixed, as for example in a gas syringe, the volume of gas increases when the temperature increases. This is because at higher temperatures, gas molecules have more kinetic energy and move faster. The higher the temperature, the greater is the force of the gas molecules on the syringe plunger. The plunger is pushed out until the pressure is balanced by the pressure of the atmosphere.

SUMMARY QUESTIONS

1. Give the names of these changes of state:
   a. liquid to gas   b. solid to liquid   c. gas to liquid
2. Describe what happens to the energy and motion of the particles when ice changes to water.
3. Explain, using the kinetic particle theory, how the volume of a gas in a gas syringe changes when the temperature decreases. The outside pressure is constant.
4. Explain, using the kinetic particle theory, how the volume of a gas in a gas syringe changes when the outside pressure decreases.

EXAM TIP

When writing about the energy in moving particles, make sure that you state kinetic energy and not just energy.

KEY POINTS

- The terms melting, boiling, condensing and freezing are used for specific changes of state.
- In melting, boiling and evaporation energy is absorbed (put in).
- In condensing and freezing, energy is released.
- The volume of a gas decreases when pressure increases because the particles are pushed closer together.
- The volume of a gas in a syringe increases when temperature increases. This is because the particles hit the walls of the syringe with more force at higher temperatures. The volume increases as the plunger moves outwards until the gas pressure in the syringe equals the atmospheric pressure.

1.3 Heating and cooling curves

LEARNING OUTCOMES

- Explain differences in physical state in terms of melting points and boiling points
- **S** Explain changes of state using the kinetic particle theory
- Interpret cooling curves and heating curves

Using melting and boiling point data

We can determine the physical state of a substance at any given temperature by comparing the temperature with its melting point and boiling point. Figure 1.3.1 shows the melting point and boiling point of water.

- At 120°C water is a gas because 120°C is above its boiling point (100°C)
- At 60°C water is a liquid because 60°C is above its melting point (0°C) but below its boiling point (100°C)
- At −15°C water is a solid because −15°C is below its melting point (0°C)

Supplement

Explaining changes of state

Energy is absorbed or released when the particles in solids, liquids or gases rearrange themselves during a change of state. We can explain the shape of a **heating** or **cooling curve** of temperature against time using ideas about the motion of the particles and energy changes.

Heating curves

PRACTICAL

A heating curve for stearic acid

A sample of solid stearic acid is heated at a constant rate using the apparatus shown. The temperature of the stearic acid is recorded every 30 seconds.

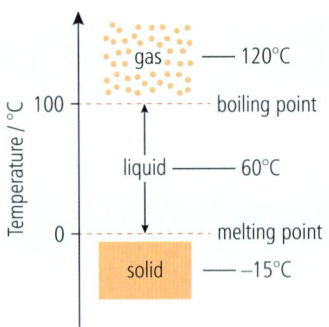

Figure 1.3.1 Water is a gas above 100°C and a solid below 0°C.

EXAM TIP

When using melting and boiling point data remember, for example, that −200°C is a lower temperature than −100°C.

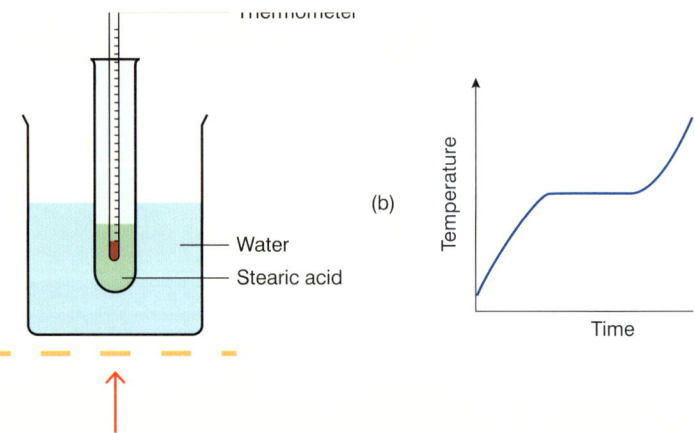

Figure 1.3.2 a apparatus and b graph of results.

Unit 1: Particle theory

The flat part of the graph in Figure 1.3.2 shows where the solid is changing to liquid. This is the melting point of stearic acid. There is no temperature rise here. So the energy supplied is not raising the temperature. The energy is being absorbed to overcome the attractive forces holding the particles of solid in position.

Cooling curves

A **cooling curve** is a graph which shows temperature against time when a substance is cooled at a constant rate. Figure 1.3.3 shows a cooling curve obtained when a gas at a temperature above its boiling point is cooled to form a solid below its melting point.

We can explain the cooling curve using ideas about the energy and motion of the particles.

- A–B: Decreasing the kinetic energy decreases the speed of the gas particles. So the temperature of the gas falls.
- B–C: The forces of attraction between the particles are strengthened. The temperature is constant here because thermal energy is released when intermolecular attractive forces are formed. The thermal energy given out during **condensation** stops the temperature from falling.
- C–D: Decreasing the kinetic energy decreases the speed of the particles in the liquid. So the temperature of the liquid falls.
- D–E: The temperature is constant because thermal energy is released when a liquid changes to a solid. The thermal energy given out during freezing stops the temperature from falling.
- E–F: Decreasing kinetic energy decreases the vibration of the particles in the solid. So the temperature decreases.

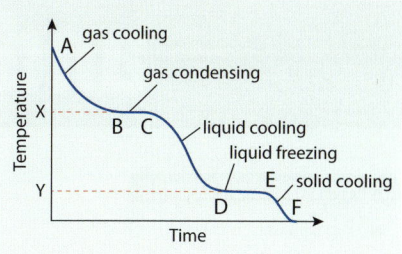

Figure 1.3.3 A cooling curve for the change of state from gas to solid showing the boiling point, X, and the melting point, Y.

KEY POINTS

- The physical state of a substance at a particular temperature can be deduced from its melting point and boiling point.
- In melting, boiling and evaporation energy is absorbed.
- In condensing and freezing, energy is released.
- The horizontal (flat) parts of cooling and heating curves show the melting point and boiling point.

SUMMARY QUESTIONS

1. Sodium chloride melts at 801°C and boils at 1413°C. Describe the physical state of sodium chloride at **a** 970°C **b** 1500°C. Explain your answers.

2. Methane condenses at −164°C and freezes at −182°C. Describe the physical state of methane at **a** −190°C **b** −150°C. Explain your answers.

3. Copy and complete using the words below:

   absorbed   flat   forces   energy   melting

   When we heat a solid, _____ is absorbed and raises the temperature of the solid. At the _____ point, the energy is _____ to overcome the attractive _____ between the particles rather than raising the temperature. That is why there is a _____ part to the heating curve.

4. Sketch a graph to show how temperature changes when a solid is heated at a constant rate to form a liquid and then a gas. Label the melting point of the liquid.

5. Explain why there is no change in temperature when a solid melts.

1.4 Solvents, solutes and solutions

LEARNING OUTCOMES

- Define the terms solvent, solute, solution and saturated solution
- State that concentration can be measured in g/dm³ or mol/dm³
- Describe the chemical tests for water
- Explain why distilled water is used in practical chemistry

Solutes, solvents and solutions

- A **solvent** is a substance which dissolves another substance. Water, ethanol and hexane are examples of solvents which are often used in chemistry.
- A **solute** is a substance which dissolves in a solvent. Salt dissolves in water. So salt is a solute. Solutes can be solids, liquids or gases.
- A **solution** is a **mixture** in which a solute is spread evenly throughout a solvent.
- An **aqueous solution** is formed when a solute dissolves in water.
- A **saturated solution** contains the maximum concentration of a solute dissolved in a solvent at a specified temperature.

When a solution is made, the solute particles are completely mixed up with the solvent particles (Figure 1.4.1). Every part of the solution has the same concentration of particles. We know that the substance has dissolved completely when we cannot see the solute any more.

Solution concentration

If we dissolve a lot of salt in 100 cm³ of water, we say that the solution is concentrated. If we dissolve only a little salt in the same amount of water, we say that the solution is dilute. We measure concentration in grams per **decimetre cubed** (written as g/dm³).

1 decimetre cubed (dm³) = 1000 cm³

We can also measure concentration in moles per decimetre cubed (written as mol/dm³). Moles are the unit used by scientists to measure 'amount of substance'.

We can use this equation to calculate **concentration** in g/dm³:

$$\text{concentration (in g/dm}^3\text{)} = \frac{\text{mass of substance (in g)}}{\text{volume of solution (in dm}^3\text{)}}$$

Example:

Calculate the concentration of a solution of magnesium chloride containing 5 g of magnesium chloride in 200 cm³ of water.

Step 1: change cm³ into dm³. 200 cm³ = $\frac{200}{1000}$ = 0.2 dm³

Step 2: use the equation for concentration: concentration = $\frac{5}{0.2}$ = 25 g/dm³

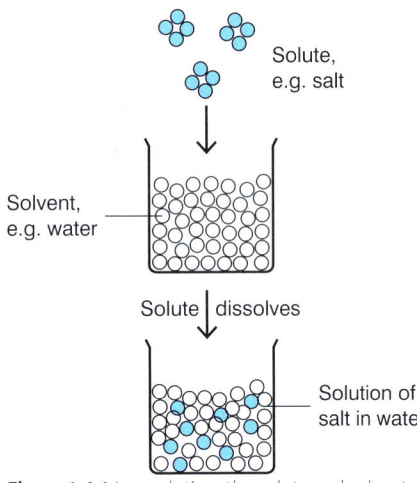

Figure 1.4.1 In a solution, the solute and solvent particles are completely mixed up.

EXAM TIP

When calculating the concentration of a solution, remember to convert cm³ to dm³. You can do this by dividing the number of cm³ by 1000.

Unit 1: Particle theory

Solubility

You can only dissolve a certain amount of substance in a solvent. When we can dissolve no more substance, we say that the solution is saturated. A **saturated solution** contains the maximum concentration of a solute dissolved in a solvent at a particular temperature. Different substances have different solubilities in a solvent. The **solubility** in g/dm^3 is the maximum number of grams of solute that can be dissolved to make 1 dm^3 of solution. For example, you can dissolve up to 340 g of potassium chloride in water to make 1 dm^3 of potassium chloride solution. We say that potassium chloride is soluble in water.

You can only dissolve about 0.3 g of iodine in water to make 1 dm^3 of iodine solution. We say that iodine is very slightly soluble in water. Its solubility is so low that we can think of iodine as being **insoluble**. Sand is a substance that does not dissolve in water at all. Sand is insoluble in water.

Water

Testing for water

There are two common chemical tests for water:

1. When we add water to white **anhydrous** copper(II) sulfate it turns blue.
2. When we add water to anhydrous cobalt(II) chloride the cobalt chloride changes from blue to pink.

Anhydrous means 'without water'. When water is added to anhydrous copper(II) sulfate or anhydrous cobalt(II) chloride, these compounds become **hydrated**. Hydrated means that the crystals have water chemically combined in their structure. This water is called **water of crystallisation**.

Water as a solvent

Water from the tap or from the ground contains soluble impurities. We use pure water in chemistry rather than tap water because these impurities can interfere with chemical experiments. For example, magnesium compounds in tap water can react with chemicals added during an experiment, such as sodium hydroxide. This can cause the solution to go cloudy. For most work in the chemical laboratory, distilled water is used because this contains very few chemical impurities.

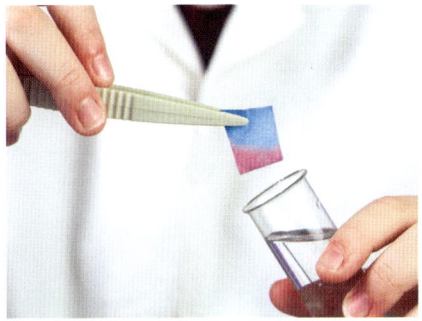

Figure 1.4.2 Blue cobalt chloride paper turns pink if water is present.

KEY POINTS

- A solvent is a substance which dissolves another substance.
- A solute is a substance which dissolves in a solvent.
- A solution is a mixture in which a solute is spread evenly throughout a solvent.
- Concentration in g/dm^3 = mass in g / volume in dm^3
- Anhydrous copper(II) sulfate and anhydrous cobalt(II) chloride are used to test for water.

SUMMARY QUESTIONS

1. Give the meaning of these terms: **a** solution **b** anhydrous **c** insoluble **d** aqueous solution
2. Explain why tap water is not used for making solutions for chemical experiments.
3. Calculate the concentration in g/dm^3 of a solution which contains 4 g of sodium hydroxide in 125 cm^3 of water.
4. Describe the colour change of anhydrous cobalt(II) chloride when an aqueous solution of sodium chloride is added.

1.5 Diffusion

LEARNING OUTCOMES
- Describe and explain diffusion using the kinetic particle theory
- **S** Describe and explain the effect of molecular mass on the rate of diffusion in gases

EXAM TIP

You may be asked to explain using the kinetic particle theory why the colour of a crystal spreads through the water in which it is placed. Don't forget to write about the particles going into solution as well as writing about diffusion.

If we put a drop of ink into some water and leave it, the colour of the ink will spread throughout the water. Why is this?

The gradual spreading out and mixing up of different particles by their random movement is called **diffusion**.

Diffusion results in the particles spreading throughout the space available. The overall direction of the movement is from where the particles are more concentrated to where the particles are less concentrated.

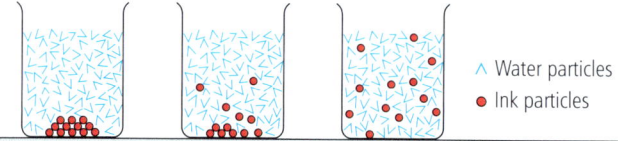

Figure 1.5.1 The colour of the ink spreads because the moving particles of ink mix with the moving water particles.

Diffusion occurs only in liquids and gases because the particles are able to move. Diffusion in gases is faster than in liquids. This is because in gases the particles move rapidly but in liquids they move less rapidly. Diffusion does not occur readily in solids because the particles are packed tightly together and, although they vibrate, they cannot move around.

Diffusion in gases

Diffusion provides evidence for the kinetic particle theory. Diffusion occurs in gases because the molecules in gases are constantly moving, colliding with each other and changing directions. This results in the gases spreading out and mixing. If we put a gas jar containing (colourless) oxygen above a gas jar of bromine vapour (brown), the molecules of the bromine vapour and oxygen gradually mix because all the particles are moving and colliding randomly.

Figure 1.5.2 Bromine gradually diffuses throughout the space available.

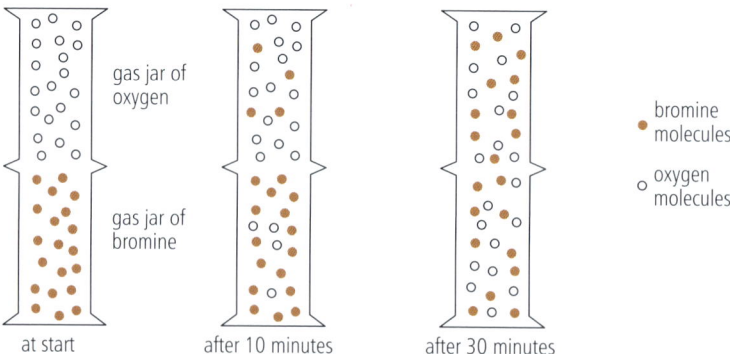

Figure 1.5.3 The molecules in the gas jars collide and move randomly. This leads to the gases mixing by diffusion.

Unit 1: Particle theory

PRACTICAL

Diffusion along a glass tube

A long glass tube is set up as shown.

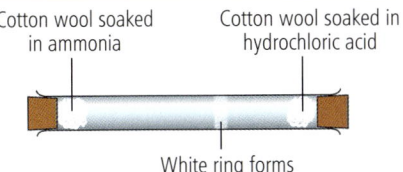

Figure 1.5.4 Apparatus to show the diffusion of gases.

Concentrated hydrochloric acid gives off fumes of a colourless gas called hydrogen chloride. Concentrated ammonia solution gives off colourless ammonia gas. These gases diffuse along the tube. After a few minutes a white ring is seen nearer one end of the tube. When the molecules of ammonia and hydrogen chloride collide with each other they react and form a white solid, ammonium chloride.

Supplement

Rate of diffusion and molecular mass

The speed at which a gas diffuses depends on the mass of its molecules. We compare the mass of molecules with each other by using their relative molecular masses, M_r. The greater the relative molecular mass, the heavier the molecule.

At the same temperature, molecules with a lower M_r move faster than molecules with a higher M_r. The white ring is nearer the hydrogen chloride end of the tube. This shows that hydrogen chloride has heavier molecules than ammonia. So hydrogen chloride diffuses more slowly than ammonia.

KEY POINTS

- Diffusion is the gradual spreading out and mixing up of different particles by random movement.
- Diffusion in liquids is slower than diffusion in gases.
- A gas with a higher molecular mass diffuses more slowly than a gas with a lower molecular mass.

SUMMARY QUESTIONS

1 Copy and complete using the words below:

different   diffusion   gases   mixed   particles   random

The kinetic particle theory states that the _____ in liquids and _____ are in constant _____ motion. When freely moving particles collide, they bounce off each other in _____ directions. If the particles are different, they get _____ up together. This process is called _____.

2 Explain the following using the kinetic particle theory:

  a A red-coloured crystal is placed at the bottom of a beaker of water. At the start of the experiment no colour is seen in the water. After two days, the red colour has spread throughout the water.

  b A bottle of perfume is opened at the front of the classroom. After a little while, you can smell the perfume at the back of the classroom.

3 The relative molecular masses of four gases are: carbon dioxide 44; methane 16; nitrogen 28; oxygen 32. Put these gases in order of their rate of diffusion, with the fastest first.

SUMMARY QUESTIONS

1 Give definitions of:
   a diffusion
   b evaporation
   c condensation

2 Match each of the words on the left with two of the statements on the right.

solid	particles close together
	particles move everywhere
liquid	particles far apart can flow but
	has a definite surface
gas	has a definite shape
	particles only vibrate

3 State whether energy is absorbed or released in these changes of state:
   a A solid changes to a liquid
   b A gas changes to a liquid

4 Match the words on the left with the definitions on the right.

anhydrous	a substance which dissolves in a solvent
hydrated	a substance which does not contain water of crystallisation
solute	a substance which dissolves a solid, liquid or gas
solvent	a substance which contains water of crystallisation

5 Describe and explain using the kinetic particle theory how the volume of a gas in a gas syringe changes when the temperature is gradually increased. Pressure is constant.

6 Use ideas about particles to explain why:
   a a balloon gets bigger when you blow into it
   b solids have a fixed shape
   c you can't squash a sealed syringe full of water

Practice questions

1 Sodium chloride is a solid which dissolves in water but not in hexane. Iodine is a solid which dissolves in hexane but not in water. Choose the mixture of substances which forms a solution.
   A a mixture of iodine and sodium chloride
   B a mixture of sodium chloride and hexane
   C a mixture of iodine and hexane
   D a mixture of iodine and water [1]

(Paper 1)

2 A solution contains 2.0 g of copper(II) sulfate dissolved in 50 cm³ of water. Choose the correct concentration of copper(II) sulfate solution.
   A 0.4 g/cm³   C 40 mol/dm³
   B 25 g/cm³   D 40 g/dm³   [1]

(Paper 2)

3 A crystal of a water-soluble red dye was placed in a beaker of water.
   (a) Describe what you would see:
       (i) after 10 minutes
       (ii) after several days. [2]
   (b) Describe the arrangement and motion of the particles in:
       (i) the crystal [2]   (ii) the water. [2]

4 (a) Name the change of state from:
       (i) solid to liquid   (ii) liquid to gas
       (iii) gas to liquid. [3]
   (b) Which two of these changes of state occur when energy is absorbed? Explain your answer using ideas about forces between particles. [2]
   (c) Describe the arrangement and motion of the particles in a gas. [2]

(Paper 3)

5 Chlorine is a green gas. A sample of chlorine is placed in a gas syringe.

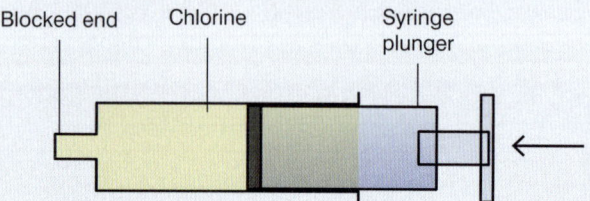

   (a) Describe and explain what happens when pressure is applied to the syringe plunger. The temperature was kept at 20 °C. [2]

(b) The syringe is warmed to at 40 °C. The pressure is constant. Describe and explain the difference in the volume of gas in the syringe. [2]

(c) Chlorine dissolves in hexane to form a solution.
  (i) Choose from the list, the word that best describes the chlorine in this solution.
      gas   saturated   solute   solvent [1]
  (ii) State the units of concentration in a solution. [1]

(d) Chlorine is a liquid at –120 °C. Describe the arrangement and motion of the particles in liquid chlorine. [2]

(Paper 3)

6 A student set up the apparatus shown.

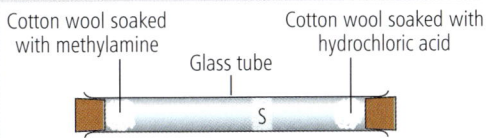

Methylamine and hydrochloric acid give off vapours which react with each other to form a white ring at point S in the tube.

(a) Use ideas about moving particles and energy to explain:
  (i) the process of evaporation from the cotton wool [2]
  (ii) how the particles of methylamine and hydrogen chloride move along the tube. [2]

(b) Describe and explain the position of the white ring, S, in the tube. [2]

(c) An ammonia molecule has about half the mass of a methylamine molecule. If ammonia is used in this experiment in place of methylamine:
  (i) predict the position of the white ring [1]
  (ii) suggest why the white ring would be in this position by referring to molecular mass. [2]

(Paper 4)

7 Sodium reacts with water to produce hydrogen gas.
(a) Describe and explain using the kinetic particle theory, the effect of decreasing the volume on the pressure of hydrogen gas in a gas syringe. The temperature stays the same. [3]

(b) Describe and explain the change in state when water changes into steam. Use the kinetic particle theory and ideas about energy changes. [3]

(c) The diagram shows the cooling curve when sodium gas is cooled to form solid sodium.

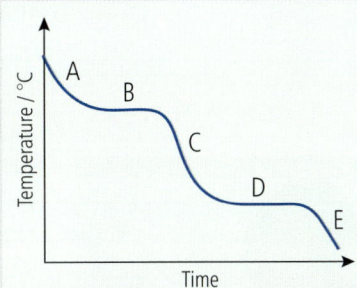

  (i) State which letter on the curve shows:
      • a liquid cooling
      • a gas changing to liquid at the boiling point. [2]
  (ii) Explain what happens to the motion and separation of the particles when a gas changes to a liquid. [2]
  (iii) Explain, in terms of energy changes, the shape of the curve between C, D and E. [3]

(Paper 4)

8 The diagram shows a porous pot, which allows the passage of gases through tiny holes in the walls of the pot.

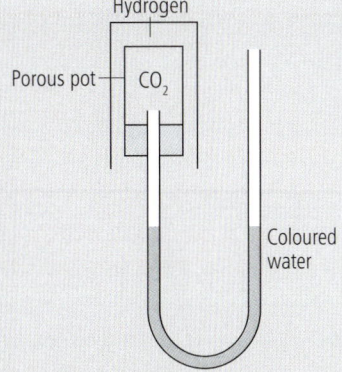

(a) Describe how this apparatus can be used to show that hydrogen has a lower molecular mass than carbon dioxide. [3]

(b) Explain your answer to part (a) using the kinetic particle theory. [5]

(Paper 4)

2.1 Apparatus for measuring

LEARNING OUTCOMES

- Name suitable apparatus for measuring time, temperature, mass and volume

Mass, time and temperature

In chemistry we often have to take accurate measurements of mass, time and temperature.

The standard unit for mass is the kilogram (shown as kg). Chemists, however, often find it more convenient to work in grams (shown as g). We usually use a top pan balance accurate to two decimal places to weigh out chemicals in the laboratory. Since many balances have a digital readout, measuring mass is fairly simple. However, you must remember to set the balance to zero with the weighing boat on it.

Sometimes we need to find out how quickly a reaction proceeds. In order to do this, we use a stop-clock. Electronic stop-clocks often read seconds to two decimal places. We can only achieve this precision, however, if we connect the stop-clock to another electronic measuring system. For example, we can use a data logger and computer.

We measure temperature in degrees **Celsius**, written as °C. Accurate thermometers can measure to of one-tenth of a degree. Very often we do not need this accuracy and a reading to the nearest degree is acceptable. When you take a thermometer reading, remember not to take the thermometer out of the liquid when you read it. If you do this it will cool down, resulting in an incorrect reading.

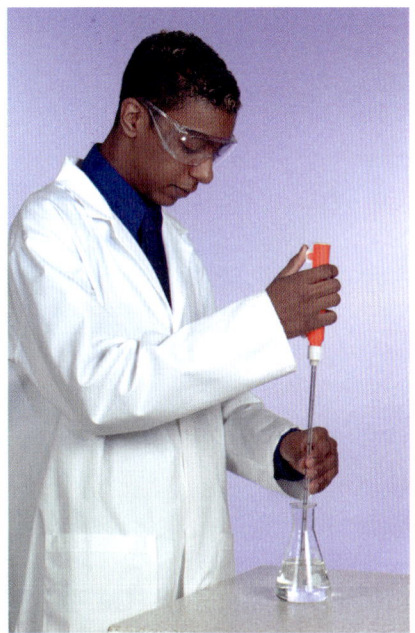

Figure 2.1.1 Using a volumetric (graduated) pipette.

Measuring volumes

Volumes of liquids

In chemistry, we generally measure volumes in centimetres cubed (written as cm^3) or decimetres cubed (dm^3). Since $1\ dm^3 = 1000\ cm^3$ we can easily change cm^3 into dm^3 by dividing by 1000.

Four pieces of apparatus are often used in chemistry for measuring volumes of liquids. These are shown in Figure 2.1.1:

There are various sizes of **measuring cylinders** and **volumetric flasks** ranging from $5\ cm^3$ to $1\ dm^3$. **Volumetric pipettes** are made in only a few sizes, the most common being $10\ cm^3$ and $25\ cm^3$. We use a **burette** to accurately deliver up to $50\ cm^3$ of liquid.

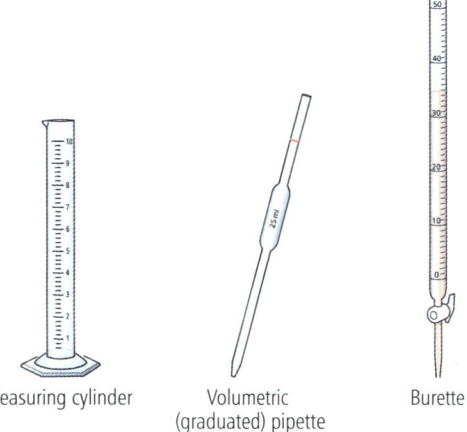
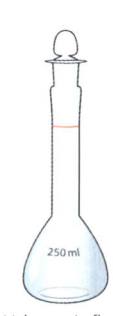

Figure 2.1.2 Apparatus for measuring volumes of liquid.

Unit 2: Separating substances

When we measure out volumes of liquids, we need to think about the accuracy required. A burette or a volumetric pipette is much more accurate than a measuring cylinder. The scale divisions on a burette can be read to the nearest 0.1 cm^3. However, a 100 cm^3 measuring cylinder might have scale divisions only every 2 cm^3.

We use a volumetric pipette to measure out a single fixed volume of liquid very accurately. A burette is used if you want to measure out volumes more accurately than by using a measuring cylinder. We also use a burette in titrations, when you are not sure of the exact volume of the solution you will be adding.

The use of a burette and volumetric pipette is further developed in Topics 6.8 and 11.5.

Measuring cylinders are most useful for making up fairly large volumes of solution or where accuracy is not as important. To make up a solution of a solid dissolved in a liquid accurately, we use a volumetric flask. We place the weighed solid in the volumetric flask, then pour in the liquid until it reaches the line marked on the glass.

Volumes of gases

We can measure volumes using either a **gas syringe** or an upturned (inverted) measuring cylinder. When using a measuring cylinder, it is completely filled with water and then turned upside down in a bowl of water. The gas collected pushes the water down – we say it **displaces** the water. A burette can also be used in this way for greater accuracy.

The use of a gas syringe and upturned measuring cylinder for the collection of gases is further developed in Topic 9.1.

Accuracy

Accurate measurements are very close to the true value. You are more likely to get accurate results for your experiments if you:

- repeat your measurements in the same way each time
- use apparatus with small scale divisions
- use the apparatus carefully.

KEY POINTS

- In chemistry, mass is measured in grams, temperature in °C and volume in cm^3 or dm^3.
- The apparatus you select for an experiment depends on the accuracy required in your experiment.
- Volumes of gases can be measured using a gas syringe or by displacement of water from a measuring cylinder.

EXAM TIP

When measuring volumes, think about the accuracy needed. A burette or volumetric pipette is far more accurate than a measuring cylinder.

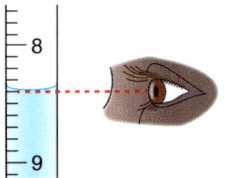

Figure 2.1.3 When reading the level of a liquid, your eye should be in line with the bottom of the meniscus (the curve in the surface of the liquid being measured).

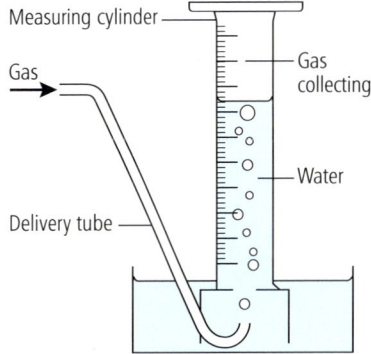

Figure 2.1.4 Measuring gas volume by displacement of water.

SUMMARY QUESTIONS

1. State what piece of apparatus you would use to:
   a. measure mass
   b. measure out 500 cm^3 of water
   c. measure out 5 cm^3 of a liquid very accurately
2. Show the units of:
   a. mass   b. volume
   c. temperature
3. State three ways you can improve the accuracy of results for an experiment.

15

2.2 Paper chromatography

LEARNING OUTCOMES

- Describe how paper chromatography separates soluble coloured substances using a suitable solvent
- Interpret chromatograms to identify unknown substances, as well as pure and impure substances
- **S** Describe the use of locating agents to identify colourless substances after chromatography
- State and use the equation $R_f = \dfrac{\text{distance travelled by substance}}{\text{distance travelled by solvent}}$

We make use of many coloured chemicals in our lives. Ink and food colourings are just two examples. These are often mixtures of several different dyes. We can demonstrate this by placing a drop of ink on a piece of filter paper. When you add a few drops of water to the ink, the colour spreads out from the ink drop and separates into several different colours.

This method of separating pigments (coloured substances) using filter paper is called paper **chromatography.** The colours separate if:

- the pigments have different solubilities in the solvent, and/or
- the pigments have different degrees of attraction for the filter paper.

These two factors determine how far the pigments move across the filter paper. If the mixture of pigments is not soluble in water, other solvents such as ethanol or propanone can be used. The type of solvent also affects how far the pigment moves across the paper.

We can get more information about the substances present in a mixture of dyes by using a special chromatography apparatus.

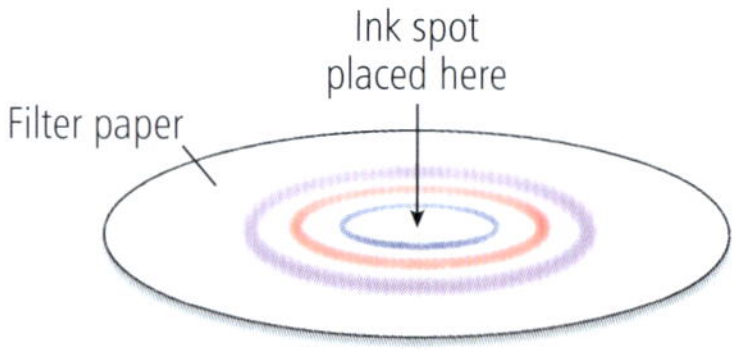

Figure 2.2.1 The blue ink contains three different dyes which are separated on the filter paper.

PRACTICAL

Making a chromatogram

1. First draw a pencil base line across a piece of chromatography paper.
2. Then place a spot of the concentrated dye mixture, M, on the base line using a very fine **pipette** or capillary tube.
3. Then put a spot of each pure dye that you think the mixture might contain, A, B and C, on the line as well. The chromatography paper is put in a jar with the solvent. Make sure that the solvent level is below the level of the spots on the base line. If not, the dye will wash off into the solvent.

As the solvent moves up the paper, the dyes in the mixture separate from each other.

EXAM TIP

When drawing chromatography apparatus, you must draw the origin line (base line) on the chromatogram so that it is above the level of the solvent.

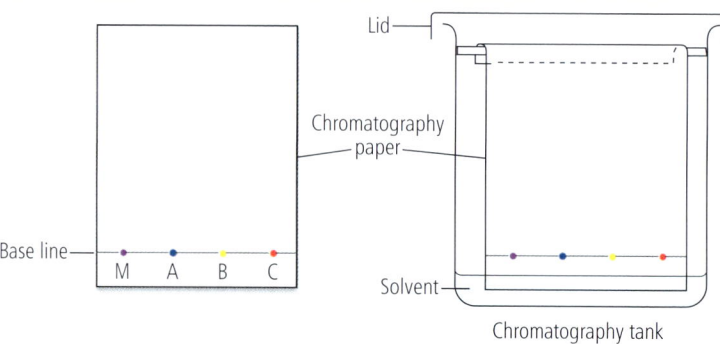

Figure 2.2.2 Apparatus for chromatography.

Unit 2: Separating substances

How can we interpret the **chromatogram** shown in Figure 2.2.3?

The mixture, M, has separated into three dyes. Two of the pure dyes have risen to the same height as two of the dyes in the mixture. So the mixture M contains dyes A and C, but does not contain dye B.

From this we can see that chromatography has two uses:
- identifying the substances in a mixture
- determining if a substance is pure or impure.

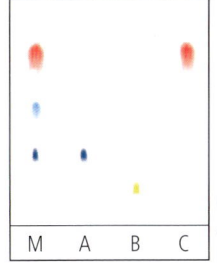

Figure 2.2.3 M contains dyes A and C, but not dye B.

Supplement

More about chromatography

Chromatography can also be used to identify colourless substances. We do this by carrying out chromatography in the same way. However, this time we mark a line near the top of the paper to show where the solvent has reached (called the **solvent front**). The chromatography paper is then dried and sprayed with a chemical called a **locating agent**. The locating agent reacts with the chemicals in the colourless spots and a coloured compound is formed. The colour is usually developed by warming the paper in an oven. Different locating agents are used for different types of compounds.

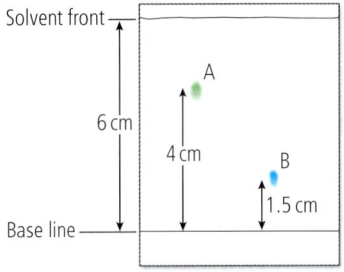

Figure 2.2.4 Values used in calculating R_f.

We can identify the substances on the chromatogram by comparing how far the spots have moved from the base line compared with the solvent front. This is called the R_f value. (R_f is short for retention factor.)

$$R_f = \frac{\text{distance travelled by substance}}{\text{distance travelled by solvent}}$$

For example, in Figure 2.2.4:

R_f of A = $\frac{4}{6}$ = 0.67

R_f of B = $\frac{1.5}{6}$ = 0.25

When R_f values have been calculated they can be compared with tables of known R_f values and the compounds present identified. R_f values vary with the solvent used.

KEY POINTS

- Chromatography is a method for separating and purifying coloured compounds using filter paper and a solvent.
- Chromatography can be used to identify compounds and to see if a substance is pure or impure.
- Locating agents are used to make colourless compounds visible on a chromatogram.
- The compounds on a chromatogram can be identified using their R_f values

SUMMARY QUESTIONS

1 Copy and complete using the words below:

filter   mixture   solubility   solvent

Chromatography can be used to separate a _____ of dyes. The _____ of the dyes in the _____ determines how far they travel up the _____ paper.

2 When setting up a chromatogram, suggest why the base line where the dyes are placed is not drawn with a pen.

3 The R_f values of four amino acids are shown:
alanine 0.38; lysine 0.14; serine 0.27; valine 0.60 Put these amino acids in order of how far they would move up the chromatography paper. Put the one that moves the greatest distance first.

2.3 Is that chemical pure?

LEARNING OUTCOMES

- Identify substances and assess their purity using melting point and boiling point data
- Interpret chromatograms to identify unknown substances and pure and impure substances
- Suggest separation and purification techniques given suitable information

Why is purity important?

In everyday life we use the idea of purity in a very inexact way. We might read 'pure orange juice' on the side of a drink bottle. However, when chemists talk about purity they mean that there is only one substance present. In orange juice there are hundreds of different compounds present so it can never be pure!

It is very difficult to make substances absolutely pure. Even distilled water is not pure. It contains tiny amounts of material from the **distillation** apparatus and dissolved gases from the air. If a substance has a small amount of unwanted substance mixed with it, the unwanted substance is called an **impurity**. Impurities in food additives or medical drugs may have harmful effects on health.

How do we know if a substance is pure?

Using chromatography

In the previous topic you saw how chromatography can be used to separate coloured substances. We can also use this method to check for purity. If there is only a single developed spot on the chromatogram, the substance is likely to be pure. This can be checked by repeating the chromatography using different solvents. If several spots are seen, the substance is impure.

We cannot use chromatography to test every substance for purity, but we can use another method – information about the substance's melting point and boiling point.

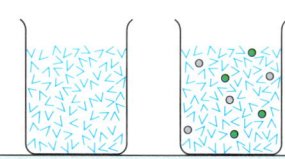

Figure 2.3.1 Pure and impure water.

Figure 2.3.2 It is important that each substance present in baby food is pure.

Using melting point and boiling point data

Most pure substances have specific melting and boiling points. We can use melting points and boiling points to identify pure substances. We can then look at tables of known melting and boiling points to name the specific substance.

Pure substances have a sharp melting point and boiling point. For example, the melting point of water is exactly 0 °C and the boiling point of water is exactly 100 °C. These can be used as a physical test for water. No other known substance has these exact melting and boiling points.

If a substance is impure:

- The melting and boiling points are not sharp – the substance melts and boils over a narrow range of temperatures. The more impurities, the wider the range of temperatures over which melting or boiling occurs.
- The boiling point is increased by impurities. For example, adding salt to water can increase the boiling point of impure water to 102 °C. The greater the amount of salt present, the greater the increase in the boiling point.

Unit 2: Separating substances

- The melting point is lowered by impurities. Adding salt to water decreases its melting point. The more salt added, the greater the decrease in melting point. This has its uses. For example, salt can be put on the roads in icy weather so that ice is less likely to form.

How do we purify mixtures?

In most chemical reactions several products are made. These products often need to be separated from unused reactants or other impurities. We can use several simple methods to purify a mixture depending on the state of the substance we want to obtain. None of these results in an absolutely pure substance but they are suitable for general school laboratory use.

We can separate an undissolved solid from a liquid or a solution by several methods. These include filtration, decanting and centrifugation. Once separated, the solid is washed to remove any solution that is trapped between the solid particles. The solid is then dried in a drying oven.

We can use the evaporation of a solvent to separate a dissolved solid from a liquid. The solution is heated in an evaporating basin. The solvent boils off leaving the solid behind. Leaving the solvent to evaporate at room temperature has the same effect. Note that if there is more than one solid in the original solution, a pure solid will not be produced.

Fractional distillation is used to separate a mixture of liquids that have different boiling points. More information on purification techniques is given in Topics 2.4, 2.5, 12.2 and 12.3.

EXAM TIP
Remember that pure substances have definite sharp melting points and boiling points. Impure substances melt and boil over a range of temperatures.

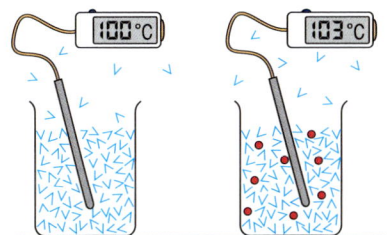

Pure water boils at 100 °C exactly

Impure water boils above 100 °C depending on the concentration of the dissolved salt

Figure 2.3.3 Boiling point of pure and impure water.

SUMMARY QUESTIONS

1. Copy and complete using the words below:

   lower   pure   range   sharp   solution

   Pure water has a _____ melting point. A _____ of salt in water melts over a _____ of temperatures and has a _____ melting point than _____ water.

2. A coloured substance, S, is analysed by chromatography. Two spots are seen on the chromatogram. State and explain what this tells you about the purity of S.

3. Sodium chloride melts at 801°C and boils at 1413°C. A sample of sodium chloride, N, contains a very small amount of potassium chloride. Suggest values for the melting point and boiling point of sample N.

4. Classify the following as either pure or impure:
   - a tap water
   - b a crystal of sulfur
   - c a mint-flavoured sweet
   - d seawater

KEY POINTS
- Chromatography can be used to show if a substance is pure or impure.
- Chromatography, melting points and boiling points can be used to identify pure substances.
- A pure substance melts and boils at definite temperatures. An impure substance melts and boils over a range of temperatures.
- Impurities lower the melting point and raise the boiling point of a substance.

2.4 Separation and purification

LEARNING OUTCOMES

- Describe and explain methods of purification by filtration and crystallisation
- Describe and explain methods of purification by the use of a suitable solvent
- Define the terms filtrate and residue

Separating a solid from a solution

Filtration

An undissolved solid can be separated from a solution or liquid by passing it through a piece of filter paper in a filter funnel. This method is called **filtration**.

- The solution which passes through the filter paper is called the **filtrate**.
- The solid that stays on the filter paper is called the **residue**.

The solid should be washed with distilled water to remove any solution left between the solid particles. The solid is then dried at either room temperature or in a drying oven. Some solids decompose when heated strongly, so it is important not to heat them too strongly. Sometimes we wash the filtrate with an organic solvent instead of distilled water. We can then just leave the solvent to evaporate.

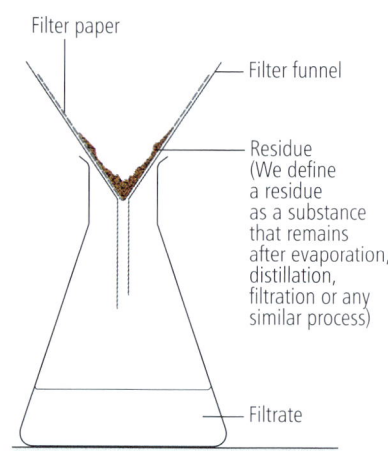

Figure 2.4.1 Filtration apparatus.

We define a residue is substance that remains after evaporation, distillation, filtration or any similar process.

Decanting and centrifugation

Decanting is pouring off a liquid from an undissolved solid. It is suitable for insoluble solids that have very heavy particles – for example, to separate sand from water.

A centrifuge is a machine which spins test tubes around and around at very high speeds. The spinning forces the solid to the bottom of the tube. You can then decant the liquid from above the solid.

Crystallisation

Crystallisation is used to obtain a crystalline solid from a solution. The solution is gently heated in an evaporating basin to concentrate it. The solvent, usually water, is evaporated until it reaches the **crystallisation point**. You can tell when this is by placing a drop of solution onto a cold tile from time to time. If crystals form quickly, the crystallisation point has been reached. The concentrated solution is then left to cool. Crystals eventually form at the bottom of the evaporating basin. These can then be filtered off or picked out and dried between pieces of filter paper.

Figure 2.4.2 We can use filtration to separate a solid from a solution.

EXAM TIP

When writing about crystallisation in exams 'heat to form a saturated solution' is the same as 'heat to the point of crystallisation'.

Unit 2: Separating substances

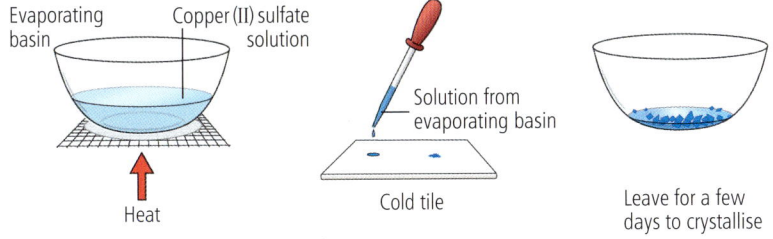

Figure 2.4.3 Crystallisation involves evaporation to the point of crystallisation. Crystals form after the solution cools down.

Solvent extraction

Solvent extraction can be used to separate two solutes dissolved in a solvent. This is especially useful if one of the solutes is **volatile** (evaporates readily). A second solvent is used to extract one of the solids from the first solvent. The second solvent must not mix with the first – we say it is **immiscible.** Two immiscible liquids will settle out into two distinct layers.

For example, in Figure 2.4.4 we have a solution of iodine in aqueous potassium iodide and we want to separate the iodine. (Remember that aqueous means dissolved in water). Iodine is soluble in aqueous potassium iodide but more soluble in a solvent called hexane. Potassium iodide is soluble in water but insoluble in hexane. We shake the solution of iodine in aqueous potassium iodide with hexane. We do this in a **separating funnel**. After shaking, the iodine moves to the hexane layer. The potassium iodide remains in the aqueous layer. To get solid iodine we: (i) run off the bottom layer of potassium iodide solution, (ii) remove the hexane layer which contains the iodine and (iii) leave the hexane layer in a fume cupboard to evaporate off to leave the iodine.

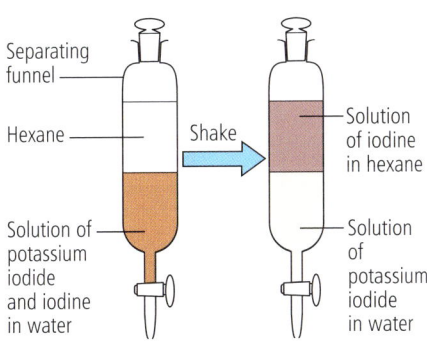

Figure 2.4.4 Separating iodine from a mixture of iodine and aqueous potassium iodide.

KEY POINTS

- Solids can be separated from solutions by filtration, decanting or centrifugation.
- In filtration, the residue is the substance trapped on the filter paper and the filtrate is the liquid which goes through the filter paper.
- Crystals are formed when a solution of a crystalline solid is partly evaporated then left to cool.
- Solvent extraction can be used to separate two solutes dissolved in a liquid.

EXAM TIP

When describing crystallisation, writing 'evaporate off some of the water and then leave to cool' is a better answer than 'heat the solution'. If you heat too much, you will get a powder and not crystals. So never write 'heat to dryness'.

SUMMARY QUESTIONS

1 Suggest suitable methods to:
   a separate crushed insoluble chalk from a mixture of chalk and water
   b get magnesium chloride *powder* from a solution of magnesium chloride in water
   c get crystals of calcium chloride from a solution of calcium chloride

2 Copy and complete using the words below:

   **filtrate   mixture   residue
   solid   trapped**

   An insoluble _____ can be separated from a liquid by filtration. When a _____ of an insoluble solid and a liquid is filtered, the solid _____ on the filter paper is called the _____ and the liquid which passes through is called the _____.

3 Describe how you can get pure, dry crystals of copper(II) sulfate from a solution of copper(II) sulfate in water.

21

2.5 More about separation and purification

LEARNING OUTCOMES

- Describe and explain methods of purification by simple distillation and fractional distillation
- Suggest separation and purification techniques given suitable information

Simple distillation

Simple distillation is used to separate a solvent from a solution.

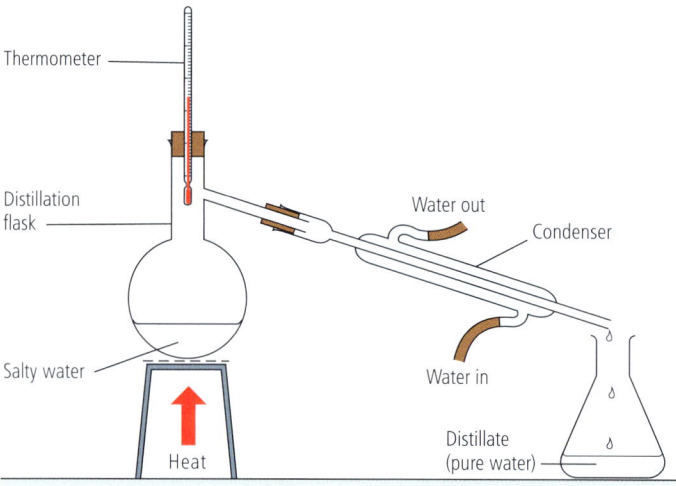

Figure 2.5.1 Simple distillation separates water from dissolved salt. The pure water is collected in the conical flask.

When a solution of salt in water is heated, the water boils and escapes as steam, leaving the salt behind as a solid. The water has a much lower boiling point than salt and readily changes to the gaseous state (steam). The steam turns back into water in the condenser. The temperature here has fallen below the boiling point of water. This is because the condenser is cold. The salt remains in the distillation flask because it has a high boiling point. You can see that distillation is a combination of two processes: evaporation and condensation.

Fractional distillation

Fractional distillation is used to separate a mixture of liquids with different boiling points. Liquids which are **miscible** (they mix with each other) are separated by this method. Fractional distillation uses a tall column in which continuous evaporation and condensation of the liquid mixture occurs. There is a range of temperatures in the column. The temperatures are higher at the bottom and lower at the top. When vaporised, the more volatile compounds in the liquid (those with lower boiling points) move further up the column than the less volatile compounds (which have higher boiling points). After a time, the most volatile compound reaches the condenser, where it changes to a liquid. This

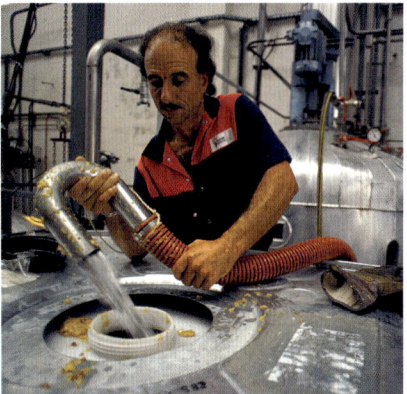

Figure 2.5.2 Jasmine oil is extracted by adding water to jasmine flowers and distilling off the oil.

EXAM TIP

When choosing a method to purify a mixture, think about the states and solubilities of each of the substances in the mixture.

Unit 2: Separating substances

is collected as the distillate. Less volatile compounds move more slowly up the column and reach the **condenser** later. They condense one at a time as **fractions** in order of increasing boiling points.

Which method of purification?

To choose the best method for purifying a mixture you must have a clear idea of how each method works. You may have to use a combination of methods to separate the mixture you want from the unwanted substances. It is often useful to know about the solubility of the substances you are dealing with.

Here is an example: describe how to separate a mixture of sand and salt to obtain pure dry samples of sand and salt crystals.

Salt is soluble in water and sand is not. So:

- Add water and stir to dissolve the salt.
- Separate the sand from the salt solution by filtration.
- Use the process of crystallisation to form the salt crystals.
- Rinse the sand with distilled water.
- Dry the sand and salt crystals separately on filter papers.

> **EXAM TIP**
> Examples of fractional distillation include petroleum fractionation (see Topic 18.4) and the purification of ethanol (see Topic 19.5).

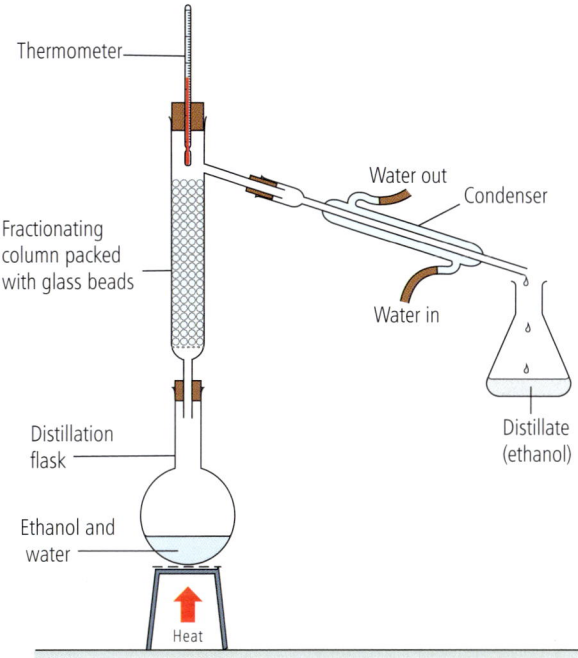

Figure 2.5.3 Fractional distillation is used to separate a mixture of liquids. The more volatile liquid, ethanol in this case, collects in the conical flask.

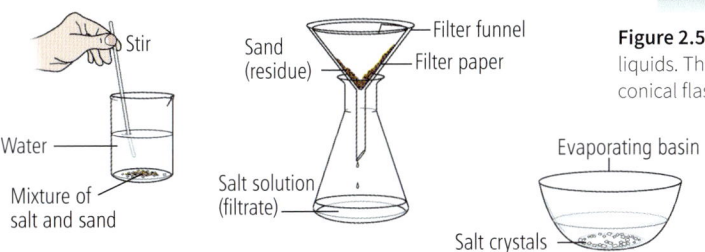

Figure 2.5.4 Separating salt from sand.

SUMMARY QUESTIONS

1. Copy and complete the paragraph using these words:

   column   distillation   lower   volatile

   Fractional _____ separates more volatile liquids from less _____ liquids. The more volatile compounds have _____ boiling points. They move further up the distillation _____.

2. Name the method or methods used to separate the following mixtures:
   a. two volatile liquids which are miscible
   b. a mixture of solid copper(II) sulfate and sand
   c. water from an aqueous solution of sodium hydroxide
   d. a mixture of aqueous amino acids of similar solubility

> **KEY POINTS**
> - Simple distillation is used to separate water from dissolved solids.
> - Fractional distillation is used to separate more volatile liquids from less volatile liquids.
> - Purification of a mixture often involves a combination of methods.

SUMMARY QUESTIONS

1 Give definitions of:
   a filtrate
   b residue
   c solvent
   d distillation

2 Give the names of the pieces of apparatus which are used for these measurements:
   a measuring time to the nearest 0.1 s
   b measuring temperature
   c measuring the volume of gas as a reaction proceeds
   d a pipette used for measuring 20.4 cm³ of liquid

3 Match each piece of apparatus on the left with the phrase on the right that describes it.

volumetric flask	can be used to measure out approximately 20 cm³ of a solution
measuring cylinder	can be used as a container to mix solutions
burette	can be used to make up 500 cm³ of a solution accurately
beaker	used to change a vapour back to a liquid
condenser	can be used to measure out 18.5 cm³ of a liquid accurately

4 Copy and complete the diagram for paper chromatography. Label your diagram.

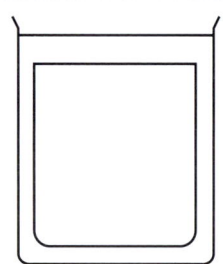

Practice questions

1 Choose the piece of apparatus that is best used for measuring out 13.5 cm³ of acid accurately. [1]

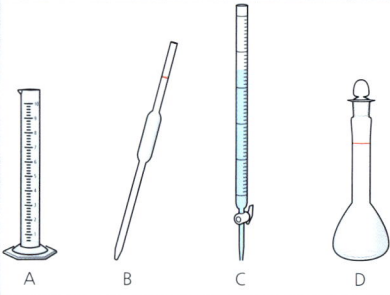

(Paper 1)

2 Choose the method that is used to separate a mixture of liquids.
   A filtration
   B crystallisation
   C evaporation
   D distillation [1]

(Paper 1)

3 Choose the correct definition of R_f.
   A $\dfrac{\text{distance travelled by solvent}}{\text{distance travelled by substance}}$
   B time taken for substance to travel to the top of the paper
   C $\dfrac{\text{distance travelled by substance}}{\text{distance travelled by solvent}}$
   D time taken for substance to reach the solvent front [1]

(Paper 2)

4 Zinc sulfide is a powder which is insoluble in water. Salt is a solid which is soluble in water.
   (a) (i) State the meaning of the term insoluble. [1]
       (ii) Describe how to obtain a dry sample of salt from a mixture of zinc sulfide and salt. [3]
   (b) Iodine is a solid that is soluble in hexane but insoluble in water. Salt is insoluble in hexane. Suggest how to obtain a sample of solid iodine from a mixture of iodine and salt using solvent extraction. [3]

(Paper 3)

5 Chromatography can be used to separate a mixture of coloured dyes.
   (a) Draw a diagram to show the apparatus used to carry out chromatography. [4]
   (b) Three different mixtures of dyes A, B and C were spotted onto a piece of chromatography paper.
   Two pure dyes, D and E, were also spotted onto the same piece of paper.

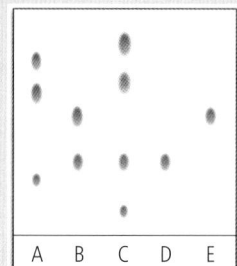

   (i) State where the dye A is placed at the start of the experiment. [1]
   (ii) State which mixture contains the greatest number of different dyes. [1]
   (iii) State which dye contains both the pure dyes D and E. [1]
   (iv) State which dye mixture contains neither dye D nor dye E. [1]

   (Paper 3)

6 The boiling points of four liquids are shown. liquid A 121°C; liquid B 22°C; liquid C 69°C; liquid D 119°C
   (a) State which liquid is most volatile. Give a reason for your answer. [2]
   (b) State the two liquids that are least easily separated by fractional distillation. Give a reason for your answer. [2]
   (c) Describe and explain how fractional distillation is used to separate the four liquids. In your answer refer to what happens in the distillation flask and in the condenser. [4]

   (Paper 3)

7 The table shows some properties of pure ethanol, P, and a sample of ethanol, S, made in the laboratory.

	melting point /°C	boiling point /°C	density g/cm^3	colour
P	−116.8	+78.6	0.79	colourless
S		+79.0	0.79	light yellow

   (a) Give two reasons how you know that sample S is impure. [1]
   (b) Predict the melting point of sample S. [1]
   (c) Predict the state of pure ethanol at −20°C. Give a reason for your answer. [2]
   (d) Cobalt(II) chloride forms hydrated crystals. Describe how to obtain a pure dry sample of cobalt(II) chloride crystals from a solution of cobalt(II) chloride in water. [3]

   (Paper 3)

8 Chromatography is used to separate a mixture of amino acids using ethanol as a solvent. The result is shown in the diagram.

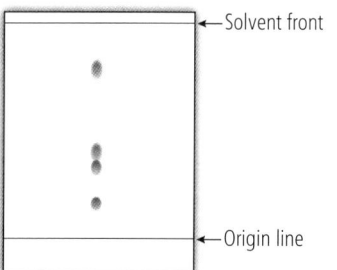

   (a) Explain why the origin line is drawn in pencil and not in ink. [1]
   (b) State how many different amino acids are present in the mixture. [1]
   (c) Copy the diagram and mark with an 'M' the amino acid which is the most soluble in ethanol. [1]
   (d) Calculate the R_f value of the amino acid 'M' using the diagram shown. [1]
   (e) Some amino acids are not easily separated by paper chromatography using ethanol. Suggest how you could separate these amino acids. [1]
   (f) Amino acids are colourless compounds. Describe how the spots of the amino acids can be made visible. [2]

   (Paper 4)

3.1 Inside the atom

LEARNING OUTCOMES

- Describe the structure of the atom
- State the relative charges and relative masses of a proton, neutron and electron
- Define proton number (atomic number)
- Describe the Periodic Table as an arrangement of elements in order of increasing proton number

Inside the atom

Every substance in our world is made up of atoms. An **atom** is the smallest uncharged particle that can take part in a chemical change. The atom is very small. If you put 5 million atoms side by side they would only measure 1mm. But even atoms themselves are made up of still smaller particles called **subatomic particles.**

At the centre of an atom is a tiny **nucleus.** The nucleus is made up of **protons** and **neutrons**. The general name given to protons and neutrons is **nucleons**. This is because they are both found in the nucleus.

Outside the nucleus, the **electrons** move around in areas which are called **electron shells**. The shells are a certain distance from the nucleus. Heavier atoms have several electron shells further and further away from the nucleus. Each of these shells can hold a certain number of electrons.

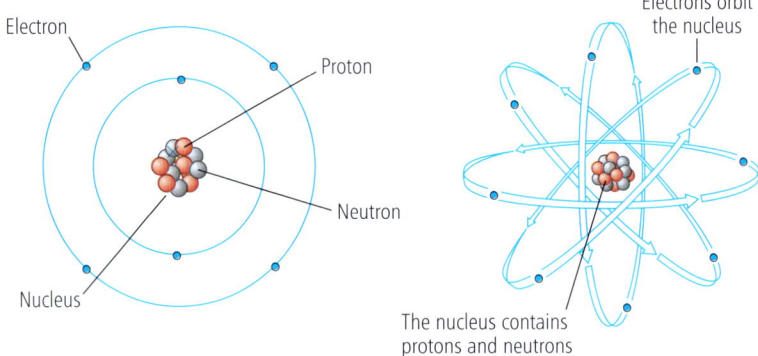

Figure 3.1.1 Two different pictures of an atom. We can draw different models of the atom but it is always difficult to show the exact position of the electrons.

The three subatomic particles have different relative charges and masses. These are shown in Table 3.1.1. Because the masses are so small, we compare their masses to the mass of the proton. That is why we use the term '**relative mass**'.

EXAM TIP

In exams, giving the value for the relative mass of an electron as $\frac{1}{2000}$ is usually acceptable.

subatomic particle	symbol	relative mass	relative charge
proton	p	1	+1
neutron	n	1	no charge
electron	e	0.000 54	−1

Table 3.1.1 Comparing protons, neutrons and electrons.

Unit 3: Atoms, elements and compounds

You can see from the table that nearly all the mass of the atom is in the nucleus. Electrons weigh hardly anything. The protons have a positive (+) charge and the electrons have a negative (−) charge. An atom has no overall charge as it is electrically neutral. This is because in any atom the number of positive protons is equal to the number of negative electrons.

The importance of proton number

The **proton number,** also called the **atomic number**, is the number of protons in the nucleus of an atom. Every atom of the same element has the same number of protons. Hydrogen atoms have one proton in their nucleus, so the proton number or atomic number of hydrogen is 1. Sodium atoms all have 11 protons, so sodium's proton number is 11.

In the **Periodic Table** of elements, the elements are arranged in order of their proton number (Topic 13.1 gives you further information about the Periodic Table).

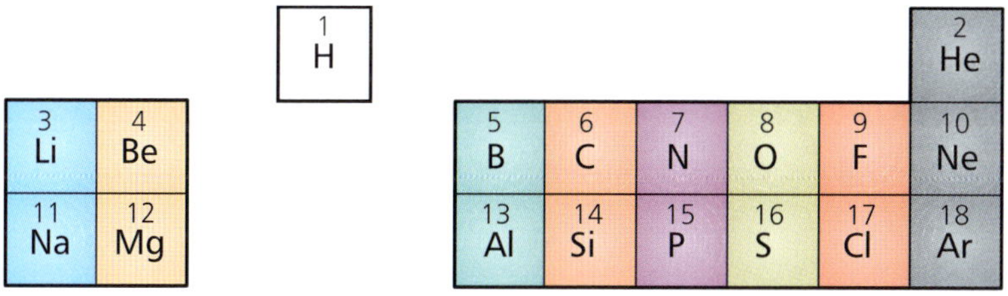

> **EXAM TIP**
>
> You will always be given a Periodic Table in Papers 1 to 4. You can use your Periodic Table to find out the number of protons in an atom.

1 H							2 He
3 Li	4 Be	5 B	6 C	7 N	8 O	9 F	10 Ne
11 Na	12 Mg	13 Al	14 Si	15 P	16 S	17 Cl	18 Ar

Figure 3.1.2 In the Periodic Table the elements are arranged in order of their proton numbers (atomic numbers).

You read the Periodic Table, line by line, starting from the top left. Notice that the proton numbers are put above the symbol for each element. The chemical properties of the elements depend on the electrons. But the number of electrons in an atom is the same as the number of protons. That is why the proton number is important.

SUMMARY QUESTIONS

1 Copy and complete using the words below:

| electrons | equal | neutrons | no |
| nucleons | positive | protons | shells |

The nucleus of an atom contains _____ and _____. The general name for protons and neutrons is _____. Protons have a _____ charge but neutrons have _____ charge. Outside the nucleus are the _____, which are negatively charged. The electrons are arranged in _____. The number of protons in an atom is _____ to the number of electrons.

2 An atom of the element neon has 10 protons and 11 neutrons. State the number of nucleons it has.

3 By referring to subatomic particles, explain why an atom of nitrogen is electrically neutral.

KEY POINTS

- The subatomic particles in atoms are protons, neutrons and electrons.
- A proton has a positive charge, an electron has a negative charge and a neutron has no charge. The protons and neutrons make up the most of the mass of an atom.
- Atoms are neutral because the number of positive protons equals the number of negative electrons.
- An electron is approximately $\frac{1}{2000}$ the mass of a proton.
- Proton number (atomic number) is the number of protons in the nucleus of an atom.
- Atoms are arranged in the Periodic Table in order of their proton number.

3.2 Isotopes

LEARNING OUTCOMES

- Define mass number, nucleon number and isotopes
- **S** Describe how isotopes have the same chemical properties because they have the same electronic configuration
- Calculate the accurate relative atomic mass from the abundances of the isotopes

How many neutrons?

Each different element has a different number of protons. The number of protons plus neutrons in the nucleus of an atom is called the **mass number** or **nucleon number.**

We show the mass number and proton number of an atom like this:

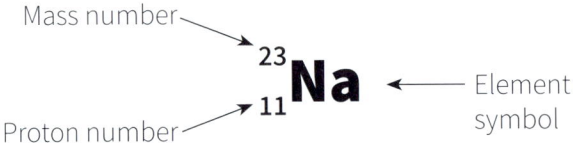

We can use the mass number and proton number to work out the number of neutrons in an atom:

Number of neutrons = mass number – proton number, *or*
Number of neutrons = mass number – atomic number

For example: sodium has 23 nucleons (protons plus neutrons) and 11 protons.

So the number of neutrons in a sodium atom is 23 – 11 = 12 neutrons.

What are isotopes?

Atoms of the same element always have the same proton number. However, in many elements some of their atoms have different numbers of neutrons. Atoms of the same element which have the same number of protons but a different number of neutrons are called **isotopes.** For example, there are three isotopes of hydrogen:

EXAM TIP

Remember that when you define isotopes the word 'atoms' is essential.

EXAM TIP

Remember that the number of protons and electrons in isotopes of a neutral atom are the same. It is only the numbers of neutrons that are different.

Figure 3.2.1 The three isotopes of hydrogen.

Here are two examples of how we interpret symbols showing mass number and proton number.

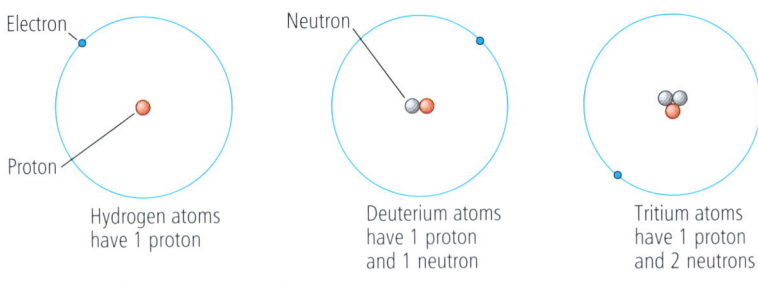

Unit 3: Atoms, elements and compounds

Supplement

Different isotopes of the same element have the same chemical properties. They react in the same way. It is the number of electrons in the outer shell of an element that determines its chemical properties. The chemical properties of isotopes are the same because they have the same electronic configuration (the electrons are arranged in their shells in the same way). However, some physical properties of isotopes such as density may be slightly different.

Accurate relative atomic masses

Many elements are mixtures of isotopes. For example, in naturally occurring chlorine about ¾ of the atoms have 18 neutrons (mass number 35) and about ¼ have 20 neutrons (mass number 37). This means that the accurate relative atomic mass of chlorine is somewhere between 35 and 37. The percentage of each isotope in naturally occurring chlorine is called its **percentage abundance.**

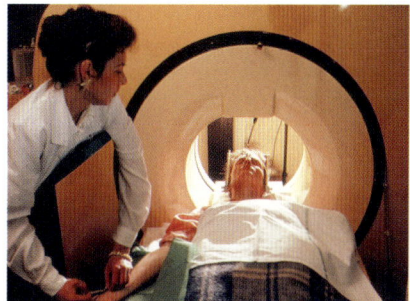

Figure 3.2.2 A patient being injected with a radioactive isotope to locate a possible tumour.

Isotope	$^{35}_{17}Cl$	$^{37}_{17}Cl$
number of neutrons	18	20
percentage of this isotope in naturally occurring chlorine	75.5%	24.5%

We calculate the accurate relative atomic mass as follows:

Step 1: multiply each mass number by its percentage abundance

$^{35}_{17}Cl = 35 \times 75.5 = 2642.5$ $^{37}_{17}Cl = 37 \times 24.5 = 906.5$

Step 2: Add up the answers to Step 1 $2642.5 + 906.5 = 3549$

Step 3: Then divide the answer to Step 2 by 100

$\frac{3549}{100}$ = **35.5** (rounded to 3 significant figures)

EXAM TIP

In exams, you are often expected to write exact definitions of important terms. You will find the definition of relative atomic mass on page 62.

KEY POINTS

- Mass number is the number of protons plus neutrons in an atom.
- Isotopes are atoms with the same number of protons but different numbers of neutrons.
- Isotopes have the same chemical properties because they have the same electronic configuration (arrangement of electrons in their shells).
- The accurate relative atomic mass of an element is calculated using the mass and percentage abundance of each of its isotopes.

SUMMARY QUESTIONS

1. Copy and complete using the words below:

   different   element   isotopes   mass   neutrons

   The _____ number of an atom is the total number of _____ and protons present. Atoms of an _____ with _____ numbers of neutrons are called _____.

2. State the number of protons, neutrons and electrons in each of these isotopes:

   a $^{235}_{92}U$ b $^{14}_{6}C$ c $^{58}_{26}Fe$

3. Write the isotopic symbols (as in question 2) for atoms of:

   a copper (atomic number 29, mass number 65)
   b iodine (53 protons and 74 neutrons)
   c boron (5 protons and 6 neutrons)

3.3 Electronic structure and the Periodic Table

LEARNING OUTCOMES

- Describe the electronic configuration (electronic structure) of the first 20 elements in the Periodic Table
- State that the number of outer shell electrons is equal to the group number in Groups I to VII
- State that the number of occupied electron shells is equal to the period number
- Explain why the noble gases are unreactive by reference to their stable outer shell of electrons

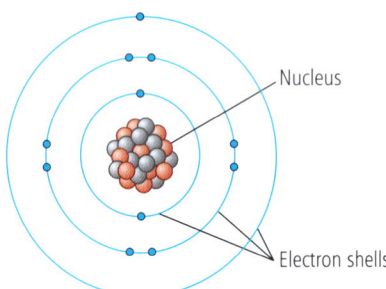

Figure 3.3.1 It is difficult to show where the electrons are. So we use a model like this one to help us.

EXAM TIP

Remember that Period 1 has a maximum of 2 electrons. It is a common error to suggest that all the shells have a maximum of 8 electrons.

Electron shells

In the electron shell model we draw the electrons as if they exist in circular orbits around the nucleus. These orbits are called electron shells or **energy levels**.

The first shell is nearest to the nucleus and can hold a maximum of two electrons. The next shell out can hold a maximum of eight electrons. When the third shell has eight electrons, the fourth shell starts filling up.

The arrangement of the electrons in shells is called the **electronic configuration**. In a sodium atom, there are two electrons in the first shell, eight electrons in the second shell and one in the third shell (see Figure 3.3.1). We have a shorthand way of writing this:

We write the electronic configuration of sodium as 2,8,1.

Arranging the electrons

The horizontal rows in the Periodic Table are called **periods**. As we move across a period, each successive element has one more electron in its outer shell. The electrons fill up the shells one by one, starting from the shell nearest to the nucleus. When one shell is full, the electrons go into the next shell out from the nucleus.

Hydrogen has one electron. This electron is in its first shell. Helium has two electrons, so these electrons are also in its first shell. Lithium has three electrons. Two of lithium's electrons are in its first shell; the third goes into the second shell because the first shell can hold only two electrons. In this way the electrons are 'put into their shells'. Figure 3.3.2 shows the electronic structures and electronic configurations of the first 20 elements.

The number of occupied electron shells is equal to the period number. So Period 1 consists of only hydrogen and helium. This is because the first shell can only hold two electrons. Period 2 elements have two occupied shells. The second (outer) shell fills up going across Period 2, starting with one electron and finishing with eight electrons. Period 3 elements have three occupied shells.

A vertical column in the Periodic Table is called a **group**. You can see that atoms of elements in the same group have the same number of electrons in their outer shell. The chemical properties of an element depend on its number of electrons in the outer shell.

Unit 3: Atoms, elements and compounds

This is why elements in some groups of the Periodic Table have similar chemical properties. You can also see that the number of electrons in the outer shell is the same as the group number. So atoms of Group I elements all have one electron in their outer shells, atoms of Group II elements have two, and so on.

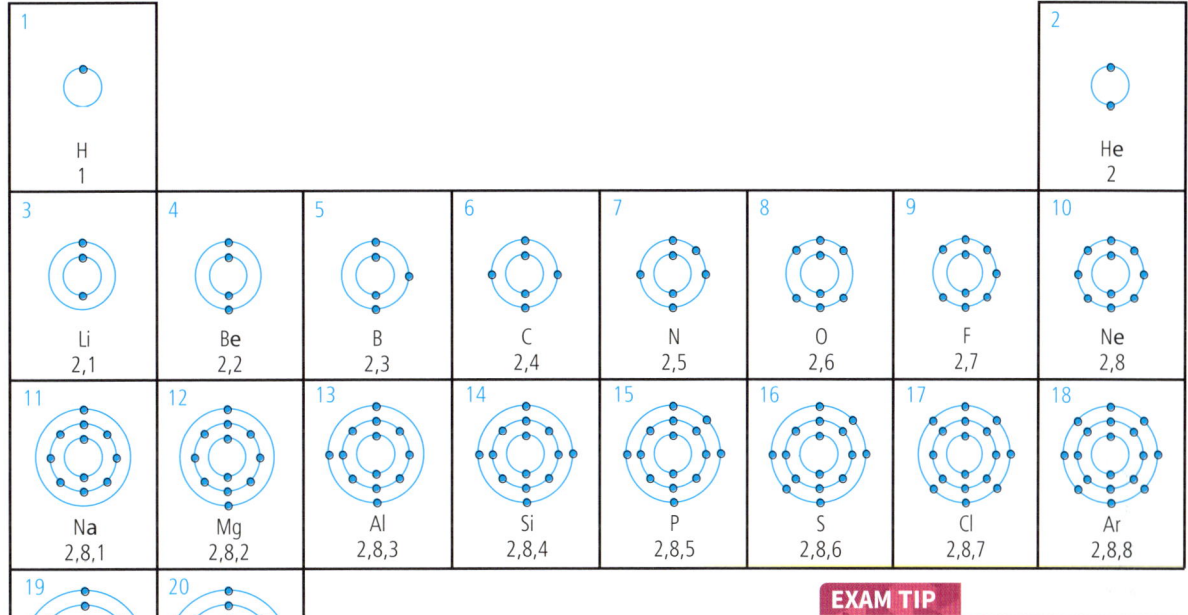

Figure 3.3.2 The electronic configuration of the first 20 elements.

The noble gas structure

When compounds are formed by combining elements, the electrons in the outer shell are transferred or shared. Group VIII elements, the **noble gases**, have a full outer shell of electrons. This type of electronic structure is very stable. The atoms cannot gain or lose electrons very easily. The stability of a full outer shell of electrons makes the noble gases very unreactive. They do not combine with other elements to form compounds. They exist as **monatomic** gases. This means that they exist as single atoms.

EXAM TIP

Make sure that you can draw the electronic configuration of the first 20 elements in shells (shown as rings containing electrons) and also write this (such as 2, 8, 1).

KEY POINTS

- Electrons in atoms are arranged in electron shells.
- The electronic configuration is the arrangement of the electrons in an atom.
- Atoms of elements in the same group in the Periodic Table have the same number of outer shell electrons.
- The chemical properties of an element depend on the number of outer shell electrons in an atom.
- The noble gases are unreactive because they have a stable outer shell of electrons.

SUMMARY QUESTIONS

1 State the number of electrons each of the following atoms have in their outer shell:
   a Group VI elements b Group II elements
   c Group VIII elements except helium
2 Write the electronic configuration in shorthand for these elements:
   a carbon b potassium c chlorine d aluminium

31

3.4 Elements, compounds and mixtures

LEARNING OUTCOMES

- Describe the differences between elements, compounds and mixtures

Elements

Every substance around us is made up of atoms. There are 118 different types of atom. A substance made up of only one type of atom is called an **element**. All the atoms in a particular element have the same number of protons. Elements cannot be split into anything simpler by chemical reactions.

Each element is given a symbol. Fluorine has the symbol F and the symbol for zinc is Zn.

Elements have particular properties which distinguish them from others. For example, silver is a shiny solid, chlorine is a green gas and bromine is a reddish-brown liquid.

The atoms of elements generally combine to form molecules or giant structures of atoms (see Topic 4.5). They rarely exist on their own. The link joining the atoms is called a **chemical bond**.

Compounds

The atoms of different elements can join to form compounds. A **compound** is a substance made up of two or more different types of atom joined together by chemical bonds.

A compound always has a fixed amount of each element in it. Water always has two hydrogen atoms for every one oxygen atom. Copper(II) oxide always contains 80% copper and 20% oxygen by mass. There are two types of compound:

- molecular compounds where atoms are bonded together. Water is an example.
- ionic compounds where many ions (atoms which have gained or lost electrons) are joined together. For example, sodium chloride.

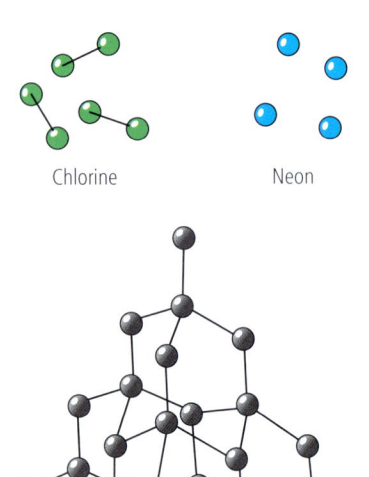

Figure 3.4.1 The structures of some elements.

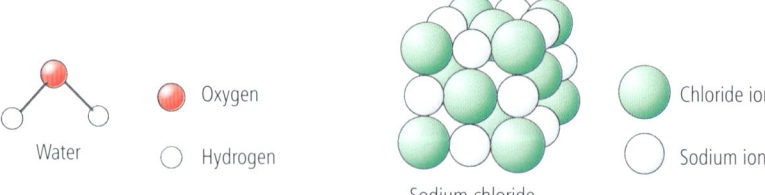

Figure 3.4.2 Water is a molecular compound of hydrogen and oxygen. Common salt is an ionic compound of sodium and chloride ions.

Compounds have different properties from the elements that they are made from. A compound of sodium and chlorine looks completely different from the elements sodium and chlorine.

Unit 3: Atoms, elements and compounds

Sodium is a silvery solid when fresh and chlorine is a green gas. But sodium chloride is a white solid. The elements in a compound cannot be separated by physical means.

PRACTICAL

Making sodium chloride from its elements

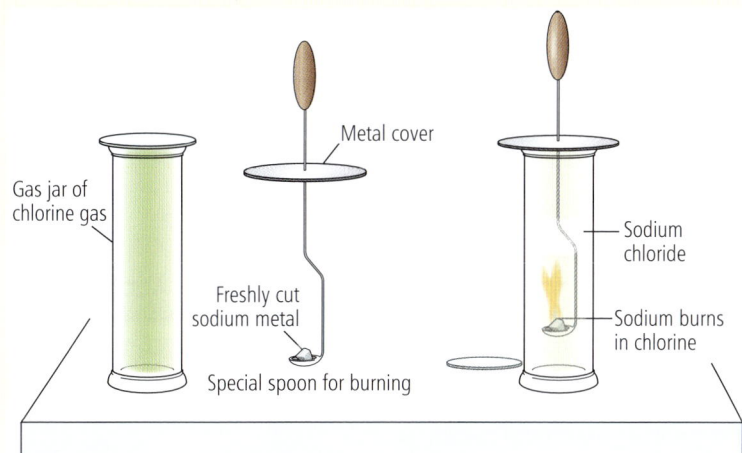

Figure 3.4.4 Burning sodium in chlorine.

Figure 3.4.3 Bromine is an element. All its atoms have the same proton number.

Mixtures

A mixture contains two or more elements or compounds that are not chemically bonded together.

- A mixture does not have a fixed amount of each element or compound in it. A mixture of iron and sulfur can contain different amounts of iron and sulfur.
- The elements or compounds in a mixture still have their characteristic properties. This is because there are no bonds between the substances in a mixture.
- We can separate the substances in a mixture by one of the physical methods described in Topics 2.4 and 2.5.

EXAM TIP

Remember that there are two parts to the definition of a compound: different elements and bonded together.

EXAM TIP

It is important that you learn the definitions for elements, compounds and mixtures and understand the differences.

SUMMARY QUESTIONS

1. Copy and complete using the words below:

    bonds   compound   different   ions

    When two or more _____ atoms join together, a _____ is formed. The atoms or _____ in a compound are held together by chemical _____.

2. Make a list of the differences between a compound and a mixture.

3. Grey-coloured iron powder was heated in reddish-brown bromine vapour. A yellowish-green powder was formed. State if the yellowish-green powder is an element, a compound or a mixture. Explain your answer.

KEY POINTS

- Elements contain only one type of atom and cannot be broken down into simpler substances.
- A compound is a substance containing two or more different types of atoms bonded together.
- Mixtures do not have a fixed composition. The substances in a mixture can be separated by physical means.

3.5 Metals and non-metals

LEARNING OUTCOMES

- Describe the general physical properties of metals and non-metals
- Describe the general chemical properties of metals and non-metals

Metals and non-metals

Most of the elements in the Periodic Table are metals. Examples include iron, sodium, tin and aluminium.

The rest are non-metals. Carbon, sulfur, chlorine and helium are examples of non-metals.

Physical properties

The best way to tell if a substance is a metal or a non-metal is to look at some of its **physical properties**. Physical properties are those such as **density,** melting point and **electrical conductivity**.

PRACTICAL

Comparing electrical conductivity

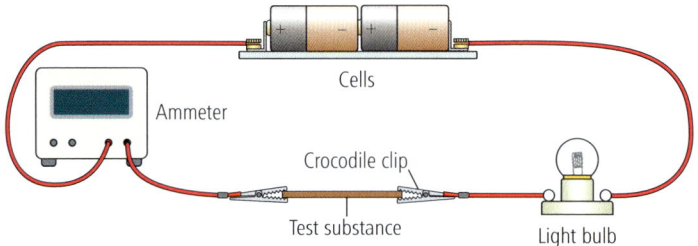

Figure 3.5.2 Comparing the electrical conductivity of solids.

1. Set up the circuit with a known substance between the crocodile clips and see if the bulb lights. If it does, the substance is an electrical conductor.
2. Repeat the experiment with different metals and non-metals.

Figure 3.5.1 Elements can be metallic or non-metallic. Can you tell which are metals and which are non-metals?

Many physical properties do not change however much substance you use. For example, the density and melting point are the same however much substance you test. In chemistry, the density of a substance is measured in grams per centimetre cubed (g/cm^3):

$$\text{density (in } g/cm^3\text{)} = \frac{\text{mass (in g)}}{\text{volume (in } cm^3\text{)}}$$

The table on the next page shows six physical differences between a metal and non-metal. There are some exceptions, but not too many.

There are also some physical properties that are less useful for telling the difference between metals and non-metals:

- Most metals have high densities. Exceptions: Group I metals and some others like gallium have low densities.

EXAM TIP

When stating the difference between a metal and a non-metal it is best to select conductivity, malleability and ductility as properties. These have fewer exceptions to the general rules.

Unit 3: Atoms, elements and compounds

- Most non-metals have low densities.
- Most metals have high melting and boiling points. Exceptions: Group I metals and mercury have relatively low melting and boiling points.
- Most non-metals have low melting and boiling points. Exceptions: carbon and silicon.
- Most metals are hard (they cannot be scratched easily) and strong. Exceptions: Group I metals, mercury, gallium.
- Many non-metals are soft and can be scratched easily. Exception: diamond.

physical property	metal	non-metal
good conductor of electricity	conducts	poor conductor (exceptions: **graphite**)
conductor of thermal energy	conducts	poor conductor (exceptions: graphite and diamond)
malleable – can be beaten into different shapes with a hammer	malleable	not malleable. Non-metals are brittle: they break easily when hit
ductile – can be drawn out into wires	ductile	not ductile. Non-metals are brittle: they break easily when a pulling force is applied
lustrous – has a shiny surface when polished	lustrous	dull surface (exceptions: graphite and iodine have shiny surfaces)
sonorous – makes a ringing sound when hit with a hard object	sonorous	not sonorous. Non-metals make a dull sound when hit with a hard object

Table 3.5.1 Comparing the general physical properties of metals and non-metals.

Chemical properties

We can also use some **chemical properties** to tell the difference between metals and non-metals. Chemical properties are to do with chemical reactions and energy changes during the reactions. Here are some differences:

- Many metal oxides are basic.
- Many non-metal oxides are acidic.
- Many metals react with acids.
- Most non-metals do not react with acids.
- When they react, metals form positive ions by losing electrons.
- When they react with metals, non-metals form negative ions by gaining electrons. (Exception: hydrogen can form positive ions.)

The last two of these points are the best chemical distinction between a metal and a non-metal.

KEY POINTS

- Metals are good conductors of electricity and thermal energy, and are malleable and ductile.
- Non-metals are poor conductors of electricity and thermal energy, and are brittle.
- Common exceptions: graphite is a non-metal that conducts electricity; most metals have high melting and boiling points, but the Group I metals and mercury have low melting and boiling points.

SUMMARY QUESTIONS

1 Copy and complete using the words below:

   conductors   ductile graphite   thermal energy low   opposite

   Metals are good _____ of electricity and _____ _____, and they are malleable and _____. Non-metals usually have properties _____ to those of metals. There are some exceptions. _____ is a non-metal that conducts electricity. Mercury is a metal that has a _____ boiling point.

2 State the names of:
   a two metals that are soft
   b a non-metal that conducts electricity

3 List four physical properties that help you to tell the difference between a metal and a non-metal.

35

SUMMARY QUESTIONS

1 For each of the following sub-atomic particles describe (i) its position in the atom (ii) its relative mass and (iii) its relative charge:

   a proton   **b** neutron   **c** electron

2 Copy and complete the table to show the number of protons, neutrons and electrons in the isotopes shown.

atom	number of protons	number of neutrons	number of electrons
$^{16}_{8}O$			
$^{207}_{82}Pb$			
$^{1}_{1}H$			
$^{37}_{17}Cl$			

3 Give definitions of the following terms:

   a nucleon number   **b** isotopes

   c mixture   **d** compound

4 A potassium ion has a mass number of 41. It has one more proton than electrons. State the number of electrons, protons and neutrons in this potassium ion. Use the Periodic Table to help you.

5 Match each word on the left with the correct description on the right.

isotopes	a substance containing two or more different elements chemically bonded together
element	atoms of the same element with different numbers of neutrons
compound	a positively charged particle in the nucleus
proton	a substance containing only one type of atom

6 Copy each of the following physical properties then write 'metal' or 'non-metal' after each one.

   a conducts thermal energy   **b** is brittle

   c is a poor electrical conductor

   d has a dull surface

   e gives a ringing sound when hit

   f cannot be drawn into wires

Practice questions

1 Choose the correct statement about the isotope $^{14}_{6}C$

   A It has 14 electrons and 6 protons.

   B It has 8 protons and 6 neutrons.

   C It has 6 electrons and 8 neutrons.

   D It has 6 protons and 14 neutrons.

(Paper 1)

2 Choose the correct statement about metals and non-metals.

   A All metals make a dull sound when hit.

   B All non-metals are ductile.

   C Mercury is a solid at room temperature.

   D Graphite is a non-metal that conducts electricity.

(Paper 1)

3 Choose the correct statement about isotopes of hydrogen.

   A The isotope $^{1}_{1}H$ has the same number of neutrons as the isotope $^{2}_{1}H$.

   B The isotope $^{1}_{1}H$ has different chemical properties to the isotope $^{2}_{1}H$.

   C An isotope of hydrogen $^{1}_{1}H$ has only one neutron.

   D The isotope $^{1}_{1}H$ has the same chemical properties as the isotope $^{2}_{1}H$.

(Paper 2)

4 Chlorine has two isotopes, $^{35}_{17}Cl$ and $^{37}_{17}Cl$.

   (a) State the meaning of the term isotope. [1]

   (b) Write down the number of neutrons in each of these isotopes of chlorine. [2]

   (c) Draw and label a diagram of an atom of chlorine to show:

   – the nucleus

   – the electron shells, with the correct number of electrons in each shell. [3]

   (d) An atom of argon has eight electrons in its outer shell. State the importance of this electronic configuration. [1]

(Paper 3)

5 Iron is a grey magnetic metal that reacts with hydrochloric acid to produce hydrogen. Sulfur is a yellow non-metal that does not react with acids.

(a) State the meaning of the term compound. [1]

(b) Describe two differences between a mixture of iron and sulfur and a compound of iron and sulfur. [2]

(c) (i) Give the electronic configuration of sulfur. [1]

(ii) Explain how this electronic configuration shows that sulfur is in Group VI of the Periodic Table. [1]

(iii) State what determines the position of an element in the Periodic Table. [1]

(d) State three differences the physical properties of iron and sulfur that have not been described in this question. [3]

(Paper 3)

6 Sodium is a soft, shiny metal with a low melting point. Chlorine is a poisonous green gas. Sodium reacts with chlorine to form sodium chloride, a white, crystalline solid with a high melting point.

(a) (i) Describe three properties shown by most metals. [3]

(ii) Describe two properties of sodium that make it an unusual metal. [2]

(b) Describe two properties of chlorine that show it is a non-metal. [2]

(c) Use the information above to explain why sodium chloride is a compound of sodium and chlorine and not a mixture of sodium and chlorine. [2]

(d) State the electronic configuration of:

(i) a sodium atom [1]

(ii) a chlorine atom. [1]

(e) An ion of calcium $^{44}_{20}Ca^{2+}$ has two fewer electrons than a Ca atom. Deduce the number of electrons in this ion. [1]

(Paper 3)

7 The electronic configuration of the atoms of five elements, A, B, C, D and E, are shown below:

A 2,8,2   B 2,8,6   C 2,1
D 2,8   E 2,6

(a) State which of these elements is in Group II of the Periodic Table. Give a reason for your answer. [1]

(b) State the two of these elements that are in the same group in the Periodic Table. Give a reason for your answer. [1]

(c) Element A is in the third period. Explain how you know this by referring to its electronic configuration [1]

(d) Element E is oxygen. An isotope of oxygen has nine neutrons.

(i) State the meaning of the term isotope. [1]

(ii) Calculate the mass number of this isotope of oxygen. [1]

(Paper 3)

8 (a) An ion of sulfur has two electrons more than the number of protons. The mass number of this ion is 36. Use this information and the information in the Periodic Table to deduce the number of protons, neutrons and electrons in this ion. [3]

(b) Explain the relationship between the electronic configuration and the position of the elements in the Periodic Table. [3]

(c) Explain why isotopes of the same element have the same chemical properties. [1]

(d) Use the information about isotopes of thallium to calculate an accurate relative atomic mass for thallium.

$^{203}_{81}Th$ abundance = 29.5%

$^{205}_{81}Th$ abundance = 70.5% [3]

(Paper 4)

4.1 Ionic bonding

LEARNING OUTCOMES

- Describe the formation of ions by electron loss or electron gain
- Describe ionic bonds as electrostatic forces of attraction between oppositely charged ions
- Describe the formation of ionic bonds between elements from Groups I and VII, including the use of dot-and-cross diagrams
- S Draw dot-and-cross diagrams for the formation of ions between other metallic and non-metallic elements

Figure 4.1.2 These crystals of sodium chloride are made up of millions of ions.

EXAM TIP

Note that you may be asked to draw dot-and-cross diagrams showing only the outer shell electrons. This is because these electrons are the ones that take part in the electron transfer (as shown in Figure 4.1.3).

EXAM TIP

When drawing the electronic configuration of an ion, make sure that the charge is shown at the top right-hand corner. Do NOT put the charge in the nucleus.

How are ions formed?

An ion is an electrically charged particle. Ions are formed when atoms gain or lose one or more electrons. We saw in Topic 3.3 that an atom with a full outer shell of electrons is stable and unreactive. It has the electronic structure of a noble gas. Most atoms do not have a full outer shell of electrons so tend to react. One way of gaining a full outer shell of electrons is by completely transferring one or more outer shell electrons from one atom to another. The diagram shows how this is done for sodium chloride.

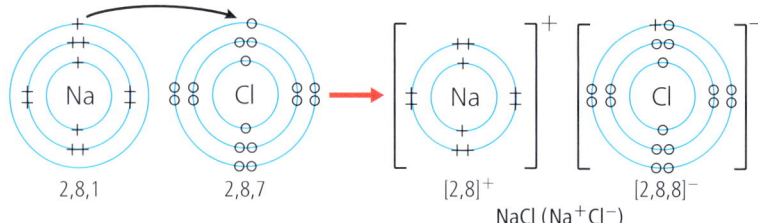

Figure 4.1.1 A sodium ion and a chloride ion are formed by the transfer of an electron from a sodium atom to a chlorine atom.

You will notice in the diagrams that the electrons are *paired*. There are four pairs of electrons in a shell of eight electrons. Pairing up the electrons like this helps to keep track of what happens when bonds are formed.

The sodium ion has one positive charge because it has 11 protons (+) in its nucleus but only 10 electrons (−). The chloride ion has one negative charge because it has 17 protons in its nucleus and 18 electrons.

This sort of diagram is called a **dot-and-cross diagram**. This does not mean that the electron transferred is any different from the others. It shows where the electrons have come from. The charge on the ion is written at the top right-hand side. Square brackets are used to show that the charge on the ion is spread evenly all over the ion.

We can write similar dot-and-cross diagrams for compounds of other Group I and Group VII elements.

Forming a stable structure

Look again at Figure 4.1.1. You will notice that the electron is transferred from a metal atom to a non-metal atom. You will also notice that the outer shells of both atoms have become complete. The sodium ion has the electronic configuration $[2,8]^+$ and the chloride ion has the configuration $[2,8,8]^-$. So both ions have a complete outer shell and have the **noble gas configuration** of eight outer electrons.

The sodium ion has the same electronic configuration as a neon atom and the chloride ion has the same electronic configuration as an argon atom. The noble gas configuration makes the ions stable. This full outer shell of 8 electrons is often called a stable **octet** of

Unit 4: Structure and bonding

electrons. Remember, however, that the first electron shell holds a maximum of two electrons, so the stable electronic configuration for a lithium ion will be [2]⁺.

In ionic compounds the positive and negative ions attract each other. The electrostatic forces of attraction operate in all directions between the oppositely charged ions and is called **ionic bonding**.

Supplement

Ions with multiple charges

We can draw dot-and-cross diagrams for many ions with 2 or 3 positive or negative charges. Look at Figure 4.1.3.

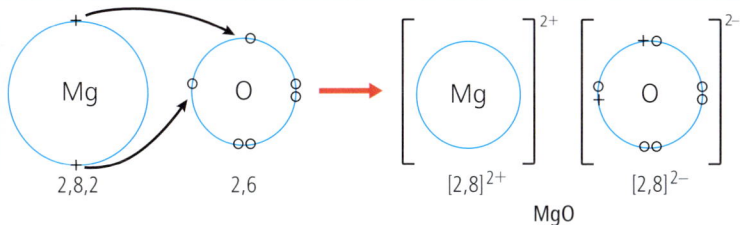

Figure 4.1.3 The formation of magnesium ions and oxide ions.

A magnesium atom has two electrons in its outer shell. An oxygen atom has six electrons in its outer shell. The two magnesium electrons are transferred to the oxygen atom to complete its outer shell. By doing so, both the magnesium ions and oxide ions have an electronic structure which is the same as the noble gas neon. So the ions have a stable electronic configuration.

Drawing the ionic structure for calcium chloride needs a little more thought: the calcium atom has two electrons in its outer shell but each chlorine atom needs only one of these to get a stable octet of electrons. So, two chlorine atoms are needed in the reaction. Each chlorine atom gains one electron.

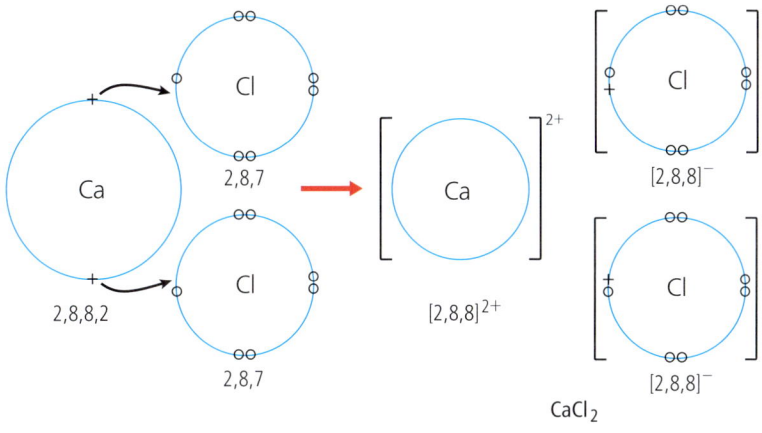

Figure 4.1.4 The formation of calcium ions and chloride ions.

KEY POINTS

- When elements in Group I and VII react, the atom of the Group I element loses the electron from its outer shell. This is transferred to the outer shell of the atom of the Group VII element.
- Ions have the electronic configuration of a noble gas with a complete outer shell of eight electrons (or two electrons for lithium).
- Dot-and-cross diagrams can be drawn to show the electronic structure of ions.
- Ionic bonding is the electrostatic force of attraction between oppositely charged ions.

SUMMARY QUESTIONS

1. Copy and complete using these words:
   **chlorine complete ions
   noble outer transferred**
   When sodium reacts with chlorine an electron from the _____ shell of sodium is _____ to the outer shell of the _____ atom. Sodium ions and chloride _____ are formed. Both these ions have the electronic configuration of a _____ gas with a _____ outer shell of electrons.

2. Write the electronic configurations in numbers for the following ions:
   a Mg^{2+} b K^+ c F^-

3. Draw dot-and-cross diagrams to show the full electronic configurations of the ions present in:
   a calcium oxide
   b aluminium chloride (Al^{3+} ion)
   c potassium sulfide (S^{2-} ion)

4.2 Covalent bonding (1): simple molecules

LEARNING OUTCOMES

- Describe the formation of a single covalent bond as a pair of electrons shared between two atoms, resulting in noble gas electron configurations
- Describe the formation of single covalent bonds in H_2, Cl_2, H_2O, CH_4, NH_3 and HCl using dot-and-cross diagrams

How are covalent bonds formed?

Think back to the formation of ions when a metal and non-metal react together. The electrons are transferred from the outer electron shell of the metal to the outer shell of the non-metal. However, when two non-metal atoms react together they share a pair of electrons. This is called **covalent bonding.**

In a hydrogen molecule (H_2) both hydrogen atoms share one electron with each other to form a pair. This pair of shared electrons between two atoms is called a **covalent bond**. It results in a noble gas configuration for each atom. We show a single covalent bond by a line joining the symbols of the atoms in the molecule, e.g. H—H.

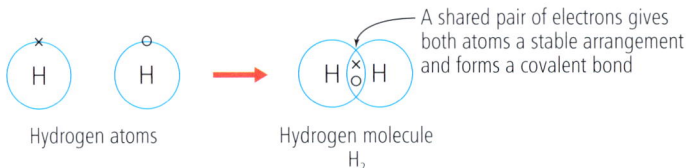

Figure 4.2.1 Atoms of hydrogen share electrons to form a single covalent bond.

The electronic structure of simple molecules

We can draw the electronic structure of simple molecules containing a single covalent bond by pairing up the electrons. In each case we aim to pair the electrons from each atom so that there is a stable octet of eight electrons around each atom. Remember though that hydrogen atoms will be stable with two electrons around them. We do not generally draw the inner shells of electrons because it is only the outer shell electrons that are involved in **bonding**.

When two atoms form covalent bonds we use a dot for the electrons from one of the atoms and a cross for the electrons from the other atom, as we did in dot-and-cross diagrams for ionic bonding. This is simply so we can see where the electrons have come from.

You will see from the examples below that not all the electrons are used in bonding. The pairs of electrons that are not used in bonding are called **lone pairs**.

EXAM TIP

When drawing dot-and-cross diagrams remember to pair up the bonding electrons in the overlap area between the atoms. Do not put them outside the area where the atoms join.

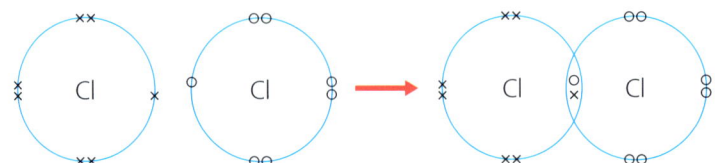

Figure 4.2.2 A chlorine molecule: the two chlorine atoms share a pair of electrons (one electron from each atom) in the covalent bond. Each atom now has eight electrons in its outer shell so has a stable electronic configuration.

Unit 4: Structure and bonding

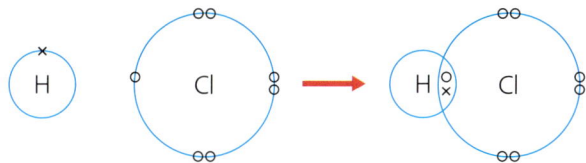

Figure 4.2.3 A hydrogen chloride molecule: hydrogen has one electron in its outer shell and chlorine has seven. One pair of electrons is shared, giving the hydrogen two electrons in its outer shell and chlorine eight.

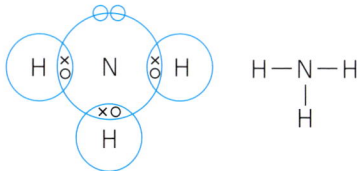

Figure 4.2.4 An ammonia molecule has 3 shared pairs of electrons and one lone pair.

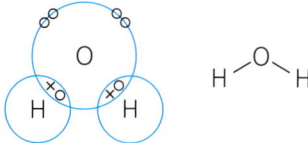

Figure 4.2.5 Water, H_2O. A hydrogen atom has one electron in its outer shell. An oxygen atom has six electrons in its outer shell. Each hydrogen atom shares a pair of electrons with the oxygen atoms to form two covalent bonds. There are four oxygen electrons remaining as two lone pairs.

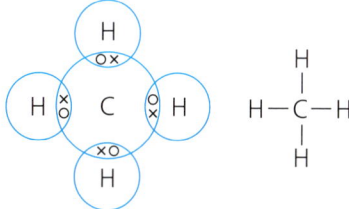

Figure 4.2.6 Methane, CH_4. Each hydrogen atom shares a pair of electrons with the carbon atom. Carbon has four electrons in its outer shell, so four covalent bonds are formed.

SUMMARY QUESTIONS

1. Copy and complete using these words:

   electrons   line   molecules   noble   non-metal

   Simple _____ have a covalently bonded structure. A covalent bond is a pair of _____ shared between two atoms that leads to a _____ gas electron configuration for each atom. A single covalent bond is shown by a single _____ between the atoms. Covalent bonds are formed between two _____ atoms.

2. Draw dot-and-cross diagrams for the following molecules. Show only the electrons in the outer shells:

   a. a fluorine molecule (fluorine is in the same group of the Periodic Table as chlorine)

   b. a hydrogen bromide molecule (bromine is in the same group of the Periodic Table as chlorine)

   c. a hydrogen sulfide molecule (sulfur is in the same group of the Periodic Table as oxygen)

3. State the number of covalent bonds formed when the following atoms combine:

   a. hydrogen and oxygen in water

   b. hydrogen and chlorine in hydrogen chloride

   c. hydrogen and carbon in methane

KEY POINTS

- A covalent bond is formed when atoms share a pair of electrons, leading to the electronic configuration of a noble gas.

- When atoms combine to form covalent bonds each atom has eight electrons in its outer shell, except for hydrogen which has two.

- The electronic configuration of simple molecular structures can be described using dot-and-cross diagrams.

- A single covalent bond is shown by a line between the symbols of the atoms. For example, H—Cl.

41

4.3 Covalent bonding (2): more complex molecules

Supplement

LEARNING OUTCOMES

- Use dot-and-cross diagrams to represent more complex covalent molecules, including N_2, O_2, C_2H_4, CH_3OH and CO_2

Molecules with three or more types of atom

Many molecules contain three or more different types of atom. For example, a molecule of methanol has four hydrogen atoms, one carbon atom and one oxygen atom. In order to draw a dot-and-cross diagram for more complex molecules it is useful to have some idea of the structure of the molecules. In methanol we know that a carbon atom has four electrons in its outer shell. These electrons must pair up with four others. They cannot pair up with all four hydrogen atoms otherwise we will have methane! One of the electrons must pair up with an oxygen atom, so the structure of methanol must be:

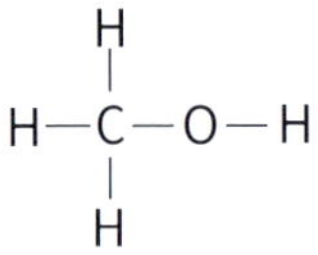

H—C—O—H (with H above and below C)

We can now draw a dot-and-cross diagram for methanol. We have three types of atom. So we need to choose another symbol other than a dot or a cross for the electrons from the third type of atom. We could use a circle, a triangle or a square. Using the same ideas as before, we pair up the electrons so that there are eight electrons in the outer shell of the carbon and oxygen atoms, and two electrons in the outer shell of the hydrogen atoms.

Figure 4.3.1 Living things contain thousands of different molecules which have covalent bonds.

× Electrons from hydrogen
• Electrons from carbon
o Electrons from oxygen

Figure 4.3.2 A dot-and-cross diagram showing the bonding in methanol.

Compounds with double and triple bonds

In order to form a stable octet of electrons, some atoms combine to form a **double bond** or a **triple bond**. A double bond is shown as a double line. For example, in oxygen the double bond is shown as O=O. In nitrogen the triple bond is shown as N≡N. Look at the following examples.

An oxygen molecule: each oxygen atom has six electrons in its outer shell. Each atom needs two electrons to complete its outer electron shell. So two pairs of electrons are shared and two covalent bonds are formed.

A nitrogen molecule: each nitrogen atom has five electrons in its outer shell. Each atom needs three electrons to complete its outer electron shell. So three pairs of electrons are shared and three covalent bonds are formed.

Carbon dioxide: each oxygen atom has six electrons in its outer shell. Each of these atoms needs two electrons to complete its outer electron shell. The carbon atom has four electrons in its outer shell so it needs four electrons to complete its outer shell. So four electrons (two pairs) are shared between the carbon atom and each of the two oxygen atoms. The structure of carbon dioxide can therefore be shown as O—C—O.

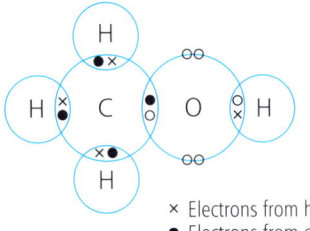

O=O

Figure 4.3.3 An oxygen molecule.

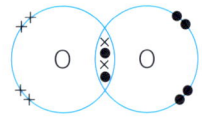

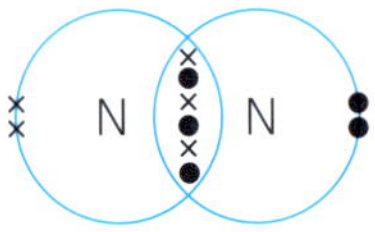

N≡N

Figure 4.3.4 A nitrogen molecule.

Unit 4: Structure and bonding

Ethene: each hydrogen atom has one electron in its outer shell. Each of these atoms needs one electron to complete its outer shell. The two carbon atoms have four electrons in their outer shells so each needs four electrons. Four bonds are formed between the hydrogen atoms and the carbon atoms, two with each carbon atom. This leaves two electrons free on each carbon atom to share with each other as a double bond.

Other molecular compounds

You may be asked to draw the electronic configuration of other simple molecules which are not specified in the syllabus. You do this by:

- deducing the electronic configuration of each atom present
- fitting the atoms together so that each atom has eight electrons around it (or two for hydrogen).

The electronic configurations for ethyne, C_2H_2, and phosphorus trichloride, PCl_3, are shown in Figure 4.3.7.

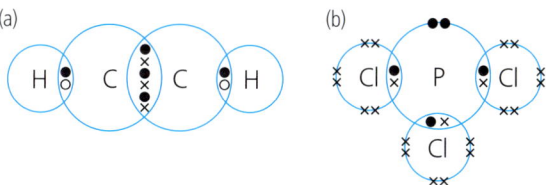

Figure 4.3.7 Dot-and-cross diagrams for (a) ethyne and (b) phosphorus trichloride.

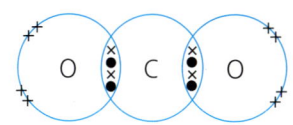

Figure 4.3.5 A carbon dioxide molecule.

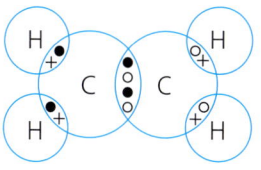

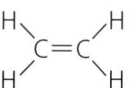

Figure 4.3.6 An ethene molecule.

EXAM TIP

When drawing the electronic structure of compounds with double and triple bonds, make sure you draw the atoms large enough to fit all the bonding electrons into the overlap areas of the atoms.

EXAM TIP

When drawing dot-and-cross diagrams for unfamiliar structures, try and think of a structure you know about which is similar. For example, PH_3 is similar to NH_3 (P and N are both in Group V) and H_2S is similar to H_2O.

SUMMARY QUESTIONS

1 Copy and complete using these words:

nitrogen   oxygen   pairs   triple   two

A double covalent bond is formed when _____ pairs of electrons are shared between two atoms. A _____ covalent bond is formed when three _____ of electrons are shared between two atoms. An _____ molecule has a double bond, and a _____ molecule has a triple bond.

2 Draw dot-and-cross diagrams for:

  a trichloromethane, $CHCl_3$ b ethene, C_2H_4

  c phosphine, PH_3

KEY POINTS

- A double covalent bond is formed when two pairs of electrons are shared between two atoms.
- A triple covalent bond is formed when three pairs of electrons are shared between two atoms.

4.4 Ionic or simple molecular?

LEARNING OUTCOMES

- Describe the properties of ionic compounds
- Describe, in terms of structure and bonding, the properties of simple molecular compounds
- **S** Describe the giant lattice structure of ionic compounds
- Explain, in terms of structure and bonding, the properties of ionic compounds and simple molecular compounds

Structure and bonding in ionic compounds and simple molecules

Substances with an ionic structure are solid at room temperature and pressure. In an ionic structure there is a regular arrangement of positive ions and negative ions. The attractive forces between the positive and negative ions are strong. The ions can only vibrate. They cannot change positions.

- Positive ions are called **cations**.
- Negative ions are called **anions**.

You can find out more about anions and cations in Topics 7.1, 7.2 and 7.3.

Substances with simple molecular structures can be either liquids, gases or low melting point solids at room temperature and pressure. In simple molecular structures which are liquids or gases, the arrangement of the molecules is irregular. The attractive forces between the molecules are weak.

Physical properties of ionic and simple molecular structures

Ionic structures:

- have high melting points and boiling points
- are good electrical conductors when molten (liquid) or in aqueous solution. Ionic structures do not conduct electricity when solid.

Simple molecular structures:

- have low melting points and boiling points
- are poor electrical conductors when solid or molten.

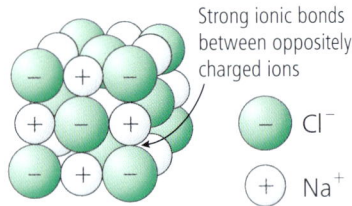

Figure 4.4.1 A giant ionic lattice structure of sodium and chloride ions.

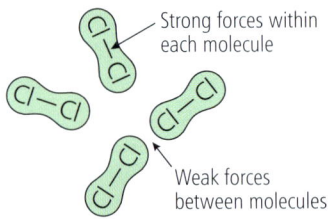

Figure 4.4.2 The forces between chlorine molecules are weak, but within the molecules the covalent bonds are strong.

EXAM TIP

Ionic compounds are generally soluble in water. Some simple molecular compounds are insoluble in water but others are either soluble in water or react with water.

Supplement

More about ions and molecules

Ionic lattices

In ionic structures, the three-dimensional network of ions is called an **ionic lattice**. An ionic lattice is defined as a regular arrangement of alternating positive and negative ions. The electrostatic attractive forces between the ions act in all directions and they are very strong. It needs a high temperature to overcome these forces. That is why **giant ionic structures** have high melting points and boiling points.

Unit 4: Structure and bonding

Intermolecular forces

The covalent bonds within molecules are very strong. However, the forces between the separate molecules are weak. These weak, attractive forces are called **intermolecular forces**.

Because the forces between molecules are weak, it only needs a little energy input to overcome these forces and get the molecules to move away from each other. This is why molecular substances have low melting points and boiling points.

Explaining some physical properties of ions and molecules

Melting points and boiling points

- Ionic compounds have high melting and boiling points because of the strong attractive forces between the ions in the giant ionic **lattice**. A high temperature is needed to weaken the strong forces of attraction enough melt the compound.
- Simple molecular compounds have low melting and boiling points because the intermolecular attractive forces between molecules are weak. A much lower temperature is needed to weaken intermolecular forces.

Electrical conductivity

- Solid ionic compounds do not conduct electricity because the ions are packed together and are not free to move around. They can only vibrate in their fixed positions in the lattice.
- Ionic compounds conduct electricity when they are molten or in aqueous solution because the ions are free to move.
- Simple molecular compounds are poor conductors of electricity. This is because they contain uncharged molecules and not charged ions.

> **EXAM TIP**
>
> Make sure you know that in a molten ionic compound or aqueous solution it is the ions that move.

> **KEY POINTS**
>
> - In giant ionic structures the attractive forces between the ions are strong. The attractive forces between molecules are weak.
> - Giant ionic structures have high melting and boiling points. Simple molecular structures have low melting and boiling points.
> - Giant ionic structures only conduct electricity when molten or in aqueous solution. Simple molecular structures are poor conductors.
> - An ionic lattice is a regular arrangement of alternating positive and negative ions.
> - Giant ionic structures only conduct electricity when molten or in aqueous solution because the ions are free to move.

SUMMARY QUESTIONS

1. Make a table of the differences between the structure, bonding and physical properties of ionic and simple molecular compounds.
2. Explain why molten sodium chloride conducts electricity but solid sodium chloride does not.
3. Copy and complete using these words:

   electrostatic   energy   forces   ionic   lattice   melting

   In giant _____ structures the _____ attractive _____ between the positive and negative ions in the ionic _____ are very strong. Ionic compounds have very high _____ points because it takes a lot of _____ to overcome these forces of attraction.

4.5 Giant covalent structures

LEARNING OUTCOMES

- Describe the giant covalent structures of graphite and diamond
- Relate the structures of graphite and diamond to their uses
- S Describe the giant covalent structure of silicon(IV) oxide
- Describe the similarity in properties between diamond and silicon(IV) oxide, related to their structures

Giant covalent structures

Many covalently bonded substances are small molecules. In some covalent structures, however, there is a network of covalent bonds throughout the whole structure. We call these structures **giant covalent structures**.

Giant covalent structures have a rigid three-dimensional network of many strong covalent bonds throughout the structure. It needs a high temperature and a lot of energy to break all these bonds. So unlike simple molecules, giant covalent structures have very high melting and boiling points. Giant covalent structures can be elements, such as carbon in the form of graphite and diamond, or compounds such as silicon (IV) oxide (silicon dioxide).

Diamond and graphite

Diamond and graphite are two different forms of carbon.

Figure 4.5.1 Diamonds owe their hardness to the large number of strong bonds in their giant covalent structure.

EXAM TIP

Giant covalent structures are sometimes described as macromolecules, but this term may give you a misleading idea about their properties. They are *not* simple molecules.

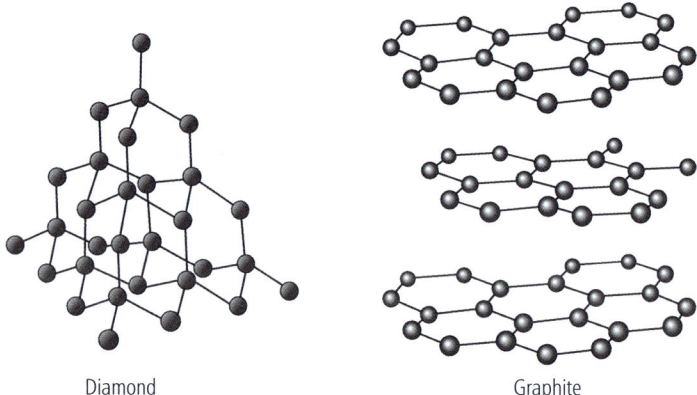

Figure 4.5.2 Diamond and graphite both have giant covalent structures.

In diamond, each carbon atom forms four covalent bonds with other carbon atoms. The carbon atoms link to form a giant covalent lattice. As well as having high melting and boiling points, diamond is very hard. This means that it can't be scratched easily. That is why diamond is used for cutting and drilling metals and glass.

Graphite is a black shiny solid. Its carbon atoms are arranged in layers. Each carbon atom is covalently bonded to three other carbon atoms. These carbon atoms are arranged in layers of hexagons – six-sided rings. The strong covalent bonding within the layers means that a high temperature and a lot of energy are needed to break the bonds. So the melting and boiling points of graphite are very high.

The bonding between the layers in graphite is weak. So the layers can slide over each other if a force is applied. This is why graphite

Unit 4: Structure and bonding

has a slippery feel and can be easily scratched. Because of this weak bonding, the layers of graphite can flake off easily. That is why graphite is used as a lubricant and in pencil 'leads'.

Graphite is used as an electrode in electrolysis. Each carbon atom in graphite is bonded to three other carbon atoms. But carbon has four electrons in its outer shell. So what has happened to the other electron from each carbon atom? The answer is that these electrons move around along the layers. They are called **delocalised electrons** because they do not belong to any particular carbon atom. Graphite conducts electricity because the delocalised electrons can move along the layers when a voltage is applied.

> **EXAM TIP**
>
> When explaining why graphite conducts electricity, make sure you state that electrons in the layers move along the layers. Just stating that 'the electrons move' suggests the electrons in the covalent bonds can move through the structure as well.

Supplement

Silicon(IV) oxide (silicon dioxide)

Silicon(IV) oxide, SiO_2, is found in sand and quartz. The silicon(IV) oxide found in quartz has a structure similar to diamond.

Each silicon atom is bonded to four oxygen atoms, but each oxygen atom is bonded to only two silicon atoms. This accounts for the formula SiO_2. Silicon(IV) oxide has a similar structure to diamond – a structure of linked tetrahedrons.

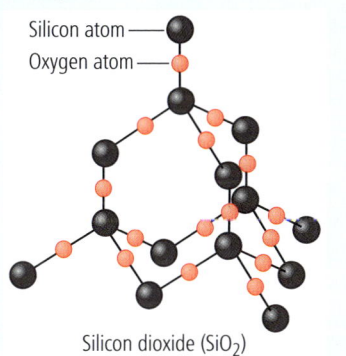

Figure 4.5.3 The structure of silicon(IV) oxide.

Silicon(IV) oxide and diamond also have similar properties:

- silicon(IV) oxide forms very hard colourless crystals and has a high melting point and a high boiling point. It needs a high temperature to break the large number of strong covalent bonds.
- silicon(IV) oxide does not conduct electricity because it has no delocalised electrons.

KEY POINTS

- Diamond and graphite have giant covalent structures.
- Diamond is used for cutting tools because of its hardness (related to the large number of strong covalent bonds in its giant covalent lattice).
- Graphite is used as an electrode because of the movement of the delocalised electrons along its layers, and as a lubricant because its layers slide over each other.
- Silicon(IV) oxide has a similar structure and properties to diamond.

SUMMARY QUESTIONS

1. Copy and complete using these words:

   **covalent   energy
   hexagons   melting
   strong   weak**

   Graphite has a giant _____ structure with the carbon atoms arranged in _____. The carbon atoms in the layers have _____ covalent bonding. The bonding between the layers is _____. Graphite has a high _____ point because it takes a lot of _____ to break the strong covalent bonds.

2. Refer to the structure of graphite to explain why it is used as a lubricant.

3. Explain why graphite conducts electricity but diamond does not.

4. Describe one similarity and one difference in the structures of diamond and silicon dioxide.

47

4.6 Metallic bonding
Supplement

LEARNING OUTCOMES

- Describe metallic bonding
- Explain in terms of structure and bonding why metals are good electrical conductors and are also malleable and ductile

Figure 4.6.2 Drawing copper out into wires depends on being able to make the layers of metal atoms slide easily over each other.

EXAM TIP

When you are drawing the structure of a metal, remember to label the particles in the lattice as positive ions. It is a common error to label them as atoms or protons.

What is metallic bonding?

Metals form a third type of giant structure. The metal atoms are packed closely together in a regular arrangement. The outer shell electrons of neighbouring atoms tend to repel each other. So the most stable state is for the outer shell to lose electrons and form a lattice of metal cations. The electrons released form a 'sea' of delocalised electrons which surrounds the lattice of positively charged ions. The positively charged metal ions are held together by their strong attraction to these delocalised electrons.

The strong electrostatic attraction between the giant lattice of positive ions and the sea of delocalised electrons is called **metallic bonding**.

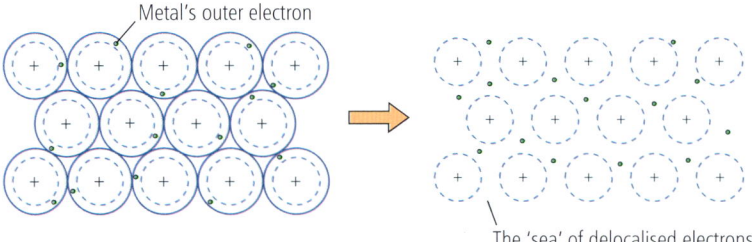

Figure 4.6.1 A metallic structure consists of positively charged metal ions surrounded by a 'sea' of delocalised electrons.

PRACTICAL

Modelling metallic structure

1. Fill a shallow dish with water and add some detergent.
2. Blow bubbles onto the surface of the water by moving the plunger of the syringe backwards and forwards until you have a single layer of lots of small bubbles.

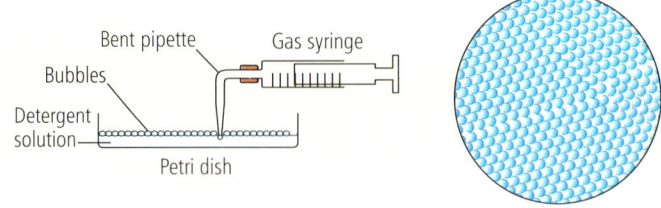

Figure 4.6.3 A bubble raft models the structure of metals.

Each bubble represents a metal atom. The bubbles are regularly arranged but in some places there are 'grain boundaries' where the directions of the layers change. The structure of metals is rather like this with irregularly shaped grains formed within the metal.

Unit 4: Structure and bonding

Explaining the properties of metals

Metals are good conductors of electricity:

When a voltage is applied to a metal, the delocalised electrons move between the layers of positive ions in the metal lattice towards the positive pole of the cell or power pack. They move towards the positive pole of the power supply because opposite charges attract.

Metals are malleable and ductile:

Substances which are **malleable** can be beaten into different shapes without breaking. Substances which are **ductile** can be drawn out into wires. We can explain the malleability and ductility of metals in the same way.

The positive ions in a metal are arranged regularly in layers. When a force is applied, the layers can slide over each other. In a metallic bond the attractive forces between the metal ions and the electrons act in any direction. So when the force is no longer applied, new bonds can easily form. This leaves the metal with a different shape.

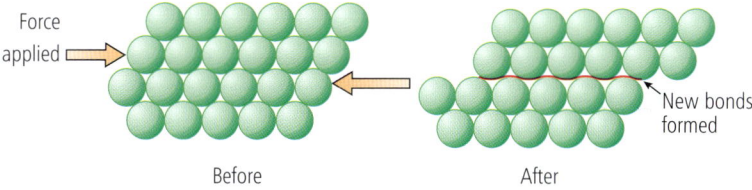

Figure 4.6.4 When a force is applied to a metal the layers slide. New bonds are formed when the force is no longer applied.

Most metals have high melting points and boiling points:

It needs a lot of energy to weaken the strong electrostatic forces of attraction between the metal ions and the delocalised electrons in the lattice. So the attractive forces can only be overcome when the temperature is high.

KEY POINTS

- Metallic bonding is the strong electrostatic attraction between the positive ions in a metal lattice and the delocalised **sea of electrons**.
- Metals conduct electricity because the delocalised electrons can move freely through the structure when a voltage is applied.
- Metals are malleable and ductile because the layers of ions can slide over each other.

EXAM TIP

It is a common error to think that conduction in metals is due to moving ions. Remember that it is only the delocalised electrons which move.

EXAM TIP

Remember that metals are malleable and ductile because the *layers* slide over each other. 'Atoms slide' is not a sufficient description.

SUMMARY QUESTIONS

1. Copy and complete using these words:

   **attractions   delocalised
   layers   malleable
   melting   positive**

   In metallic bonding the _____ metal ions are held together by a sea of _____ electrons. The strong electrostatic _____ between the ions and the electrons gives many metals their strength and accounts for their high _____ point. Metals are _____ because when a force is applied the _____ of ions can slide over each other.

2. Explain why metals conduct electricity.

3. Draw a diagram to show the structure of a metal. Label your diagram.

49

SUMMARY QUESTIONS

1. Match each type of structure on the left to the correct description on the right.

simple molecular	the solid conducts electricity
giant ionic	has a high melting point but doesn't conduct electricity
giant covalent	has a high melting point and conducts electricity when it dissolves in water
metallic	has a low melting point and does not conduct electricity

2. State whether the following pairs of elements form ionic or covalent compounds when they combine:

   a sodium and chlorine

   b hydrogen and oxygen

   c hydrogen and chlorine

   d lithium and chlorine

3. Copy and complete using words from the list:

   **eight   electrons   molecules   pair
   shares   shell   stable**

   A covalent bond is a _____ of shared electrons between two atoms. When chlorine atoms combine to form chlorine _____, each chlorine atom _____ a pair of _____ to form a covalent bond. Each chorine atom now has an outer electron _____ with _____ electrons. This makes the chlorine molecule _____.

4. The table shows some properties of different substances.

substance	melting point / °C	conductivity of solid	conductivity of liquid	solubility in water
A	−56	does not conduct	does not conduct	insoluble
B	610	does not conduct	conducts	soluble
C	−70	does not conduct	does not conduct	insoluble
D	2310	conducts	conducts	insoluble
E	680	does not conduct	conducts	soluble

   Classify each of these substances as metals, giant ionic structures or simple molecular structures.

Practice questions

1. Choose the correct statement about the electrons in sodium chloride.

   A Sodium shares two electrons with chlorine.

   B Sodium ions have eight electrons in their outer shells.

   C Chloride ions have seven electrons in their outer shells.

   D Sodium ions have one electron in their outer shells.

   (Paper 1)

2. Choose the correct statement about the outer shell electrons of the sulfur atom in hydrogen sulfide, H_2S.

   A Two lone pairs and one bonding pair.

   B One lone pair and two bonding pairs.

   C One lone pair and three bonding pairs.

   D Two lone pairs and two bonding pairs.

   (Paper 2)

3. Potassium chloride, KCl, has a giant ionic structure. Methane, CH_4, is a simple molecule.

   (a) Give two differences in the physical properties of potassium chloride and methane. [2]

   (b) Draw a dot-and-cross diagram to show the full electronic configuration of a potassium ion. [1]

   (c) State the meaning of the term ionic bond. [2]

5. Draw dot-and-cross diagrams for:

   a a sodium ion and a chloride ion

   b ammonia (NH_3)

   c hydrogen

6. Name the type of structures represented in each of these diagrams:

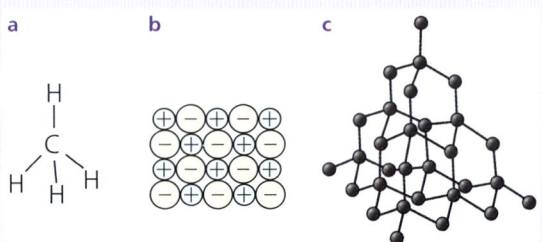

50

(d) Draw a dot-and-cross diagram of methane. [1]

(Paper 3)

4 The diagram shows the structures of diamond and graphite.

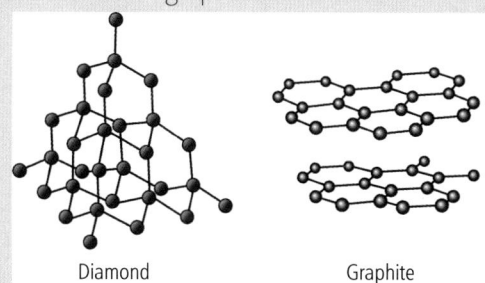

Diamond   Graphite

(a) Describe two ways in which these structures are similar. [2]
(b) Describe two differences between these structures. [2]
(c) State the name given to these types of structure. [1]
(d) Explain why graphite is used as a lubricant by referring to its structure. [2]
(e) Diamond is used in jewellery. State and explain one other use of diamond. [2]

(Paper 3)

5 Ethene, C_2H_4, is a covalent compound. Calcium oxide, CaO, is an ionic compound.
(a) Draw a dot-and-cross diagram for ethene. Show only the outer electron shells. [2]
(b) (i) Explain why ethene has a low melting point. [2]
   (ii) State one other property of ethene that demonstrates that it is a simple molecule. [1]
(c) Draw dot-and-cross diagrams for the ions present in calcium oxide. Show all the electron shells. [4]
(d) Calcium oxide has an ionic lattice structure. Explain the meaning of the term ionic lattice. [2]
(e) Explain why calcium oxide conducts electricity when molten but not when solid. [2]

(Paper 4)

6 The structures of carbon dioxide, diamond and silicon dioxide are shown below.

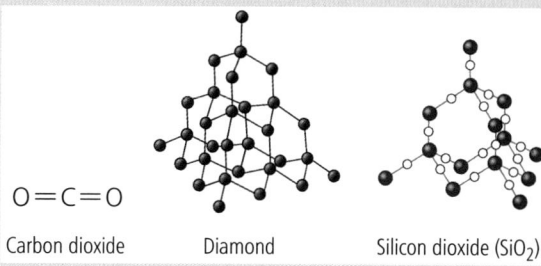

Carbon dioxide   Diamond   Silicon dioxide (SiO_2)

(a) (i) Give the physical state of carbon dioxide at room temperature. Explain your answer. [2]
   (ii) Give the physical state of diamond at room temperature. Explain your answer. [2]
(b) Draw a dot-and-cross diagram for carbon dioxide, showing only the outer shell electrons. [2]
(c) (i) Describe two ways in which the structure of diamond and silicon dioxide are similar. [2]
   (ii) Describe one difference in the structure of silicon dioxide and diamond. [1]
(d) State the formula for silicon dioxide. [1]
(e) Describe two physical properties of silicon dioxide that are similar to diamond. Suggest how each property is related to the structure of silicon dioxide. [4]

(Paper 4)

7 Zinc is a metal and sulfur is a non-metal.
(a) State three differences in the physical properties of zinc and sulfur. [3]
(b) Draw a labelled diagram to show the metallic bonding in zinc. [2]
(c) (i) Explain why zinc is malleable. [2]
   (ii) Explain why zinc conducts electricity. [2]
(d) A compound of sulfur and chlorine has the formula SCl_2.
   Draw a dot-and-cross diagram for a SCl_2 molecule. Show only the outer shell electrons. [3]

(Paper 4)

5.1 Chemical formulae

LEARNING OUTCOMES

- Deduce the formula of a simple compound from the relative number of atoms present

Figure 5.1.1 John Dalton produced a list of symbols for elements in 1808 but they are different from those we use today.

EXAM TIP

When writing symbols containing two letters, make sure that the second letter is clearly lower case. Cl is correct for chlorine. CL is wrong.

EXAM TIP

Take care when writing the second atom in a formula. Co2 is not acceptable for carbon dioxide and neither is H²o for water. The symbol for oxygen is always a capital O and numbers are always shown as subscripts (written below the line).

Chemical symbols

Every element has its own chemical symbol. For example, sulfur is S, oxygen is O and bromine is Br. Some symbols are not as obvious. For example, iron is Fe and potassium is K. If you do not know the symbol for an element, you can look it up in the Periodic Table. Notice that the first letter in a symbol is always a capital letter. If there is a second letter, it is always a small letter (lower case).

Formulae

We can work out the **formula** of a compound by knowing the combining powers of the elements it contains. You can often work out the combining power of an element from its group in the Periodic Table.

Group	I	II	III	IV	V	VI	VII	VIII
Combining power	1	2	3	4	3	2	1	0

The combining powers of the elements in Group I to III are the same as their group number. The combining powers of the non-metals on the right-hand side of the Periodic Table in Groups V to VIII are found by taking the group number away from 8. So nitrogen in Group V has a combining power of 3 (as 8 − 5 = 3).

Deducing the formula of an ionic compound

Magnesium is in Group II so it has a combining power of 2.

Chlorine is in Group VII so it has a combining power of 1.

You can use the following method to find the formula:

- write down the symbols in the same order as in the name of the compound
- write down the combining power underneath each element
- swap the numbers around
- cancel the numbers to get the smallest whole number ratio, if necessary.

	Example 1	Example 2	Example 3
What is the formula of:	magnesium chloride	aluminium oxide	calcium oxide
Combining power	Mg Cl 2 1	Al O 3 2	Ca O 2 2
Formula	$MgCl_2$	Al_2O_3	CaO

Note that:

- you cannot predict the combining power of the transition elements (see Topic 13.5)

Unit 5: Formulae and equations

- you cannot combine metals with other metals to form a compound.
- using combining powers for compounds made of two non-metals does not always work. For example some elements form more than one oxide e.g. SO_2 and SO_3.

Simple rules for naming compounds

Naming compounds of two elements

If a compound contains a metal and a non-metal, the metal is put first and the ending of the non-metal changes to 'ide'. For example, the compound of chlorine and magnesium is named magnesium chloride; the compound of calcium and oxygen is calcium oxide.

When a compound contains hydrogen, this generally comes first, for example hydrogen bromide. If a compound does not contain hydrogen the non-metal with the lower group number comes first, for example nitrogen dioxide. Nitrogen is in Group V, so it comes before oxygen which is in Group VI.

If both non-metals are in the same group then the one further down the group comes first, for example sulfur dioxide. Sulfur is lower than oxygen in Group VI, so it comes first.

Some compounds which have been known for a long time are called by their common names. For example, H_2O is water and NH_3 is ammonia.

Naming compounds with three elements

You should be able to recognise the following groups which contain oxygen as well as another element.

OH	hydroxide	Example: sodium hydroxide, NaOH
NO_3	nitrate	Example: magnesium nitrate, $Mg(NO_3)_2$
SO_4	sulfate	Example: calcium sulfate, $CaSO_4$
CO_3	carbonate	Example: sodium carbonate, Na_2CO_3

Using brackets

When writing formulae for compounds containing groups such as OH and NO_3 we must keep these groups together. In calcium nitrate, $Ca(NO_3)_2$, there are two nitrate ions. The subscript 2 to the right of the brackets shows this.

Note that in one formula unit of $Ca(NO_3)_2$ there are:

$2 \times 1 = 2$ N atoms and

$2 \times 3 = 6$ O atoms

> **KEY POINTS**
> - Each chemical element has a symbol.
> - When naming a compound containing two elements the second name often changes its ending to –ide.
> - The formula of a simple compound can be worked out from the combining powers of the elements present.

> **SUMMARY QUESTIONS**
> 1 Write the name of the compounds formed by combining:
>   a bromine and sodium
>   b magnesium and oxygen
>   c hydrogen and iodine
> 2 Work out the formulae of:
>   a magnesium bromide
>   b potassium chloride
>   c barium oxide
>   d magnesium nitride
>   e gallium oxide
>   f sodium sulfide
> 3 Name the following compounds:
>   a $MgBr_2$ b AlI_3
>   c K_2SO_4 d $Sr(NO_3)_2$

5.2 Working out the formula

LEARNING OUTCOMES

- Describe molecular formula as the number and type of different atoms in one molecule
- Deduce the formula of a simple compound from the relative number of atoms present in a model or diagram
- Draw and interpret the displayed formula of a molecule
- **S** Describe empirical formula as the simplest whole number ratio of the different atoms or ions in a compound
- Determine the formula of an ionic compound from the number and charges of the ions present

EXAM TIP

When deducing the molecular formula from models, make sure that you write down each atom only once, e.g. C_3H_6 is correct but $C_2H_4CH_2$ is incorrect.

Working out formulae from diagrams

If we are shown the structure of a molecule showing all atoms and bonds, we can work out its molecular formula by counting the number and type of each atom.

Molecular formula is the formula showing the number and type of different atoms in one molecule.

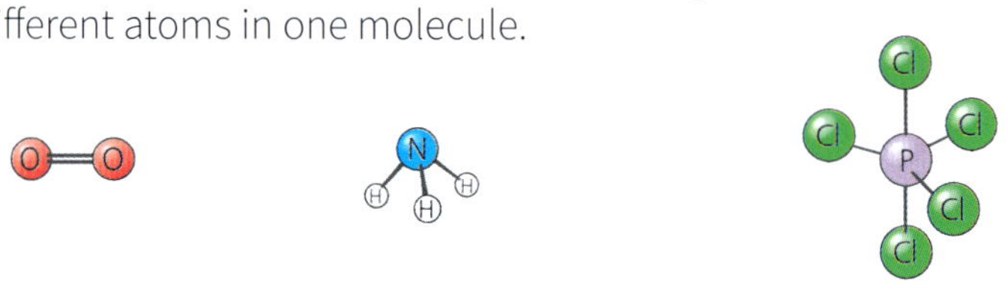

Oxygen (2 O atoms) formula O_2

Ammonia (1 N atom and 3 H atoms) formula NH_3

Phosphorus(V) chloride (1 P atom and 5 Cl atoms) formula PCl_5

Figure 5.2.1 The structures of oxygen, ammonia and phosphorus(V) chloride.

Displayed formula

The **displayed formula** shows all of the atoms and all of the bonds. Here are some examples:

The displayed formula is:

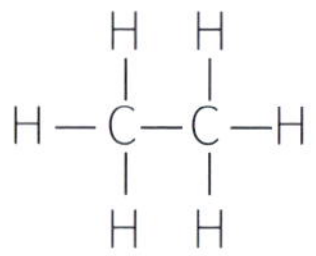

The molecular formula is C_2H_6.

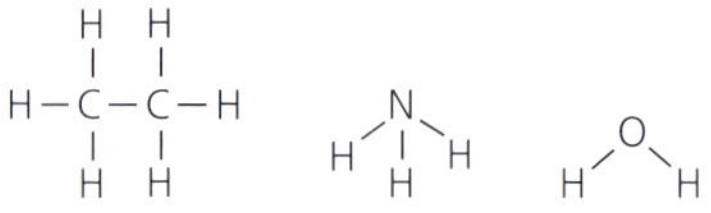

ethane   ammonia   water

Figure 5.2.2 Examples of displayed formulae.

Supplement

Empirical formula

Empirical formula is the formula showing the simplest whole number ratio of the different atoms or ions in a compound.

Look at the structure of ethane, C_2H_6, above.

The molecular formula is C_2H_6 because it has 2 C atoms and 6 H atoms.

The empirical formula (simplest whole number ratio) is CH_3.

The formula of an ionic compound is always an empirical formula.

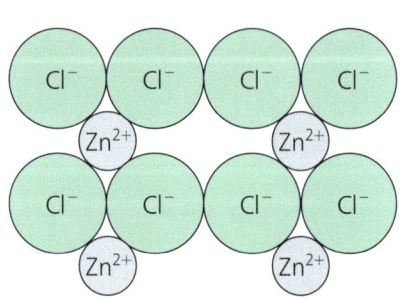

Figure 5.2.3 The empirical formula of zinc chloride is $ZnCl_2$.

Unit 5: Formulae and equations

You can work out the formula of an ionic compound from a diagram, by counting the number of positive ions and negative ions and then finding the simplest ratio.

The diagram of zinc chloride shows 4 zinc ions and 8 chloride ions. If we divide each by four, we get: $Zn = 1$ and $Cl = 2$. So the formula is $ZnCl_2$.

Deducing formulae from charges on the ions

In an ionic compound the number of positive charges is balanced out by the number of negative charges. The total charge is zero. We can work out the formula for a compound if we know the charges on the ions. Figure 5.2.3 shows how the charges on some common ions are related to the position of the element in the Periodic Table.

Group I	II		III	IV	V	VI	VII	VIII none
		H^+						
Li^+	Be^{2+}					O^{2-}	F^-	none
Na^+	Mg^{2+}		Al^{3+}			S^{2-}	Cl^-	none
K^+	Ca^{2+}	Transition metals						none

Figure 5.2.3 Ions formed by some elements in the Periodic Table.

You work out the formula for an ionic compound knowing the charge on each ion in a similar way to that used in Topic 5.1 e.g. $Al^{3+} = +3$. So you need 3 Cl^- ions to balance so that there is no overall charge. So the formula for aluminium chloride = $AlCl_3$.

Some ions contain more than one type of atom. These are called compound ions.

NH_4^+	OH^-	NO_3^-
ammonium ion	hydroxide ion	nitrate ion
SO_4^{2-}	CO_3^{2-}	HCO_3^-
sulfate ion	carbonate ion	hydrogencarbonate ion

The formulae of compounds formed from these ions are found in the same way as above, by balancing out the charges on the ions (see Figure 5.2.4).

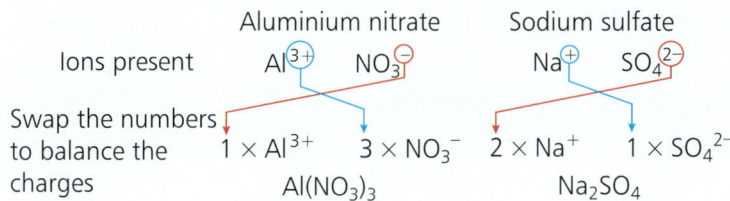

Figure 5.2.4 Formula from charges on the ions.

SUMMARY QUESTIONS

1. Write the molecular formula for each of the following compounds:
   a. N≡N
   b. O=C=O
   c. displayed formula of ethene (H₂C=CH₂)

2. Name these compounds:
   a. K_2SO_4
   b. $Mg(NO_3)_2$
   c. $Cu(OH)_2$

3. Write the formula of each of these compounds by balancing the charges on the ions:
   a. calcium carbonate
   b. magnesium nitrate
   c. ammonium chloride
   d. aluminium sulfate

KEY POINTS

- The molecular formula is the number and type of different atoms in one molecule.
- The displayed formula shows all atoms and all bonds in a molecule.
- The formula of a simple compound can be deduced from the relative number of atoms or ions present in a model.
- Empirical formula is the simplest whole number ratio of the different atoms or ions in a compound.
- The formula of an ionic compound can be worked out using the charges on the ions.

5.3 Chemical equations

LEARNING OUTCOMES
- Construct word equations and balanced symbol equations

EXAM TIP

Never write '+ heat' on the left or right of an equation. Heat is not a substance.

EXAM TIP

When balancing an equation you must never change the formula of any of the reactants or products.

EXAM TIP

You should know that the following molecules of gas are diatomic: hydrogen, H_2; nitrogen, N_2; oxygen, O_2; fluorine, F_2; chlorine, Cl_2; bromine, Br_2; and iodine, I_2.

EXAM TIP

When balancing symbol equations you must not change any of the formulae. Always balance by putting large numbers in front of the formulae. For example, balancing CaO by making it into CaO_2 is wrong. It should be 2CaO.

Word equations

We show chemical reactions by **chemical equations**. The simplest type of equation is a **word equation**. This shows the **reactants** on the left and the **products** on the right. The reactants are the substances that we add at the start and are changed during the reaction. The products are the substances formed by the reaction. The arrow shows that the reaction goes from left to right – from reactants to products. For example:

$$\text{magnesium} + \text{oxygen} \longrightarrow \text{magnesium oxide}$$
$$\text{reactants} \qquad\qquad \text{product}$$

Any conditions such as heating or adding catalysts are written over the arrow.

$$\text{nitrogen} + \text{hydrogen} \xrightarrow{\text{catalyst}} \text{ammonia}$$

Symbol equations

Atoms cannot be formed from nothing and they cannot be destroyed in a chemical reaction. So there must be the same number of each type of atom on both sides of the equation. We balance an equation by counting the number of each type of atom in the reactants and in the products. If the numbers are equal on both sides of the arrow, the equation is balanced.

This equation is balanced: $\quad Fe + S \longrightarrow FeS$

There is one atom of iron and one atom of sulfur on each side of the equation.

This equation is *not* balanced: $\quad Mg + O_2 \longrightarrow MgO$

There are two atoms of oxygen on the left but only one on the right. In order to balance an equation you need to follow these steps.

Step 1: Write down the formulae for the reactants and products. For example:

$H_2 + O_2 \longrightarrow H_2O$

Some gaseous elements exist as **diatomic** molecules. They have two atoms per molecule.

Step 2: Count up the atoms on each side of the arrow. You can use coloured dots beneath to help you.

$$H_2 + O_2 \longrightarrow H_2O$$

There are two oxygen atoms on the left of the arrow but only one on the right.

Step 3: Balance the atoms on each side by putting a number in front of one of the reactants or products. The number in front multiplies all the way through the molecule. So $2H_2O$ has $(2 \times 2) = 4$ hydrogen atoms and $(2 \times 1) = 2$ oxygen atoms.

Unit 5: Formulae and equations

$$H_2 + O_2 \longrightarrow 2H_2O$$

Now the oxygen atoms are balanced on both sides.

But the hydrogen atoms have become unbalanced. There are 2 atoms of hydrogen in the reactants but 4 atoms of hydrogen in the products.

Step 4: Keep balancing the equation in this way until you get the same numbers of all the atoms on both sides. This is done here by putting 2 in front of H_2.

$$2H_2 + O_2 \longrightarrow 2H_2O$$

The equation is now balanced.

Supplement

More symbol equations

When brackets are used, the small number at the bottom right of the brackets multiplies the atoms inside the brackets. So $Mg(NO_3)_2$ has 1 Mg, (1 × 2) = 2; 2 × 1 = 2 N and (3 × 2) = 6 O. If we wrote $2Mg(NO_3)_2$ we would have twice as many of each of these atoms: 2 Mg, 4 N and 12 O.

Some examples are:

Example 1: aluminium + water ⟶ aluminium + hydrogen
chloride   hydroxide   chloride

$AlCl_3 + H_2O \longrightarrow Al(OH)_3 + HCl$

Balance the Cl: $AlCl_3 + H_2O \longrightarrow Al(OH)_3 + 3HCl$

Balance the OH: $AlCl_3 + 3H_2O \longrightarrow Al(OH)_3 + 3HCl$

Example 2: calcium   +   nitric ⟶   calcium + water
hydroxide   acid   nitrate

$Ca(OH)_2 + HNO_3 \longrightarrow Ca(NO_3)_2 + H_2O$

Balance the nitrate (NO_3): $Ca(OH)_2 + 2HNO_3 \longrightarrow Ca(NO_3)_2 + H_2O$

The H is not balanced: there are 4 H on the left but only 2 on the right.

Balance the H: $Ca(OH)_2 + 2HNO_3 \longrightarrow Ca(NO_3)_2 + 2H_2O$

KEY POINTS

- We write reactants on the left of an equation and product on the right.
- In a symbol equation there is the same number of each type of atom on each side of the equation.
- Equations are balanced by writing numbers in front of particular reactants or products.

SUMMARY QUESTIONS

1. Write word equations for the following reactions:

   a $2Na + 2H_2O \longrightarrow 2NaOH + H_2$

   b $Mg + ZnSO_4 \longrightarrow Zn + MgSO_4$

   c $CuO + H_2SO_4 \longrightarrow CuSO_4 + H_2O$

2. Write balanced symbol equations for the following reactions:

   a $Na + Cl_2 \longrightarrow NaCl$

   b $Cl_2 + H_2 \longrightarrow HCl$

   c $Na + O_2 \longrightarrow Na_2O_2$

3. Write balanced symbol equations for the following reactions:

   a chlorine + potassium bromide ⟶ bromine + potassium chloride

   b copper(II) chloride, $CuCl_2$ + sodium hydroxide ⟶ copper hydroxide, $Cu(OH)_2$ + sodium chloride

   c phosphorus + chlorine ⟶ phosphorus trichloride, PCl_3

5.4 More about equations

LEARNING OUTCOMES

- Construct symbol equations with state symbols
- S Construct balanced ionic equations
- Deduce the symbol equation for a reaction given relevant information

Using state symbols

We use **state symbols** in equations to show if a substance is a solid, liquid, gas or dissolved in water. The state symbols used are:

(s) solid   (l) liquid   (g) gas

(aq) aqueous solution (solute dissolved in water)

State symbols are written after the formula of each reactant and product:

$$Zn(s) + H_2SO_4(aq) \longrightarrow ZnSO_4(aq) + H_2(g)$$

When water or other liquids are reactants or products they can be liquid or gas, depending on the conditions:

at room temperature: $MgO(s) + 2HCl(aq) \longrightarrow MgCl_2(aq) + H_2O(l)$

at above 100 °C: $CO(g) + H_2O(g) \longrightarrow CO_2(g) + H_2(g)$

Supplement

Ionic equations

Many chemical reactions involve ionic compounds. When ionic compounds dissolve in water, the ions separate:

$$MgCl_2(s) + aq \longrightarrow Mg^{2+}(aq) + 2Cl^-(aq)$$

Notice how we write the separate ions. The $2Cl^-$ shows that there are twice as many chloride ions as magnesium ions in solution.

You can often tell if a substance forms ions. Examples are:

- halides, nitrates, hydroxides and sulfates of Group I metals
- acids such as hydrochloric acid, HCl, sulfuric acid, H_2SO_4, and nitric acid, HNO_3
- ammonium compounds – these contain the ammonium ion, NH_4^+.

In most ionic reactions, only some of the ions take part in the reaction. The ones that do not are called **spectator ions**.

An **ionic equation** is a symbol equation that shows only those ions and molecules that react. We can change an ordinary symbol equation into an ionic equation in the following way:

Step 1: Write down the balanced equation with the state symbols.

$$BaCl_2(aq) + Na_2SO_4(aq) \longrightarrow 2NaCl(aq) + BaSO_4(s)$$

Step 2: Write down all the ions present in the equation. Any reactant or product that is an insoluble solid, a liquid or a gas is not split into ions.

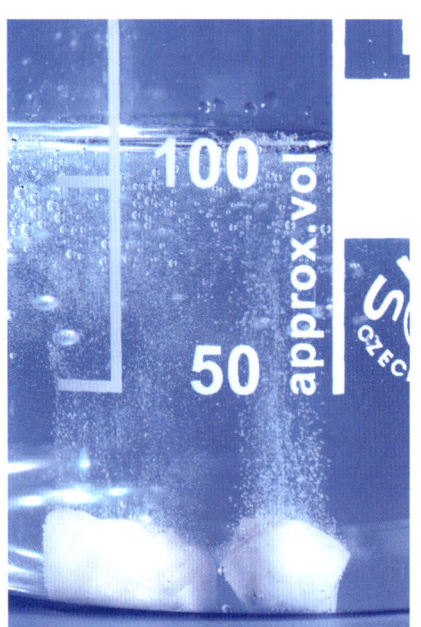

Figure 5.4.1 The products of a reaction may have a different state from the reactants.

EXAM TIP

Most questions about ionic equations will make the physical states of the reactants and products clear.

Unit 5: Formulae and equations

$Ba^{2+}(aq) + 2Cl^-(aq) + 2Na^+(aq) + SO_4^{2-}(aq) \longrightarrow BaSO_4(s) + 2Cl^-(aq) + 2Na^+(aq)$

Step 3: Cross out the ions that are the same on both sides of the equation. These are the spectator ions.

$Ba^{2+}(aq) + 2\cancel{Cl^-}(aq) + 2\cancel{Na^+}(aq) + SO_4^{2-}(aq) \longrightarrow BaSO_4(s) + 2\cancel{Cl^-}(aq) + 2\cancel{Na^+}(aq)$

Step 4: Only write down the reactants and products that are left to get the ionic equation.

$Ba^{2+}(aq) + SO_4^{2-}(aq) \longrightarrow BaSO_4(s)$

Another example

Aqueous chlorine reacts with aqueous sodium bromide to form aqueous sodium chloride and aqueous bromine. Write an ionic equation for this reaction.

Step 1: $Cl_2(aq) + 2NaBr(aq) \longrightarrow Br_2(aq) + 2NaCl(aq)$

Step 2: $Cl_2(aq) + 2Na^+(aq) + 2Br^-(aq) \longrightarrow Br_2(aq) + 2Na^+(aq) + 2Cl^-(aq)$

Step 3: $Cl_2(aq) + 2\cancel{Na^+}(aq) + 2Br^-(aq) \longrightarrow Br_2(aq) + 2\cancel{Na^+}(aq) + 2Cl^-(aq)$

Step 4: $Cl_2(aq) + 2Br^-(aq) \longrightarrow Br_2(aq) + 2Cl^-(aq)$

If two solutions are added together and a precipitate (solid) is formed you can simplify the method. All you have to do is to write down the ions that go to make up the precipitate as the reactants, and the formula of the precipitate as the product:

symbol equation: $FeCl_3(aq) + 3NaOH(aq) \longrightarrow Fe(OH)_3(s) + 3NaCl(aq)$

ionic equation: $Fe^{3+}(aq) + 3OH^-(aq) \longrightarrow Fe(OH)_3(s)$

In this reaction we know that the iron must be Fe^{3+}, because each of the three Cl^- ions has one negative charge to balance the three positive charges on the Fe^{3+} ion.

> **EXAM TIP**
>
> When writing ionic equations do not try to make molecules such as Cl_2, Br_2 and H_2O into ions.

SUMMARY QUESTIONS

1. **a** Copy and complete the equation for the reaction of zinc with hydrochloric acid to include state symbols. $ZnCl_2$ is soluble in water.

    $Zn + 2HCl \longrightarrow ZnCl_2 + H_2$

    b Give the state symbol which has not been used in this equation.

2. Write ionic equations for the following reactions:

    a $AgNO_3(aq) + KBr(aq) \longrightarrow KNO_3(aq) + AgBr(s)$

    b $CuCl_2(aq) + 2NaOH(aq) \longrightarrow Cu(OH)_2(s) + 2NaCl(aq)$

    c $Br_2(aq) + 2KI(aq) \longrightarrow I_2(aq) + 2KBr(aq)$

> **KEY POINTS**
>
> - We can add the state symbols (s), (l), (g) or (aq) after the formula of each reactant and product.
> - Ionic equations are simplified symbol equations showing only those ions that react and the product(s) of their reaction.
> - Ions that are present but do not take part in a reaction are called spectator ions.

SUMMARY QUESTIONS

1 Write word equations for the reactions of:
   a magnesium with chlorine
   b potassium with oxygen
   c sodium with bromine
   d carbon with oxygen

2 Write the formulae for:
   a sodium chloride
   b magnesium oxide
   c aluminium oxide
   d calcium chloride
   e aluminium chloride

3 Match the names with the symbols:

   oxide Cl_2
   hydrogen O_2
   calcium O^{2-}
   chlorine H^+
   chloride Ca^{2+}
   oxygen Cl^-

4 Balance these equations:
   a $Ca + O_2 \longrightarrow CaO$
   b $SO_2 + O_2 \longrightarrow SO3$
   c $Na + Cl_2 \longrightarrow NaCl$
   d $C_4H_8 + O_2 \longrightarrow CO_2 + H_2O$
   e $Fe_2O_3 + CO \longrightarrow Fe + CO_2$
   f $PbO + C \longrightarrow Pb + CO_2$

5 Write word equations for these reactions:
   a $Zn + CuSO_4 \longrightarrow ZnSO_4 + Cu$
   b $4Fe + 3O_2 \longrightarrow 2Fe_2O_3$
   c $NH_4Cl + NaOH \longrightarrow NH_3 + NaCl + H_2O$
   d $Cu(NO_3)_2 + Mg \longrightarrow Cu + Mg(NO_3)_2$

S 6 Change these equations into ionic equations:
   a $Cl_2(aq) + 2NaBr(aq) \longrightarrow Br_2(aq) + 2NaCl(aq)$
   b $FeCl_2(aq) + 2KOH(aq) \longrightarrow Fe(OH)_2(s) + 2KCl(aq)$
   c $Mg(s) + 2HCl(aq) \longrightarrow MgCl_2(aq) + H_2(g)$

Practice questions

1 Gallium is in Group III of the Periodic Table. Oxygen is in Group VI. The formula of gallium oxide is:
   A Ga_2O_3 C Ga_3O_2
   B Ga_3O_6 D GaO_2
   (Paper 1)

2 Part of the structure of magnesium chloride is shown.

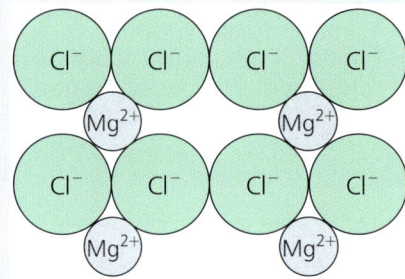

The correct empirical formula for magnesium chloride is:
   A Mg_2Cl C Mg_8Cl_4
   B Mg_4Cl_8 D $MgCl_2$
   (Paper 2)

3 The formulae for some halogens and halogen compounds are shown below.

   NaCl   KBr Br_2 Cl_2 I_2 LiBr   KI

   (a) State which of these are compounds. [1]
   (b) Write the names of (i) NH_4Br (ii) Na_2SO_4 [2]
   (c) Copy and complete the equation for this reaction:
       $Cl_2 + $ ___ $KBr \longrightarrow 2KCl + $ ___ [2]
   (d) Name these ions:
       (i) I^- (ii) O^{2-} (iii) NO_3^- [3]
   (Paper 3)

4 Magnesium reacts with hydrochloric acid to form magnesium chloride, and also reacts with sulfuric acid to produce magnesium sulfate.
   (a) Give the formula for:
       (i)   hydrochloric acid
       (ii) sulfuric acid
       (iii) magnesium sulfate. [3]
   (b) Magnesium also reacts with very dilute nitric acid.

(i) Copy and complete the equation for this reaction:

$Mg(s) + \ldots HNO_3(aq) \longrightarrow Mg(NO_3)_2(aq) + H_2(g)$ [1]

(ii) Write a word equation for this reaction. [2]

(iii) State the meaning of the symbols (s), (aq) and (g). [2]

(c) Magnesium reacts with oxygen to form magnesium oxide, MgO. Write a symbol equation for this reaction. [2]

(Paper 3)

5 Calcium carbonate reacts with hydrochloric acid to form calcium chloride, carbon dioxide and water.

(a) Copy and complete the equation for this reaction:

$CaCO_3 + \_\_\_ HCl \longrightarrow CaCl_2 + \_\_\_ + \_\_\_$ [3]

(b) When heated, calcium carbonate decomposes:

$CaCO_3 \longrightarrow CaO + CO_2$

Write a word equation for this reaction. [1]

(c) The 'model equation' below describes the combustion of methane.

[Diagram of methane combustion: CH₄ + 2 O=O → O=C=O + 2 H-O-H]

(i) Write a balanced symbol equation for this reaction using molecular formulae. [2]

(ii) State the meaning of the term molecular formula. [1]

(Paper 3)

6 This question is about some compounds containing nitrogen.

(a) Nitrogen reacts with hydrogen to make ammonia gas, NH_3.
Write a balanced equation for this reaction, including state symbols. [3]

(b) When ammonia is bubbled through a solution of hydrochloric acid, a solution of ammonium chloride is formed.
Write a balanced equation for this reaction including state symbols. [2]

(c) Aqueous ammonia contains ammonium ions, NH_4^+, and hydroxide ions, OH^-. When a few drops of aqueous ammonia reacts with aqueous copper(II) chloride, $CuCl_2$, the products are a precipitate of copper(II) hydroxide, $Cu(OH)_2$, and aqueous ammonium chloride.
Construct the ionic equation for this reaction. [2]

(d) Nitric acid reacts with calcium hydroxide to form calcium nitrate and water.
Write a balanced equation for this reaction. [2]

(Paper 4)

7 This question is about sodium and some compounds of sodium.

(a) When sodium metal reacts with water the products are hydrogen gas and a solution of sodium hydroxide.
Write a balanced equation for this reaction, including state symbols. [3]

(b) An aqueous solution of sodium chloride reacts with an aqueous solution of silver nitrate, $AgNO_3$. The products are a precipitate of silver chloride, AgCl, and a solution of sodium nitrate.

(i) Write a full symbol equation for this reaction, including state symbols. [2]

(ii) Convert the full symbol equation to an ionic equation. [1]

(Paper 4)

8 When iron(III) oxide, Fe_2O_3, and aluminium are heated together at a high temperature, iron and aluminium oxide are formed.

(a) Suggest the meaning of the symbol (III) in iron(III) oxide. [1]

(b) Write a balanced symbol equation for this reaction. [2]

(c) Iron reacts with oxalic acid. The structure of oxalic acid is shown.

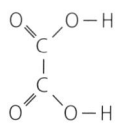

(i) Give the name of the type of formula shown by this structure. [1]

(ii) Deduce the empirical formula of oxalic acid. [1]

(iii) Define empirical formula. [1]

(Paper 4)

6.1 Masses and molecules

LEARNING OUTCOMES

- Define relative atomic mass and relative molecular mass
- Calculate relative molecular mass and relative formula mass

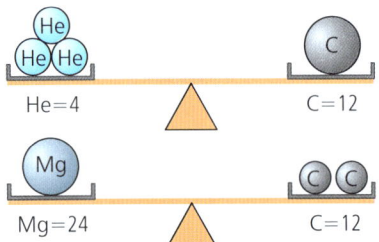

Figure 6.1.1 Three helium atoms have about the same mass as one carbon atom. One magnesium atom has the same mass as two carbon atoms.

EXAM TIP

When writing definitions remember to focus on the key words, e.g. average mass, isotopes, carbon-12.

EXAM TIP

In your exam always use the definitions used in the syllabus. You may see slightly different definitions of relative atomic mass in different books.

Relative atomic mass

The atoms of different elements have different masses. So if we want to know what quantities of reactants to use to make a certain amount of product, we have to know how heavy one atom is compared with another. We call this their relative mass.

The mass of a single atom is so small that you cannot weigh it using even the most accurate balance. To overcome this problem, scientists weigh a huge number of atoms and then compare them with the mass of the same number of 'standard' atoms. Scientists have chosen to use an isotope of carbon called carbon-12 (^{12}C) as their standard. This is given a value of 12 units.

The mass of other atoms is then found by comparing their mass with the same number of carbon-12 atoms. The mass found is called the **relative atomic mass**. The symbol for relative atomic mass is A_r. Note that there are no units for relative atomic mass.

Relative atomic mass is defined as:

the average mass of the isotopes of an element compared to $1/12^{th}$ the mass of an atom of carbon-12.

The reason why we use the average mass of the atoms is to take account of naturally occurring isotopes.

We can also use the values of the relative atomic masses for the mass of ions.

We can use relative atomic masses to see how heavy one atom is compared with another. For example the A_r of sulfur is 32, the A_r of copper is 64 and the A_r of oxygen is 16. This means that one sulfur atom has the same mass as two oxygen atoms. One copper atom has the same mass as two sulfur atoms or four oxygen atoms.

Relative molecular mass

Relative molecular mass is the sum of the relative atomic masses of all the atoms shown in the formula of a molecule. To calculate relative molecular mass all we have to do is to add all the relative atomic masses together. The symbol for relative molecular mass is M_r. Here are two examples.

Example 1

Calculate the relative molecular mass of methane.

formula	CH_4
atoms present	1 carbon + 4 hydrogen
adding A_r values	$(1 \times A_r \text{ carbon}) + (4 \times A_r \text{ hydrogen})$
M_r	$(1 \times 12) + (4 \times 1) = 16$

Unit 6: Chemical calculations

Example 2

Calculate the relative molecular mass of phosphorus(V) oxide, P_2O_5.

formula	P_2O_5
atoms present	2 phosphorus + 5 oxygen
adding A_r values	$(2 \times A_r$ phosphorus$) + (5 \times A_r$ oxygen$)$
M_r	$(2 \times 31) + (5 \times 16) = 142$

Relative formula mass

Relative formula mass is the sum of the relative masses of all the ions shown in the formula of an ionic compound.

In order to find the relative formula mass we follow exactly the same method used for relative molecular mass. We can apply the term relative formula mass to all compounds, molecular or ionic. Example 3 shows how to calculate the relative formula mass of calcium hydroxide, $Ca(OH)_2$. In this case we also have to consider the use of the brackets: $(OH)_2$ means that there are 2 O's and 2 H's.

Example 3

Relative formula mass of calcium hydroxide

formula	$Ca(OH)_2$
atoms present	1 calcium + 2 oxygen + 2 hydrogen
adding A_r values	$(1 \times A_r$ calcium$) + (2 \times A_r$ oxygen$) + (2 \times A_r$ hydrogen$)$
relative formula mass	$(1 \times 40) + (2 \times 16) + (2 \times 1) = 74$

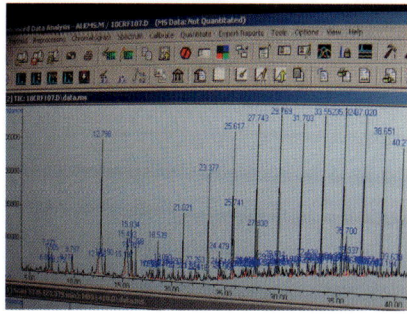

Figure 6.1.2 Mass spectrometry is one way of finding the relative molecular mass of a molecule.

> **EXAM TIP**
>
> If a formula has brackets, first work out the atomic masses inside the brackets then multiply by the number outside the brackets. Finally, add the atomic masses which were not bracketed.

> **KEY POINTS**
>
> - Relative atomic mass is the average mass of the isotopes of an element compared to 1/12th the mass of an atom of carbon-12.
> - The relative molecular mass is the sum of the relative atomic masses in the formula of a molecule.
> - Relative formula mass is the sum of the relative 'atomic' masses of all the ions shown in the formula of an ionic compound.

SUMMARY QUESTIONS

1. a Deduce the number of helium atoms that have the same mass as one atom of sulfur.

   b Deduce the number of lithium atoms that have the same mass as one atom of iron.

2. Calculate the relative formula mass of:

   a potassium chloride, KCl   b chlorine(VII) oxide, Cl_2O_7

   c calcium nitrate, $Ca(NO_3)_2$

3. Calculate the relative formula mass of $CuSO_4 \cdot 5H_2O$ (Hint: find the formula mass of $CuSO_4$ and $5H_2O$ separately then add them together).

6.2 Chemical calculations and the mole

LEARNING OUTCOMES

- Be able to calculate **reacting masses** using simple proportions
- **S** Define the mole and the Avogadro constant
- Use the mole in simple calculations to find relative molecular mass and relative atomic mass
- Use the mole and the Avogadro constant to calculate the number of particles present

Chemical calculations using simple proportion

We can work out how much product we get from a given amount of reactant by simple proportion. We can do this very easily if we have information about the mass of product made by a given amount of reactant. For example:

A student obtains 48 g of magnesium sulfate from 9.6 g of magnesium. Calculate the maximum mass of magnesium sulfate that can be obtained from 1.2 g of magnesium.

9.6 g magnesium gives 48 g magnesium sulfate

so 1.2 g magnesium gives $\frac{1.2}{9.6} \times 48 = 6$ g magnesium sulfate

Supplement

The Avogadro constant and the mole

If we have 6.02×10^{23} atoms, ions or molecules of a substance, we have an amount that can be easily weighed. This number of atoms, ions or molecules is called the **Avogadro constant**.

The amount of substance containing the Avogadro number of particles is called a mole of that substance.

6.02×10^{23} atoms of hydrogen, H, have a mass of 1 gram. Its $A_r = 1$.

6.02×10^{23} molecules of oxygen, O_2, have a mass of 32 grams ($2 \times A_r$). Its $M_r = 32$.

- The amount of substance that contains 6.02×10^{23} atoms, ions or molecules is called a **mole** (shown as mol). The mole has no units.
- the Avogadro constant is the number of atoms, ions or molecules in one mole of substance. Its units are /mol (per mole).

Using the mole

You can find the amount of substance in moles using this formula:

$$\text{amount of substance (mol)} = \frac{\text{mass (g)}}{\text{molar mass (g/mol)}}$$

The amount of substance can refer to moles of atoms, moles of molecules or moles of ions.

Simple mole calculations

Example 1

Calculate the number of moles of water in 4.5 grams of water. (M_r of water = 18)

1 mol water = 18 g

So 4.5 g of water is $\frac{4.5}{18} = 0.25$ mol

EXAM TIP

It is important to make clear what types of particles you are referring to. If you just state 'moles of oxygen' it is not clear whether it is moles of oxygen atoms or moles of oxygen molecules. A mole of oxygen molecules (O_2) contains 6.02×10^{23} molecules but it contains twice as many atoms: 1.204×10^{24} atoms.

Unit 6: Chemical calculations

Example 2

Calculate the mass of sodium hydroxide in 0.5 moles of sodium hydroxide.
(M_r of NaOH = 40)

mass = number of moles × mass of 1 mole (formula mass)
= 0.5 × 40 = 20 g NaOH

Example 3

0.6 moles of ammonia has a mass of 10.2 g. Calculate the relative molecular mass, M_r, of ammonia.

$$M_r = \frac{mass}{moles} = \frac{10.2}{0.6} = 17$$

Relative atomic mass can be calculated in a similar way when given suitable data.

Moles and numbers of particles

We can calculate the number of specified particles in a given mass of substance when given the value of the Avogadro constant (6.02×10^{23}/mol).

Example

Calculate the number of molecules of methane in 4 g of methane, CH_4.

(M_r of CH_4 = 16)

Step 1: Calculate the number of moles:
$$\frac{4}{16} = 0.25 \text{ mol}$$

Step 2: Multiply the number of moles by the Avogadro constant:
$0.25 \times (6.02 \times 10^{23}) = 1.505 \times 10^{23}$

A practical method for deducing the relative molecular mass of a gas is given in Topic 6.5.

SUMMARY QUESTIONS

1 In the reaction $Zn + 2AgNO_3 \rightarrow Zn(NO_3)_2 + 2Ag$
 6.750 g of silver are formed from 10.625 g of silver nitrate. Calculate the mass of silver nitrate needed to get 54.00 g of silver.

2 Copy and complete using the words below:

 amount   mole   molecules

 A mole is the _____ of substance that contains 6.02×10^{23} atoms, _____ or ions. The Avogadro constant is the number of atoms, molecules or ions in one _____ of a substance.

3 Calculate the number of moles of:
 a sodium hydroxide in 8 g of sodium hydroxide, NaOH
 b water in 5.4 g of water
 c aluminium oxide in 12.75 g of aluminium oxide, Al_2O_3.
 (A_r values: Na = 23, O = 16, H = 1, Al = 27)

KEY POINTS

- Reacting masses can be calculated using simple proportions when given suitable information.
- One mole is the amount of substance that contains 6.02×10^{23} atoms, molecules or ions.
- The Avogadro constant is the number of atoms, molecules or ions in one mole of substance.
- Number of moles
$$= \frac{\text{mass of substance}}{\text{mass of 1 mole of substance}}$$

65

6.3 More chemical calculations

Supplement

LEARNING OUTCOMES
- Calculate stoichiometric reacting masses
- Calculate percentage composition by mass

Figure 6.3.1 Bleach is used in some swimming pools to kill harmful bacteria. Getting the quantities right involves careful calculation! Adding too much bleach can cause burns to the skin and eyes.

EXAM TIP

When doing calculations, put the relative formula masses or moles below the appropriate reactants or products in the symbol equation, so that you can see which reactants or products are relevant. Be sure to take the stoichiometry of the equation into account.

Reacting masses

We often want to find out what mass of one reactant we need to add to another so that they react exactly, and there is no waste. To do this we need to know the ratio in which the reactants combine. The ratio of each reactant and product in a balanced equation is called the **stoichiometry** of the equation.

In the equation:

$2Mg + O_2 \rightarrow 2MgO$

the stoichiometry of the reaction is two moles (of magnesium atoms) to one mole (of oxygen molecules) to form two moles (of magnesium oxide).

We can use relative formula masses to work out the minimum mass of oxygen needed to react completely with a given mass of magnesium. Remember that one mole is the relative formula mass in grams.

Look at this question:

Calculate the mass of oxygen needed to react exactly with 12 g of magnesium.

A_r values: Mg = 24, O = 16

There are two ways of doing this.

Method 1: using relative formula masses

Step 1: Write the balanced equation:

$2Mg + O_2 \rightarrow 2MgO$

Step 2: Calculate the amounts using A_r values, taking into account the stoichiometry of the reaction:

$2Mg + O_2 \rightarrow 2MgO$
$2 \times 24 \quad\quad 32$
$48\,g \quad\quad\quad 32\,g$

Step 3: Use simple proportion:

12 g Mg reacts with $\frac{12}{48} \times 32 = 8\,g\,O_2$

Method 2: using moles

Step 1: Write the balanced equation

$2Mg + O_2 \rightarrow 2MgO$

Step 2: Calculate the moles of Mg: $\frac{12}{24} = 0.50$ mol

66

Unit 6: Chemical calculations

Step 3: Calculate the moles of O_2 taking the stoichiometry of the balanced equation into account:

moles O_2 is ½ moles Mg = 0.25 mol

Step 4: Calculate the mass of O_2 (mass = moles × formula mass) = 0.25 × 32 = 8 g

Percentage composition by mass

We can use the formula of a compound, A_r and M_r to calculate the **percentage composition by mass** of a particular element present in a compound. It is useful to know this, as it helps us, for example, to compare the amount of nitrogen in different fertilisers or to work out how much metal we can obtain from a metal ore.

$$\% \text{ composition by mass} = \frac{\text{mass of a particular element in one mole of a compound}}{\text{mass of one mole of that compound}} \times 100$$

Since the relative formula mass is equal to the **molar mass** in grams, we can rewrite this as:

% composition by mass

$$= \frac{\text{relative atomic mass of an atom of a particular element} \times \text{number of these atoms in one formula unit}}{\text{relative formula mass of compound}} \times 100$$

Example

Calculate the percentage by mass of iron in 1 mole of iron oxide, Fe_2O_3.

A_r values: Fe = 56, O = 16

Step 1: Calculate mass of iron in 1 mol Fe_2O_3.

There are two moles of iron in every mole of iron oxide. So mass of iron = 2 × 56 g = 112 g.

Step 2: Calculate M_r of Fe_2O_3.

It is (2 × 56 g) + (3 × 16 g) = 160 g.

Step 3: Apply the relationship for % by mass: $\frac{112 \text{ g}}{160 \text{ g}} \times 100 = 70\%$

> **EXAM TIP**
>
> In % by mass calculations make sure that you check the correct number of atoms of the type asked for.

> **KEY POINTS**
>
> - Stoichiometry is the ratio of each reactant and product in a balanced equation.
> - When calculating reacting masses, we need to use the relative formula masses of each reactant and their stoichiometric ratios.
> - Percentage composition by mass is the mass of a particular element in one mole of a compound × 100/mass of 1 mole of the compound.

SUMMARY QUESTIONS

1. Aluminium reacts with iron(III) oxide.

    $2Al + Fe_2O_3 \rightarrow Al_2O_3 + 2Fe$

    Calculate the mass of iron(III) oxide needed to react exactly with 18.9 g Al.

2. Ethane burns in oxygen to produce carbon dioxide and water.

    $2C_2H_6 + 7O_2 \rightarrow 4CO_2 + 6H_2O$

    Calculate the mass of ethene needed to react exactly with 15.68 g O_2.

3. Calculate the percentage by mass of nitrogen in NH_4NO_3.

6.4 Amount of product
Supplement

LEARNING OUTCOMES
- Calculate stoichiometric reacting masses
- Apply the terms limiting reactant and excess reactant

How much product?

We often need to know how much product we can get from a given amount of reactant. There are two ways of doing this: we can use relative formula masses and simple proportion or we can work out the number of moles of reactant and product. These two methods are shown for the following question:

Calculate the mass of water formed when 4 g of methane is completely burned in excess oxygen.

A_r values: C = 12, H = 1, O = 16

There are two ways of doing this.

Method 1: using relative formula masses

$$CH_4 + 2O_2 \rightarrow CO_2 + 2H_2O$$

M_r values: 12 + (4 × 1)   2 × [(2 × 1) + 16]

16 g methane produces 36 g water

4 g methane produces $\frac{4 \times 36}{16}$ = 9 g water

EXAM TIP

When answering questions about reacting masses and moles of product formed, work only with the reactants and/or products asked for in the question. Ignore all the other reactants and products.

Method 2: using moles

$$CH_4 + 2O_2 \rightarrow CO_2 + 2H_2O$$

moles $CH_4 = \frac{4}{16}$ = 0.25 mol

mole ratio of methane : water = 1 : 2

So moles of water formed = 2 × 0.25 = 0.5 mol

mass of water in g = moles × M_r of water

So mass of water = 0.5 × 18 = 9 g

The method used to calculate the amount of product formed is the same as used for reacting masses in Topic 6.3

Figure 6.4.1 For methane to burn completely the oxygen has to be in excess.

EXAM TIP

The limiting reactant is the reactant that is NOT in excess. It has the smaller number of moles after you have taken into account the ratio in which the reactants combine.

Limiting reactants

When we carry out a reaction we sometimes use an **excess** of one of the reactants. The reactant that is not in excess is called the **limiting reactant**. The reaction stops when the limiting reactant is used up. You can calculate which reactant is limiting by calculating the reactant that has the lower number of moles. You must, however, take into account the stoichiometry of the reaction. If you just work out the number of moles from the masses given, it does not take into account the fact that twice as much of one of the reactants might be used up compared with the other.

Figure 6.4.2 shows how the particles of calcium carbonate and hydrochloric acid decrease in number as the reaction proceeds.

The equation for the reaction is:

$$CaCO_3 + 2HCl \rightarrow CaCl_2 + CO_2 + H_2O$$

Unit 6: Chemical calculations

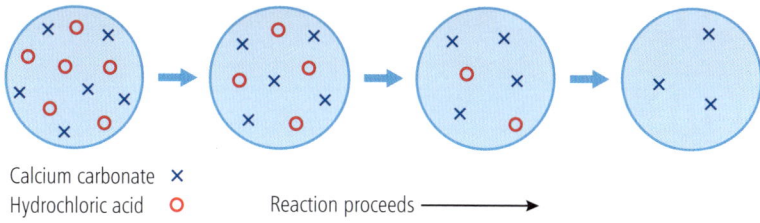

Calcium carbonate ×
Hydrochloric acid ○   Reaction proceeds ⟶

Figure 6.4.2 Model of reactant particles being used up. The hydrochloric acid is the limiting reactant because it gets used up before the calcium carbonate.

The equation shows that for each mole of calcium carbonate that reacts, two moles of hydrochloric acid are converted to products. So in this case, although we had the same number of particles of calcium carbonate and hydrochloric acid to start with, the hydrochloric acid gets used up first. So hydrochloric acid is the limiting reactant.

Calculations involving limiting reactants

The example shows a typical problem involving a limiting reactant.

1.2 g of magnesium reacts with a solution containing 2.74 g of hydrochloric acid. Show by calculation which is the limiting reactant.

A_r values: Mg = 24, Cl = 35.5, H = 1

Step 1: Calculate the number of moles of each reactant.

Mg = $\frac{1.2}{24}$ = 0.05 mol   HCl = $\frac{2.74}{(1 + 35.5)}$ = 0.075 mol

Step 2: Take into account the stoichiometry of the equation

$$Mg + 2HCl \rightarrow MgCl_2 + H_2$$

1 mol of magnesium reacts with 2 mol of HCl

Step 3: Calculate number of moles to react completely.

To react completely, 0.05 mol magnesium will need to react with 2 × 0.05 = 0.1 mol of HCl.

But we have only 0.075 mol of HCl, so HCl is the limiting reactant and magnesium is in excess.

From this type of calculation, you can also find out by how much one reactant is in excess. In the example above all the hydrochloric acid was used up.

Hydrochloric acid used up = 0.075 mol

From the equation: 2 mol of HCl reacts with 1 mol of Mg.

So 0.075 mol HCl will react with $\frac{0.075}{2}$ = 0.0375 mol Mg

Excess Mg = mol Mg at start – mol Mg reacted

= 0.05 – 0.0375 = 0.0125 mol Mg in excess

KEY POINTS

- The limiting reactant is the one that is not in excess.
- We can work out which reagent is limiting by comparing the number of moles of each reactant, taking into account the stoichiometry of the equation.
- We can use the chemical equation and relative formula masses to calculate the mass of product formed from a given amount of reactant.

SUMMARY QUESTIONS

1 Calculate the mass of copper(II) sulfate formed when 2.5 g of copper(II) oxide reacts with excess sulfuric acid. The equation for the reaction is:

$$CuO + H_2SO_4 \rightarrow CuSO_4 + H_2O$$

A_r values: Cu = 63.5, O = 16.0, S = 32.0

2 In the reaction:

$$Mg + 2CH_3CO_2H \rightarrow (CH_3CO_2)_2Mg + H_2$$

2.4 g of magnesium is reacted with 6 g of ethanoic acid, CH_3CO_2H.

Show, by calculation, the reagent that is in excess.

A_r values: Mg = 24, C = 12, H = 1, O = 16

6.5 Gas volume calculations

Supplement

LEARNING OUTCOMES

- Apply the molar gas volume in calculations involving gases
- Convert cm^3 to dm^3
- Calculate the relative atomic mass or relative molecular mass of a substance

Gas volume calculations

Look at the two flasks in Figure 6.5.1. Each has a volume of 1 dm^3. They are each filled with a different gas and the volume of the gas is measured accurately.

Each flask contains the same number of moles. This means that the same volume of gas has the same number of moles.

At room temperature and pressure the volume of one mole of any gas is 24 dm^3 or 24 000 cm^3. This is called the **molar gas volume**. Room temperature and pressure (shown as **r.t.p.**) is 20 °C and 1 atmosphere pressure.

We can use the fact that one mole of gas occupies 24 dm^3 to do chemical calculations for reactions where gases are produced.

volume of gas (in dm^3) = number of moles of gas × 24

Example 1

Calculate the volume of 0.2 moles of carbon dioxide at r.t.p.

1 mol occupies 24 dm^3. So 0.2 mol occupies 0.2 × 24 = 4.8 dm^3.

Example 2

Calculate the mass of carbon dioxide present in 60 cm^3 of carbon dioxide.

$M_r [CO_2] = 44$

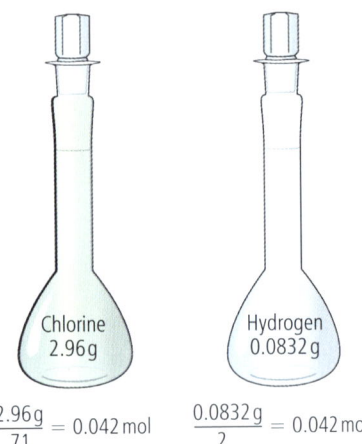

$\frac{2.96g}{71} = 0.042\,mol$ $\frac{0.0832g}{2} = 0.042\,mol$

Figure 6.5.1 There are the same number of moles of gas in each flask.

PRACTICAL

Deducing the volume of one mole of gas

1. Put 0.10 g of magnesium ($A_r = 24$) in the flask and add excess hydrochloric acid.
2. Record the maximum volume of gas produced.

The volume produced is 100.0 cm^3. The equation for the reaction is:

$Mg + 2HCl \rightarrow MgCl_2 + H_2$

The equation shows us that one mole of hydrogen is produced when one mole of magnesium reacts.

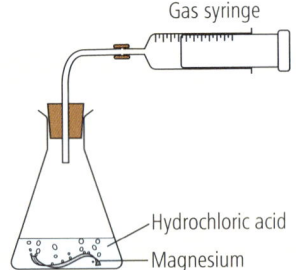

Figure 6.5.2 Apparatus for following the progress of a reaction where a gas is released.

Moles of magnesium used = $\frac{0.1}{24}$ = 0.004166 mol = moles of hydrogen

So the volume of *one mole* of hydrogen must be = 100.0 × $\frac{1}{0.004166}$ = 24 000 cm^3 (24 dm^3)

Unit 6: Chemical calculations

Step 1: convert the volume of gas given in cm^3 to dm^3

$60\ cm^3 = 0.06\ dm^3$ (as $1cm^3 = 0.001\ dm^3$)

Step 2: rearrange the expression between moles, volume of gas and $24\ dm^3$

$$\text{moles } CO_2 = \frac{\text{volume of gas (in } dm^3)}{24} = \frac{0.06}{24} = 2.5 \times 10^{-3}\text{ moles}$$

Step 3: use the expression

$$\text{mass (g)} = \text{moles} \times M_r$$
$$= 2.5 \times 10^{-3} \times 44 = 0.11\text{ g } CO_2$$

Example 3

Calculate the volume of carbon dioxide that is produced when 2.8 g of butene burns in excess air. The equation for the reaction is:

$$C_4H_8(g) + 6O_2(g) \rightarrow 4CO_2(g) + 4H_2O(l)$$

M_r [butene] = 56

Step 1: calculate the number of moles of butene = 2.8/56 = 0.05 moles

Step 2: use the stoichiometry of the reaction.

From the equation, 1 mole of butene produces 4 moles of carbon dioxide.

So 0.05 moles of butene will produce $0.05 \times 4 = 0.2$ moles of carbon dioxide.

Step 3: Calculate the volume of 0.2 moles of carbon dioxide, knowing that 1 mole of gas occupies $24\ dm^3$.

0.2 moles of carbon dioxide gas will occupy $0.2 \times 24\ dm^3 = 4.8\ dm^3$.

Deducing relative molecular mass or relative atomic mass using gas volumes

We can use the apparatus in Figure 5.6.3 to deduce the value of M_r or A_r of some compounds or elements.

Example

0.05 g of magnesium is added to excess hydrochloric acid. The volume of hydrogen produced is $50\ cm^3$. Deduce the relative atomic mass of magnesium. It is known that 1 mole of magnesium produces 1 mole of hydrogen gas.

Step 1: convert cm^3 hydrogen to dm^3 hydrogen: $50/1000 = 0.05\ dm^3$

Step 2: calculate moles of hydrogen using molar gas volume: $\frac{0.05}{24} = 2.083 \times 10^{-3}\text{ mol}$

Step 3: use simple proportion to calculate the mass in 1 mole of magnesium.

$$\frac{1}{2.083 \times 10^{-3}} \times 0.05\text{ g}$$

So the relative atomic mass of magnesium is 24.0 (to 3 significant figures).

> **EXAM TIP**
>
> If you are given gas volumes in cm^3, remember to convert the volume to dm^3 by dividing by 1000, e.g. $25\ cm^3 = 0.025\ dm^3$.

> **KEY POINTS**
>
> - One mole of any gas has a volume of $24\ dm^3$ at room temperature and pressure.
> - For reactions involving gases, the molar gas volume can be used to calculate reacting masses.
> - Relative atomic mass and relative molecular mass can sometimes be calculated using the molar gas volume.

> **SUMMARY QUESTIONS**
>
> 1 Copy and complete using the words below:
>
> | all | mole | pressure |
> | room | | volume |
>
> The _____ of one _____ of a gas is $24\ dm^3$ at _____ temperature and _____. This is the same for _____ gases.
>
> 2 a Calculate the volume occupied by 11 g of carbon dioxide gas at r.t.p.
>
> b Calculate the mass of nitrogen, N_2, in $1200\ cm^3$ of nitrogen gas at r.t.p.
>
> 3 When 0.167 g of calcium carbonate is added to excess hydrochloric acid, $40\ cm^3$ of carbon dioxide is produced. Calculate the M_r of calcium carbonate. One mole of calcium carbonate reacts to produce one mole of carbon dioxide.

6.6 Yield and purity

Supplement

LEARNING OUTCOMES
- Calculate percentage yield
- Calculate percentage purity

Percentage yield

When you carry out a chemical reaction in the laboratory not all the reactants are changed to the products you want. This is because there may be other reactions happening at the same time or the reaction doesn't go to completion. The word **yield** describes how much of a particular product you can get from the reactants when the reaction is complete. The yield can be given in moles or grams. It is more useful to talk about **percentage yield**:

$$\text{percentage yield} = \frac{\text{experimental yield}}{\text{theoretical yield}} \times 100$$

The **experimental yield** is the amount of a particular product we get in an experiment. Experimental yield is sometimes called the actual yield. We don't know what this will be until we have weighed the product.

The **theoretical yield** (or predicted yield) is found by using relative formula masses together with the equation for the reaction to calculate the maximum amount of product we can get from given amounts of reactants.

Example

A student reacts 9.0 g of aluminium powder with excess chlorine. The mass of aluminium chloride collected is 35.6 g. Calculate the percentage yield.

A_r values: Al = 27, O = 16

Step 1: Calculate the theoretical yield.

$$2Al + 3Cl_2 \rightarrow 2AlCl_3$$

1 mole of aluminium produces 1 mole of aluminium chloride.

Using formula masses: 27 g of Al produces $27 + (3 \times 35.5) = 133.5$ g of $AlCl_3$

Theoretical yield for 9.0 g of Al = $\frac{9}{27} \times 133.5 = 44.5$ g of $AlCl_3$

Step 2: Apply the expression for percentage yield

$$\% \text{ yield} = \frac{\text{experimental yield}}{\text{theoretical yield}} \times 100 = \frac{35.6}{44.5} \times 100 = 80\%$$

EXAM TIP

Always show your working in calculations. If you make an error at the start, such as using an incorrect value of M_r, you can still you can still gain marks for the rest of the calculation.

EXAM TIP

When applying the expression for percentage yield, make sure that you are using the same units (moles or grams).

PRACTICAL

How much copper(II) sulfate can we get from malachite?

Put a known amount of crushed malachite (copper carbonate ore) in a beaker. This is treated as shown.

$$\% \text{ yield} = \frac{\text{actual mass of copper(II) sulfate}}{\text{predicted mass of copper(II) sulfate}} \times 100$$

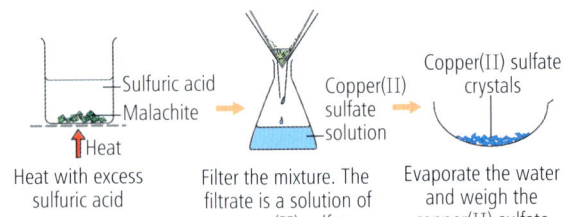

Figure 6.6.1 Extracting copper(II) sulfate from malachite.

Unit 6: Chemical calculations

Percentage purity

When we carry out a chemical reaction, the product may still contain very small amounts of impurities mixed with it. These impurities may be small amounts of unreacted starting material, other products formed or additional products caused by unwanted reactions. You can work out **percentage purity** in a similar way to percentage yield, using either moles or mass in grams:

$$\% \text{ purity} = \frac{\text{mass of pure product}}{\text{mass of impure product}} \times 100$$

Figure 6.6.2 Silicon chips used for computers have to be very pure.

Example

A chemist made 60.0 g of aspirin. Chemical analysis showed that this sample contained 58.5 g of pure aspirin and 1.5 g of impurities. Calculate the percentage purity of this aspirin sample.

Using the expression for % purity:

$$\% \text{ purity} = \frac{58.5}{60.0} \times 100 = 97.5\% \text{ pure}$$

(The impurities make up (100 − 97.5)% = 2.5% of the product).

SUMMARY QUESTIONS

1. Copy and complete using the words below:

   equation   masses   product   theoretical   yield

   The amount of _____ made in a chemical reaction is called the _____. We can use relative formula _____ and the stoichiometry of the _____ to calculate the _____ yield.

2. When 11.50 g of sodium reacts with excess chlorine, 22.30 g of sodium chloride is made.

   $2Na + Cl_2 \rightarrow 2NaCl$

   A_r values: Na = 23, Cl = 35.5

   Calculate:

   a the theoretical yield of sodium chloride

   b the % yield

3. A 26.5 g sample of impure paracetamol contains 24.5 g of pure paracetamol.

   Calculate the % purity of this paracetamol.

KEY POINTS

- Percentage yield
  $$= \frac{\text{experimental yield}}{\text{theoretical yield}} \times 100$$
- The theoretical yield is found using relative atomic masses and the stoichiometry of the equation.
- Percentage purity
  $$= \frac{\text{mass of pure product}}{\text{mass of impure product}} \times 100$$

6.7 Empirical formula and molecular formula

Supplement

LEARNING OUTCOMES
- Calculate empirical formula
- Calculate molecular formula

Calculating the empirical formula

In Topic 5.2 we defined empirical formula as the simplest whole number ratio of the different atoms or ions in a compound. For example, the empirical formula for hydrogen peroxide, H_2O_2, is HO. The empirical formula of butene, C_4H_8, is CH_2.

PRACTICAL

Finding the formula of magnesium oxide

1. Burn a weighed amount of magnesium (0.6 g) in a crucible. The magnesium reacts with the oxygen in the air to form magnesium oxide.

2. When the apparatus has cooled, weigh the magnesium oxide formed. In this experiment it was 1.0 g.

1.0 − 0.6 = 0.4 g = mass of oxygen in magnesium oxide

moles Mg = $\frac{0.6}{24}$ = 0.025 mol

moles O = $\frac{0.4}{16}$ = 0.025 mol

The numbers of moles of magnesium and oxygen atoms are equal, so the empirical formula of magnesium oxide is MgO.

EXAM TIP

Remember that the formula for an ionic compound is always its empirical formula.

If we know the masses of each element (or the % by mass) that combine to form a compound, we can work out its empirical formula.

Worked example 1: using masses

A compound of tin (Sn) and chlorine (Cl) contains 29.75 g of tin and 35.50 g of chlorine. Calculate the empirical formula of tin chloride.

A_r values: Sn = 119, Cl = 35.5

	Sn	Cl
Step 1: calculate the number of moles	$\frac{29.75}{119}$ = 0.250 mol	$\frac{35.50}{35.5}$ = 1.00 mol
Step 2: divide each by the lowest number of moles	$\frac{0.250}{0.250}$ = 1	$\frac{1.00}{0.250}$ = 4

Step 3: write the ratio (Sn:Cl) and formula 1 Sn : 4 Cl = $SnCl_4$

Figure 6.7.1 We can find the formula of magnesium oxide by weighing magnesium and the product magnesium oxide.

EXAM TIP

When calculating empirical formulae, make sure that between Steps 1 and 2 you don't round up the figures. This often leads to errors.

Unit 6: Chemical calculations

Worked example 2: using percentage by mass

A compound of carbon and hydrogen contains 80% carbon and 20% hydrogen by mass. Calculate the empirical formula of this compound.

A_r values: C = 12, H = 1

	C	H
Step 1: divide % by A_r values	$\frac{80}{12} = 6.67$	$\frac{20}{1} = 20$
Step 2: divide each by the lowest number of moles	$\frac{6.67}{6.67} = 1$	$\frac{20}{6.67} = 2.99$
Step 3: write the ratio and formula	1 C : 3 H	= CH_3

In some empirical formula calculations you may have to include an additional step at the start. Look at the following question:

A compound of carbon, hydrogen and oxygen contains 37.5% carbon and 12.5% hydrogen by mass. Calculate the empirical formula.

In this question the first step is to calculate the % of oxygen:

% oxygen = 100 – (37.5 + 12.5) = 50% oxygen.

Deducing the molecular formula

Molecular formula is the formula showing the number and type of different atoms in one molecule. For example, ethane has the empirical formula CH_3 but its molecular formula is C_2H_6. To find the molecular formula we need to know:

- the empirical formula of the compound
- the relative formula mass of the compound.

Worked example

A compound has the empirical formula CH_2. Its relative formula mass is 84. Deduce its molecular formula.

A_r values: C = 12, H = 1

Step 1: find the empirical formula mass of CH_2 $12 + (2 \times 1) = 14$

Step 2: divide relative formula mass by empirical formula mass $\frac{84}{14} = 6$

Step 3: multiply the empirical formula by the number calculated in Step 2 to get the molecular formula of the compound: $6 \times CH_2 = C_6H_{12}$

KEY POINTS

- Empirical formula is calculated using the mass in grams or % by mass of the elements present and their relative atomic masses to get the simplest ratio of moles of the elements in a compound.
- Molecular formula is calculated using the empirical formula and the relative molecular mass of a compound.

SUMMARY QUESTIONS

1. Copy and complete using the words below:

   atomic   dividing   empirical   lowest   mass   ratio

   The _____ formula shows the simplest whole number _____ of atoms in a compound. It is found by _____ the _____ of each element by its relative _____ mass and then dividing by the _____ number of moles to get the ratio of atoms.

2. A compound contains 26.67% carbon and 2.22% hydrogen by mass. The rest of the compound is oxygen. Calculate its empirical formula. A_r values: C = 12, H = 1, O = 16

3. Calculate the molecular formula of an oxide of phosphorus which has an empirical formula P_2O_5 and a relative molecular mass of 284. A_r values: O = 16, P = 31

6.8 Titrations

Supplement

LEARNING OUTCOMES

- Calculate solution concentration and volumes of solutions using data provided
- Use experimental data from a titration to calculate the concentration of a solution

EXAM TIP

Mole calculations involving concentrations are easier if you change cm^3 to dm^3 and then use the formula concentration = number of moles ÷ volume of solution in dm^3.

Figure 6.8.1 Titrations give you valuable practice in calculating the concentration of solutions.

EXAM TIP

You may be asked to give your answers to a certain number of significant figures (SF's): The number 23.64 quoted to 3 SF's is 23.6; to 2 SF's it is 24 and to 1 SF it is 20. Zeros before a value are not SF's, e.g. 0.0024 is to 2 SF's.

Solution concentration

In Topic 1.4 we defined the concentration of a solution in terms of mass of substance per dm^3. For many chemical calculations we need to express the concentration in terms of mol/dm^3.

$$\text{concentration (in mol/dm}^3\text{)} = \frac{\text{amount of substance in moles}}{\text{volume of solution in dm}^3}$$

There are three points to note about this equation:

- you will have to convert volumes of solution in cm^3 into dm^3.
- you may be given the amount of solute dissolved in grams. You need to convert this into moles.
- you will also need to be able to rearrange the equation so that you can calculate moles or volume (see worked examples below).

Worked examples

Example 1

Calculate the mass of potassium hydroxide in 20 cm^3 of a solution of concentration 0.4 mol/dm^3.

A_r values: K = 39, O = 16, H = 1

Step 1: convert cm^3 to dm^3 20 cm^3/1000 = 0.02 dm^3

Step 2: find number of moles using concentration and volume

moles KOH = concentration (mol/dm^3) × volume (dm^3)
= 0.4 × 0.02 moles = 0.008 moles

Step 3: calculate mass from (moles × M_r)
0.008 × (39 + 16 + 1) = 0.448 g

Example 2

Calculate the volume, in cm^3, of a solution of sodium hydroxide, NaOH, containing 0.8 g of sodium hydroxide of concentration 0.40 mol/dm^3.

A_r values: Na = 23, O = 16, H = 1

Step 1: convert grams of NaOH to moles:

$$\frac{0.8}{(23 + 16 + 1)} = 0.02 \text{ mol NaOH}$$

Step 2: calculate volume in dm^3,

using volume (dm^3) = $\frac{\text{moles}}{\text{concentration (mol/dm}^3\text{)}} = \frac{0.02}{0.4} = 0.05$ dm^3

Titrations

We can find the concentration of alkali needed to completely react with an acid using a procedure called an acid–base **titration**. In a titration, the concentration of one of the solutions is known

Unit 6: Chemical calculations

accurately. We put this solution into the burette. We put a measured volume of the other solution into the titration flask and add an indicator.

We then run the solution from the burette into the flask until the indicator changes colour. The change in colour shows that the reaction is complete.

We use the burette reading together with the known concentration of the solution in the burette to find the unknown concentration of the solution in the flask.

Worked example

25 cm³ of potassium hydroxide of unknown concentration is put into a flask and an acid-base indicator added. The indicator changes colour when 30 cm³ of dilute hydrochloric acid of concentration 0.20 mol/dm³ has been added from the burette. Calculate the concentration of the potassium hydroxide in mol/dm³.

$$KOH + HCl \rightarrow KCl + H_2O$$

Step 1: calculate the moles of hydrochloric acid used

moles HCl = concentration × volume (in dm³)

$$= 0.20 \times \frac{30}{1000} = 0.0060 \text{ mol HCl}$$

Step 2: look at the equation to find the ratio of KOH to HCl; in this case it is 1:1.

So moles HCl = moles KOH = 0.006 mol KOH

Step 3: change moles of KOH to concentration in mol/dm³ (note that 25 cm³ = 0.025 dm³)

$$\text{concentration (mol/dm}^3\text{)} = \frac{\text{number of moles of solute}}{\text{volume of solution (dm}^3\text{)}}$$

$$= \frac{0.0060}{0.025} = 0.24 \text{ mol/dm}^3 \text{ KOH}$$

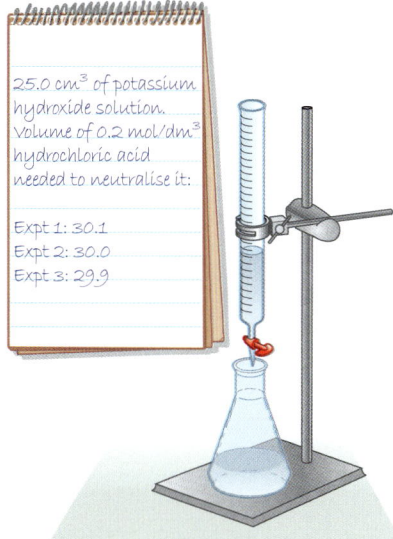

25.0 cm³ of potassium hydroxide solution. Volume of 0.2 mol/dm³ hydrochloric acid needed to neutralise it:

Expt 1: 30.1
Expt 2: 30.0
Expt 3: 29.9

Figure 6.8.2 Titration apparatus.

KEY POINTS

- The concentration of a solution is the number of moles of solute dissolved in 1 dm³ of solution.
- An acid–base titration is a method used to calculate the concentration of a solution whose concentration is not known.

SUMMARY QUESTIONS

1. Copy and complete using the words below:

   **acid   concentration   moles
   stoichiometric   titration   volume**

   We find the concentration of an alkali by carrying out a _____ with an _____ of known _____. The first step in the calculation is to use the concentration and _____ of the acid to find the number of _____ of acid used. We then use information from the _____ equation and volume of alkali to calculate the concentration of alkali.

2. The equation shows the reaction of sulfuric acid with sodium hydroxide:

   $$H_2SO_4 + 2NaOH \rightarrow Na_2SO_4 + 2H_2O$$

   25 cm³ of sulfuric acid of concentration 0.2 mol/dm³ reacted with exactly 10 cm³ of sodium hydroxide. Calculate:

   a the number of moles of sulfuric acid present
   b the number of moles of sodium hydroxide reacting
   c the concentration of the sodium hydroxide in mol/dm³

SUMMARY QUESTIONS

1 Calculate the relative formula mass of:
   a Al_2O_3
   b Na_2SO_4
   c $Al_2(SO_4)_3$
   d PCl_5

2 Match the words on the left with the phrases on the right.

Avogadro number	the reagent that is not in excess
molar gas volume	the number of specified particles in one mole of those particles
limiting reagent	the sum of the relative atomic masses
relative molecular mass	the volume occupied by one mole of any gas

3 Calculate the number of moles of:
   a bromine molecules in 2.4 g of bromine, Br_2
   b iron atoms in 9.6 g of iron oxide, Fe_2O_3
   c chloride ions in 79.17 g magnesium chloride, $MgCl_2$

4 Calculate the volume in cm³ at r.t.p. of:
   a 0.3 moles of neon
   b 2.5 moles of carbon dioxide
   c 0.03 moles of ammonia

5 Calculate the mass of:
   a 0.5 moles of hydrogen chloride, HCl
   b 0.2 moles of calcium nitrate, $Ca(NO_3)_2$
   c 3 moles of phosphorus(V) chloride, PCl_5

6 Calculate the moles of gas in:
   a 1680 cm³ of carbon dioxide
   b 960 cm³ of nitrogen
   c 2.4 dm³ of neon

7 Calculate the concentration in mol/dm³ of the following solutions:
   a 0.5 moles of hydrochloric acid in 500 cm³ of solution
   b 0.15 moles of magnesium chloride in 250 cm³ of solution
   c 5.85 g of sodium chloride in 2 dm³ of solution
   d 4.9 g of sulfuric acid in 250 cm³ of solution

Practice questions

1 Choose the best definition of relative atomic mass.
   A The mass of 12 atoms of the isotopes of an element compared with the mass of an atom of carbon-12.
   B The average mass of a molecule of the isotopes of an element compared with the mass of a molecule of carbon-12.
   C The mass of the isotopes of an element compared with the mass of carbon-14.
   D The average mass of the atoms of the isotopes of an element compared with the mass of an atom of carbon-12.

(Paper 1)

2 Choose the correct number of particles of CO_2 present in 240 cm³ of CO_2. Avogadro constant = 6.02×10^{23}/mol.
   A 6.02×10^{21}
   B 6.02×10^{25}
   C 1.802×10^{23}
   D 1.802×10^{21}

(Paper 2)

3 Iron reacts with hydrochloric acid to form iron(II) chloride, aqueous $FeCl_2$, and hydrogen.
   (a) Write a balanced equation for this reaction. [3]
   (b) When 28 g of iron reacts with excess hydrochloric acid, 63.5 g of iron(II) chloride and 1 g of hydrogen are formed.
      (i) Calculate the mass of iron(II) chloride formed from 7 g of iron. [1]
      (ii) Calculate the mass of iron that will be needed to produce 10 g of hydrogen. [1]

(Paper 3)

4 3.2 g of iron is added to 50 cm³ of 1.0 mol/dm³ sulfuric acid.

$Fe + H_2SO_4 \rightarrow FeSO_4 + H_2$

   (a) Calculate the number of moles of sulfuric acid present. [1]
   (b) Show by calculation that iron is in excess. [2]

(c) Calculate the mass of iron remaining after the reaction is complete. [2]
(d) Calculate the maximum volume of hydrogen formed in the reaction at r.t.p. [2]
(e) Calculate the theoretical yield of iron(II) sulfate in grams. [2]
(f) The actual yield of iron(II) sulfate was 7.22 g. Calculate the percentage yield. [2]

(Paper 4)

5 A solution of potassium hydroxide of unknown concentration was titrated with sulfuric acid.

$2KOH + H_2SO_4 \rightarrow K_2SO_4 + 2H_2O$

(a) Describe how to carry out an acid–alkali titration. [4]
(b) It required 15 cm³ of 0.05 mol/dm³ sulfuric acid to neutralise 25 cm³ of potassium hydroxide. Calculate:
  (i) the number of moles of sulfuric acid in the titration [1]
  (ii) the concentration of the aqueous potassium hydroxide. [2]
  (iii) the mass of potassium hydroxide dissolved in 25 cm³ of solution. [1]

(Paper 4)

6 Compound Z contains 0.96 g of carbon, 0.16 g of hydrogen and 2.84 g of chlorine. No other elements are present.

(a) Calculate the empirical formula of Z. [3]
(b) The relative molecular mass of Z is 99. Calculate the molecular formula of Z. [2]
(c) A different compound of carbon, hydrogen and chlorine was made by reacting 8.1 g of chlorine with 1.6 g of methane.

$CH_4 + Cl_2 \rightarrow CH_3Cl + HCl$

  (i) Show by calculation which is the limiting reactant. [2]
  (ii) Calculate the theoretical yield of CH_3Cl. [2]
  (iii) The actual yield of CH_3Cl was 3.79 g. Calculate the percentage yield. [2]
(d) CH_3Cl is a gas. Calculate the volume of 5.05 g of CH_3Cl at r.t.p. [2]
(e) Calculate the percentage by mass of hydrogen in CH_3Cl. [2]

(Paper 4)

7 Calcium carbonate reacts with hydrochloric acid:

$CaCO_3 + 2HCl \rightarrow CaCl_2 + CO_2 + H_2O$

(a) Calculate the number of moles of hydrochloric acid required to react exactly with 5 g of calcium carbonate. [2]
(b) Calculate the volume of 0.5 mol/dm³ hydrochloric acid that will react exactly with 5 g of calcium carbonate. [2]
(c) Calculate the volume of carbon dioxide produced at r.t.p. when 35 g of calcium carbonate react with excess hydrochloric acid. [3]
(d) A student crystallised the calcium chloride produced in this reaction. The percentage yield was only 90%.
  (i) Give two possible reasons why the yield was not 100%. [2]
  (ii) Describe the meaning of percentage yield. [1]

(Paper 4)

8 Ammonium nitrate, NH_4NO_3, and ammonium sulfate, $(NH_4)_2SO_4$, are used as fertilisers.

(a) (i) Calculate the percentage by mass of nitrogen in $(NH_4)_2SO_4$. [3]
  (ii) The percentage by mass of nitrogen in NH_4NO_3 is 35%. Calculate the percentage by mass of nitrogen in a mixture containing 4 parts of NH_4NO_3 and 1 part of $(NH_4)_2SO_4$. [2]
(b) NH_4NO_3 can be made by titrating aqueous ammonia with nitric acid.

$NH_3 + HNO_3 \rightarrow NH_4NO_3$

50 cm³ of a solution of ammonia was neutralised by 15 cm³ of nitric acid of concentration 2.0 mol/dm³.

Calculate the concentration of the solution of ammonia:
  (i) in mol/dm³ [2]
  (ii) in g/dm³. [2]

(Paper 4)

7.1 Conductors and electrolysis

LEARNING OUTCOMES

- Describe electrical conductors and electrical insulators
- Describe the reasons for the use of copper and steel-cored aluminium for electrical cables, and why we use plastics and ceramics as insulators
- Define electrolysis
- Identify the anode, cathode and electrolyte in an electrolytic cell

Electrical conductors

Electrical **conductors** are substances which have a low resistance to the passage of electricity. In other words, they allow electricity to flow through them easily. All metals, as well as graphite, conduct electricity. They are good electrical conductors because some of their electrons are able to move throughout the structure when a voltage is applied. We describe these free-moving electrons as 'mobile electrons'.

Copper is commonly used in electrical wiring and in thicker electricity cables because it is a very good conductor of electricity. It is also more ductile than many other metals so can be easily stretched out into wires.

We use very thick wires in the high-voltage power lines used to transfer electricity over long distances. This is because thick wires do not transfer as much thermal energy to the air as thin wires. High-voltage power lines are made from aluminium cables with a **steel** core in the middle.

- Aluminium is used because it is a good conductor of electricity. It also has a low density so that the wires hanging between pylons over long distances do not break under their own weight. Aluminium is also resistant to corrosion.
- The steel core gives the cables additional strength to stop them sagging and breaking.

Electrical insulators

An electrical **insulator** is a substance that resists the flow of an electric current. Insulators do not conduct electricity. This is because they do not have mobile electrons or mobile ions. Examples of insulators include plastics and ceramic materials. Ceramics are often made by heating clay.

Plastics such as PVC are useful insulators, not only because they do not conduct electricity. They are also flexible and non-biodegradable. This makes them useful for covering electrical wires so that we do not get an electric shock. Plastics are less useful as insulators where large electric currents are used. The thermal energy of the electric current can easily melt the plastic.

Ceramics are useful insulators because

- they do not conduct electricity
- they have very high melting points so do not melt when large electric currents flow.

EXAM TIP

It is a common error to think that the steel in an electricity cable just conducts electricity. It is there to strengthen the cable.

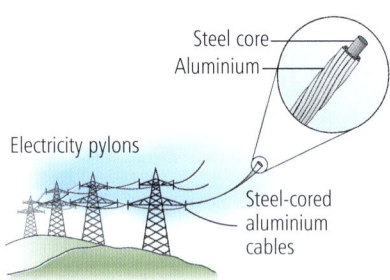

Figure 7.1.1 A steel-cored aluminium cable.

Figure 7.1.2 Overhead power lines have to be strong as well as good conductors of electricity.

Unit 7: Electricity and chemistry

We use ceramics in high-voltage electricity towers to keep the wires from touching the metal pylons or from touching each other.

Electrolysis

Electrolysis is the **decomposition** (breaking down) of an ionic compound when molten or in an aqueous solution by the passage of an electric current.

Electrolysis can only happen when the ionic compound is molten (liquid) or when it is dissolved in water.

The important parts of an electrolysis apparatus are shown in Figure 7.1.3. We call this an electrolysis cell. The voltage is provided by a direct current power supply such as batteries or a power pack.

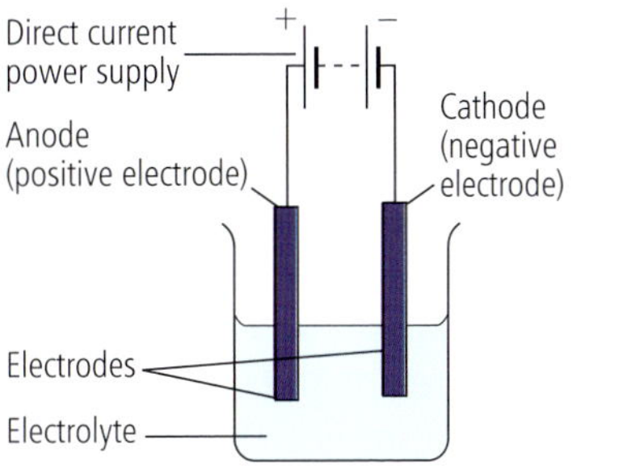

The molten or aqueous ionic compound that conducts electricity and decomposes during electrolysis is called the **electrolyte**.

Figure 7.1.3 The key parts of an electrolysis cell.

The **electrodes** are rods that carry the electric current to and from the electrolyte. Two properties that electrodes must have are:

- they are good electrical conductors
- they are **inert**, so that they do not react with the electrolyte or the products of electrolysis.
- Graphite or platinum are commonly used as electrodes.

The positive electrode is called the **anode**. The negative electrode is called the **cathode**.

KEY POINTS

- Steel-cored aluminium cables are used in high-voltage power lines because aluminium is a good conductor and steel strengthens the cables.
- Insulators such as plastics and ceramics prevent an electric current flowing.
- Copper is used in electrical wiring because it is a very good conductor of electricity.
- Electrolysis is the decomposition of an ionic compound when molten or in an aqueous solution by the passage of an electric current.
- Electrodes are rods that carry the electric current to and from the electrolyte.
- The positive electrode is called the anode and the negative electrode is called the cathode.

EXAM TIP

Note that the positive pole of the battery is shown as a long line and the negative pole as a short line.

EXAM TIP

Exam questions often ask you to draw and/or label electrolysis apparatus. Make sure that you are able to label the anode, cathode and electrolyte.

SUMMARY QUESTIONS

1 Copy and complete using the words below:

**electricity   melt
resistance   thicker   thin**

Conductors have a low _____ to the passage of _____. The _____ an electrical wire, the greater is the amount of current that can flow through it. A _____ wire will heat up when an electric current passes through it. It may even _____.

2 State the meaning of the term electrical insulator.

3 State the meaning of the terms:
   a electrolysis
   b decomposition
   c anode

81

7.2 The products of electrolysis

LEARNING OUTCOMES

- Describe the electrode products and observations made during electrolysis
- State which electrode metals and non-metals form at during electrolysis
- Predict the products formed in the electrolysis of a given molten binary compound

The electrolysis of molten lead(II) bromide, $PbBr_2(l)$

PRACTICAL

The electrolysis of molten lead(II) bromide

When the lead(II) bromide is heated it begins to melt and at that point the lamp lights. This shows that an electric current is flowing because the ions in the molten electrolyte are able to move to the electrodes. The electric current decomposes lead(II) bromide. Grey beads of molten lead form at the cathode. At the anode, reddish-brown bromine gas is formed.

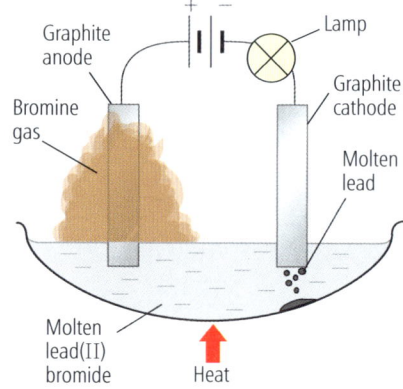

Figure 7.2.1 Electrolysis of molten lead(II) bromide

Figure 7.2.2 A series of electrolysis cells being used to extract aluminium.

The electrolysis of dilute sulfuric acid, $H_2SO_4(aq)$

PRACTICAL

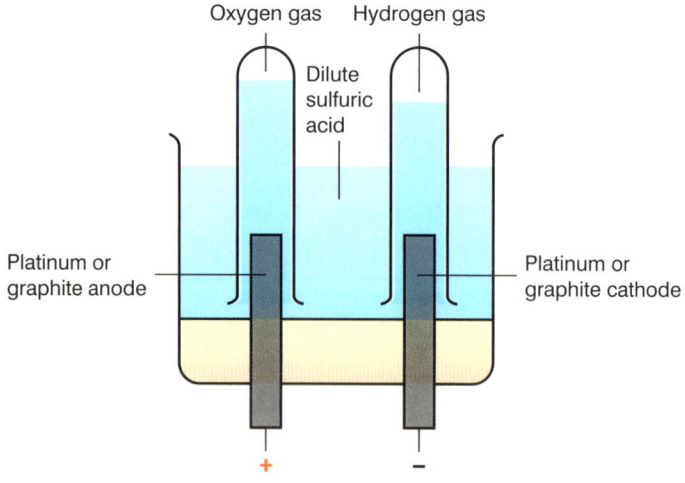

Figure 7.2.3 Electrolysis of dilute sulfuric acid.

When dilute sulfuric acid is electrolysed, bubbles of gases are seen at both the anode and cathode. The gases can be collected as shown. Oxygen is produced at the anode and hydrogen is produced at the cathode.

EXAM TIP

Remember that the products of electrolysis are always either *m*etals or *m*olecules. It is a common error to suggest that the products are ions.

Unit 7: Electricity and chemistry

Predicting the products of electrolysis

Molten binary ionic compounds

The equation for the electrolysis of molten lead(II) bromide shows that it is broken down into its elements, lead and bromine.

$$PbBr_2(l) \longrightarrow Pb(l) + Br_2(g)$$

Does the same happen with other molten binary ionic compounds? (Binary means that a compound contains only two elements.) Table 7.2.1 gives more examples of electrolysis of molten compounds using inert electrodes.

compound electrolysed	product at the cathode (− electrode)	product at the anode (+ electrode)
aluminium oxide	aluminium	oxygen
copper bromide	copper	bromine
sodium chloride	sodium	chlorine
zinc chloride	zinc	chlorine

Table 7.2.1 The products formed from the electrolysis of molten ionic compounds.

You can see a pattern in the table. When the electrolyte decomposes, a metal is formed at the cathode (negative electrode) and a non-metal forms at the anode (positive electrode). So we can easily predict the products of electrolysis of a molten ionic compound.

Dilute sulfuric acid

Dilute sulfuric acid decomposes during electrolysis to form hydrogen and oxygen. The reaction can also be thought of as the electrolysis of water. The water breaks down to hydrogen ions, H^+, and hydroxide ions, OH^-. The aqueous hydrogen ions form hydrogen gas at the cathode. The aqueous hydroxide ions break down at the anode to make oxygen gas.

> **EXAM TIP**
>
> Remember the general pattern that in electrolysis metals or hydrogen are formed at the cathode and non-metals, other than hydrogen, are formed at the anode.

> **KEY POINTS**
>
> - When a molten ionic compound is electrolysed, a metal is formed at the negative electrode and a non-metal is formed at the positive electrode.
> - When lead(II) bromide is electrolysed, a reddish-brown gas is formed at the anode and a silvery-grey metal is formed at the cathode.
> - When dilute sulfuric acid is electrolysed, bubbles of oxygen are formed at the anode and bubbles of hydrogen are formed at the cathode.

> **SUMMARY QUESTIONS**
>
> 1. Predict the products formed at the anode and cathode when the following molten compounds are electrolysed using inert electrodes:
>   a. sodium fluoride
>   b. copper(II) iodide
>   c. zinc bromide
>
> 2. Describe the observations and products formed at each electrode during the electrolysis of:
>   a. molten lead(II) bromide
>   b. dilute sulfuric acid
>   c. concentrated hydrochloric acid (chlorine formed at the anode)

7.3 Electrolysis of aqueous solutions

LEARNING OUTCOMES

- Describe the electrode products and observations made during the electrolysis of concentrated aqueous sodium chloride
- S Predict and identify the electrolysis products of aqueous halides in dilute or concentrated aqueous solution

EXAM TIP

Remember that the observations at the electrodes are what you *see* (e.g. yellow-green gas/bubbles) and are *not* the name of the substance formed.

EXAM TIP

Remember that when *aqueous* sodium chloride is electrolysed, hydrogen is formed at the cathode. But when *molten* sodium chloride is electrolysed, sodium is formed at the cathode.

Electrolysing concentrated aqueous sodium chloride

When we electrolyse concentrated aqueous solutions of ionic metal chlorides, we find that chlorine is produced at the positive electrode. This is the same result at the anode that we get with the molten compounds. However, we do not get a metal at the negative electrode. We get hydrogen gas at the cathode instead.

PRACTICAL

Electrolysing a concentrated solution of sodium chloride, NaCl(aq)

This experiment is done in a fume cupboard because chlorine gas is toxic (poisonous).

An electric current is passed through the concentrated aqueous solution of sodium chloride. Bubbles of colourless hydrogen gas are seen at the cathode (negative electrode) and bubbles of yellow-green chlorine gas are seen at the anode (positive electrode). The gases are collected in upturned test tubes.

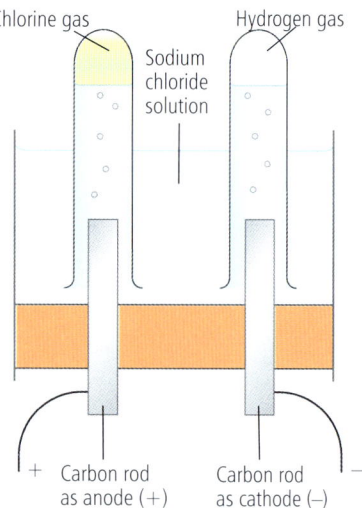

Figure 7.3.1 When concentrated sodium chloride is electrolysed, chlorine forms at the anode and hydrogen at the cathode.

Supplement

Explaining the results

In concentrated solutions of ionic compounds in water we have a greater variety of ions than in molten ionic compounds. Water is a weak electrolyte. It has a very low concentration of hydrogen ions, H^+, and hydroxide ions, OH^-.

$$H_2O \rightleftharpoons H^+ + OH^-$$

So in an aqueous solution of sodium chloride we have these ions: Na^+, H^+, Cl^-, OH^-. So why is hydrogen given off at the cathode instead of sodium?

The answer lies in the ion **discharge series**. This is similar to the metal reactivity series. The lower down the series, the more likely it is that the ion will be discharged (changed from an ion to an atom or molecule at the electrode). The order of this series is:

Unit 7: Electricity and chemistry

for positive ions: $\underrightarrow{Na^+ \; Mg^{2+} \; Al^{3+} \; H^+ \; Cu^{2+}}$

more likely to be discharged

for negative ions: $\underrightarrow{SO_4^{2-} \; NO_3^- \; OH^- \; Cl^- \; Br^- \; I^-}$

more likely to be discharged

When a concentrated aqueous solution of sodium chloride is electrolysed, hydrogen rather than sodium is discharged at the negative electrode. This is because hydrogen is lower in the discharge series (reactivity series).

Electrolysis of dilute aqueous solutions

When dilute aqueous solutions are electrolysed we can usually predict the electrode products from the discharge / reactivity series. The table shows some examples.

aqueous solution	ions present	product at cathode	product at anode	ions remaining	change to the electrolyte
potassium iodide	K^+, I^-, H^+, OH^-	hydrogen	iodine	K^+, OH^-	becomes alkaline
copper(II) chloride	Cu^{2+}, Cl^-, H^+, OH^-	copper	chlorine	H^+, OH^-	colour of solution fades
dilute sulfuric acid	SO_4^{2-}, H^+, OH^-	hydrogen	oxygen	SO_4^{2-}	water used up

Electrolysis of aqueous sodium chloride

The product at the cathode when aqueous solutions of reactive metals are electrolysed using inert electrodes is usually hydrogen. With sodium chloride, the product discharged at the anode varies with the concentration.

- concentrated NaCl(aq): anode product is chlorine
- moderately concentrated NaCl(aq): anode product is chlorine and oxygen
- very dilute NaCl(aq): anode product is mainly oxygen with a little chlorine.

SUMMARY QUESTIONS

1. Copy and complete using the words below:

   **bubbles   cathode
   concentrated   green
   hydrogen   ions**

   A _____ aqueous solution of sodium chloride contains sodium, chloride, _____ and hydroxide _____. When this solution is electrolysed, _____ chlorine gas is seen at the anode and _____ of hydrogen are seen at the _____.

2. State the products formed at the anode (+) and cathode (−) when the following are electrolysed:

   a aqueous magnesium bromide

   b concentrated aqueous sodium chloride.

   c very dilute aqueous sodium chloride

3. State the *observations* at each electrode when concentrated aqueous solutions of these compounds are electrolysed using inert electrodes:

   a copper(II) chloride

   b concentrated hydrochloric acid

   c sodium iodide

KEY POINTS

- When concentrated aqueous solutions of salts of reactive metals are electrolysed, hydrogen rather than a metal is formed at the cathode.
- When concentrated aqueous sodium chloride is electrolysed, chlorine is formed at the anode and hydrogen at the cathode.
- When aqueous bromides or iodides are electrolysed the halogens are discharged at the anode.
- The anode product formed when aqueous chlorides are electrolysed depends on the concentration of the solution.

7.4 Explaining electrolysis
Supplement

LEARNING OUTCOMES

- Describe the charge transfer during electrolysis
- Construct ionic half-equations for the reactions at the anode and cathode

From ions to atoms

Figure 7.4.1 shows how charge is transferred during electrolysis.

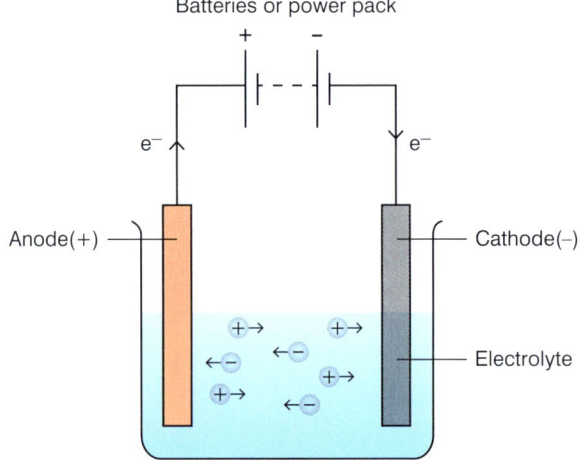

Figure 7.4.1 The transfer of charge during electrolysis.

- Electrons, e⁻, move in the external circuit from the negative pole of the battery to the cathode and towards the positive pole of the battery from the anode.
- Electrons are lost from the cathode and gained at the anode.
- The movement of the ions in the electrolyte completes the electrical circuit.

When an ionic compound is molten or aqueous, the ions are free to move around within the liquid. When an electric current is applied, the cations (positive ions) move towards the cathode and anions (negative ions) move towards the anode.

When the ions reach the electrodes they gain or lose electrons. A reaction where electrons are lost is an **oxidation** reaction. A reaction where electrons are gained is a **reduction** reaction.

When lead(II) bromide is electrolysed using inert electrodes, reduction occurs at the cathode and oxidation at the anode.

At the cathode, lead ions in the electrolyte take electrons from the external circuit and become lead atoms. This is a reduction reaction because lead ions gain electrons:

$$Pb^{2+}(l) + 2e^- \longrightarrow Pb(l)$$

This type of equation, which shows what is happening at only one of the electrodes, is called an **ionic half-equation**. In this case two electrons are added to the Pb^{2+} ion, balancing its 2+ charge.

Figure 7.4.2 The first person to explain electrolysis was Michael Faraday, who worked on this about 200 years ago.

EXAM TIP

Remember that ANions move towards the ANode, and CAtions move towards the CAthode because opposite charges attract.

Unit 7: Electricity and chemistry

At the anode, bromide ions in the electrolyte lose electrons to the anode and become bromine molecules. This is an oxidation reaction because bromide ions lose electrons. The ionic half-equation for this reaction is:

$2Br^-(l) \longrightarrow Br_2(g) + 2e^-$

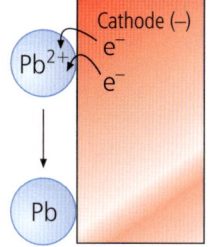
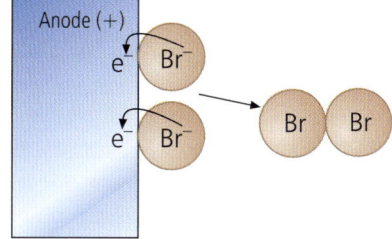
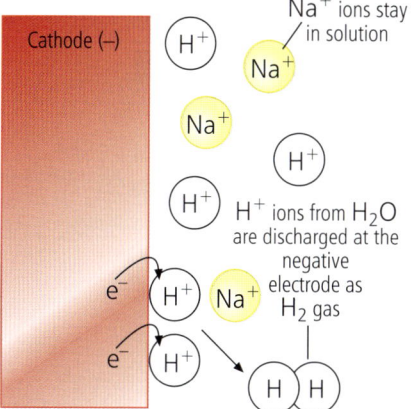

Figure 7.4.3 At the cathode, metal ions gain electrons. At the anode, non-metal ions lose electrons.

Figure 7.4.4 Hydrogen ions rather than sodium ions are discharged at the cathode from aqueous sodium chloride solution.

Similar **half-equations** can be written for other molten electrolytes. The metal ions gain electrons and the non-metal ions lose electrons.

What happens in aqueous solutions?

In aqueous solutions of metal salts or acids, H^+ and OH^- ions are present as well as the ions from the salt or acid. The less reactive element is discharged at the cathode. In a solution of sodium chloride, hydrogen ions, rather than sodium ions, gain electrons. Hydrogen gas is produced:

$2H^+(aq) + 2e^- \longrightarrow H_2(g)$

When sulfuric or nitric acid or solutions of sulfates or nitrates are electrolysed, oxygen is formed at the anode. The hydroxide ions in the water, rather than the sulfate or nitrate ions, lose electrons. Oxygen gas is produced:

$4OH^-(aq) \longrightarrow O_2(g) + 2H_2O(l) + 4e^-$

KEY POINTS

- Charge transfer during electrolysis involves (i) movement of electrons in the external circuit and oxidation-reduction reactions at the electrodes (ii) movement of ions in the electrolyte.
- Oxidation (electron loss) takes place at the anode and reduction (electron gain) takes place at the cathode.
- Ionic half-equations show the gain or loss of electrons when atoms or molecules are formed from ions.

SUMMARY QUESTIONS

1 Copy and complete using the words below:

   **anode   cathode   gas   hydrogen   lose
   oxygen   positive**

   During electrolysis, _____ ions move towards the _____ and negative ions move towards the _____. At the cathode, metal or _____ ions gain electrons and form metal atoms or hydrogen _____. At the anode, non-metal ions _____ electrons and form halogens or _____.

2 Write ionic half-equations for these reactions:

   a lead ions producing lead, Pb

   b hydrogen ions producing hydrogen molecules

   c hydroxide ions producing oxygen and water

3 Write half-equations for the anode and cathode reactions when these compounds are electrolysed:

   a molten zinc iodide

   b dilute aqueous potassium sulfate

7.5 Purifying copper
Supplement

LEARNING OUTCOMES

- Identify the electrode products, and describe the observations made, during the electrolysis of aqueous copper(II) sulfate using graphite electrodes and copper electrodes
- Describe the purification of copper by electrolysis

Refining copper

Thousands of tonnes of copper are used every year to make electrical wiring and pieces of electrical equipment. Copper extracted from copper ore is too brittle to be drawn into wires easily and the impurities reduce its electrical conductivity. So it needs to be refined (made purer).

Copper is refined by electrolysis. An impure strip of copper is connected to the positive end of a power supply. This forms the anode. A thin strip of pure copper is connected to the negative end of the power supply. This forms the cathode. The electrolyte is a solution of copper(II) ions, usually copper(II) sulfate solution.

Figure 7.5.1 Copper-plated sheets being removed from an electrolysis cell containing copper(II) sulfate electrolyte.

EXAM TIP

Remember OIL RIG: *O*xidation *Is* *L*oss of electrons; *R*eduction *Is* *G*ain of electrons.

EXAM TIP

Remember that in electrolysis the electrodes are usually inert (graphite or platinum). If the anode is not inert, it may react and decrease in size.

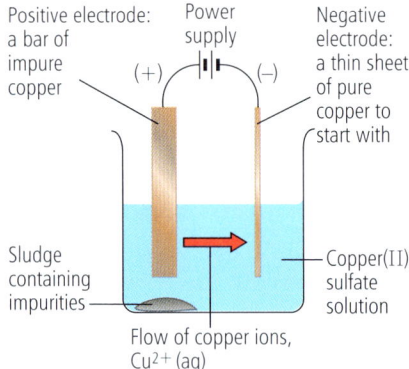

Figure 7.5.2 Copper is made purer by electrolysis.

At the impure copper anode (the positive electrode), copper atoms are oxidised by loss of electrons and form copper ions. These go into solution as part of the electrolyte:

$$Cu(s) \longrightarrow Cu^{2+}(aq) + 2e^-$$

At the pure copper cathode (the negative electrode), copper ions from the electrolyte are reduced by gaining electrons. The copper ions form copper atoms. These copper atoms are deposited on the strip of pure copper:

$$Cu^{2+}(aq) + 2e^- \longrightarrow Cu(s)$$

As the electrolysis proceeds the cathode becomes thicker as it gains more and more copper. After a time the cathode of pure copper is removed and replaced by a new one.

The anode loses mass and the impurities fall to the bottom of the electrolysis cell as 'anode sludge'. Other valuable metals, such as gold and platinum, can be extracted from the impurities in this 'anode sludge'.

The overall result of this electrolysis is that pure copper is transferred from the anode to the cathode.

Unit 7: Electricity and chemistry

Changing the electrodes

We can also electrolyse copper(II) sulfate solution using inert electrodes (graphite or platinum).

The type of electrodes used influences the reaction occurring at the anode.

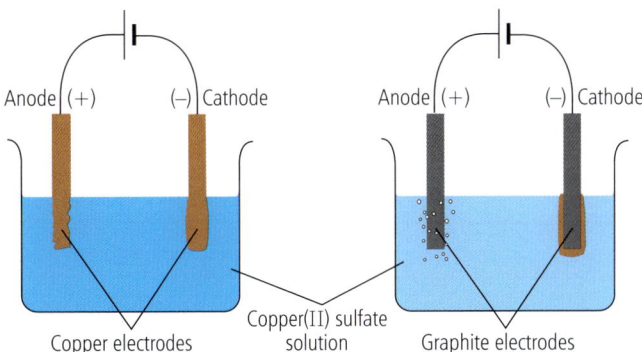

Figure 7.5.3 The reaction at the anode during the electrolysis of copper(II) sulfate solution depends on the type of electrodes used.

The ions present in an aqueous solution of copper(II) sulfate are $Cu^{2+}(aq)$, $SO_4^{2-}(aq)$, $H^+(aq)$ and $OH^-(aq)$.

Electrolysis with inert electrodes:

- At the anode:

The anode cannot lose electrons because it is inert. Hydroxide ions rather than sulfate ions are discharged. This is because hydroxide ions are lower in the discharge series. Oxygen gas bubbles off:

$4OH^-(aq) \longrightarrow O_2(g) + 2H_2O(l) + 4e^-$

- At the cathode:

Copper ions rather than hydrogen ions are discharged because Cu^{2+} is lower in the discharge series than H^+. Copper metal is deposited:

$Cu^{2+}(aq) + 2e^- \longrightarrow Cu(s)$

- The electrolyte gradually loses its blue colour. This is because the blue copper ions in solution are turning to copper atoms at the cathode but are not being replaced in the solution at the anode.

Electrolysis with copper electrodes:

- At the anode:

Because the copper anode is not inert, it loses electrons and copper ions go into solution. The anode gets smaller.

$Cu(s) \longrightarrow Cu^{2+}(aq) + 2e^-$

- At the cathode:

Copper ions rather than hydrogen ions are discharged because they are lower in the discharge series.

$Cu^{2+}(aq) + 2e^- \longrightarrow Cu(s)$

- The electrolyte remains the same deep blue colour. This is because as copper ions are removed from the solution at the cathode, they are replaced in solution by copper ions formed at the anode.

KEY POINTS

- Copper is purified by using an impure copper anode and a pure copper cathode.
- Electrolysis of copper(II) sulfate solution using inert electrodes produces copper at the cathode and oxygen at the anode.
- Electrolysis of copper(II) sulfate solution using copper electrodes produces copper at the cathode. The anode reacts as Cu atoms are converted to Cu^{2+} ions which go into solution.

SUMMARY QUESTIONS

1 Copy and complete using the words below:

**anode   cathode
electrolysed   electrons
gain**

When a solution of copper(II) sulfate is _____ using copper electrodes the copper atoms at the _____ lose _____ to form copper ions. At the _____ the copper ions _____ electrons and turn into copper atoms.

2 Write ionic half-equations for the reactions at the anode and cathode during the electrolysis of copper(II) sulfate solution using:

a platinum electrodes

b copper electrodes

7.6 Electroplating

LEARNING OUTCOMES

- Know that metal objects are electroplated to improve their appearance and resistance to corrosion
- Describe how metals are electroplated
- S Construct ionic half-equations for reactions at the anode and cathode

Electroplating metals

Electroplating is used to put a thin layer of one metal on top of another metal using electrolysis. We electroplate metals in the following way.

- We connect the object to be electroplated to the negative pole of the power supply. It becomes the cathode. The object can be anything which is made of metal, e.g. a spoon or a small metal statue. The object to be electroplated must be very clean so that the metal which is to cover it does not flake off.
- The plating metal is connected to the positive pole of the power supply. It becomes the anode.
- The electrolyte is a solution of an ionic compound of the plating metal. For example, if you want to plate an object with silver, you can use silver cyanide (which contains Ag^+ ions) as the electrolyte.

PRACTICAL

Electroplating with copper

Figure 7.6.1 Plating articles with silver makes them more attractive.

When you turn on the electric current, the steel ring starts getting covered with a thin layer of copper. The copper anode gradually gets smaller as the copper is transferred from the anode to form a thin layer on the steel cathode. The depth of colour of the copper(II) sulfate electrolyte does not change. This is because Cu^{2+} ions removed from solution to form Cu atoms at the cathode are continually being replaced by Cu^{2+} ions going into solution at the copper anode.

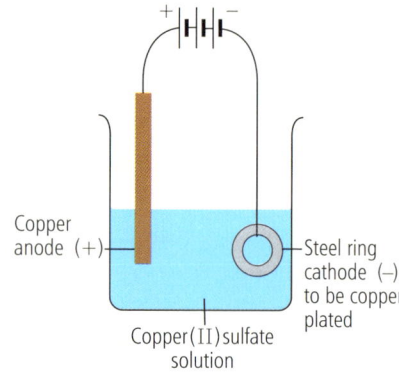

Figure 7.6.2 Electroplating apparatus.

EXAM TIP

Remember that the Plating metal is the Positive electrode (anode) and the object to be electroplated is the cathode.

Supplement

How electroplating works

Electroplating works in the same way as metal **refining** using electrolysis. Figure 7.6.1 shows what happens when an object is electroplated with silver.

At the anode the silver atoms lose electrons. They become silver ions which go into solution.

$Ag(s) \longrightarrow Ag^+(aq) + e^-$

Unit 7: Electricity and chemistry

The silver ions move to the cathode. At the cathode the silver ions gain electrons to become silver atoms which form a thin layer of silver on the surface of the object to be plated.

$$Ag^+(aq) + e^- \longrightarrow Ag(s)$$

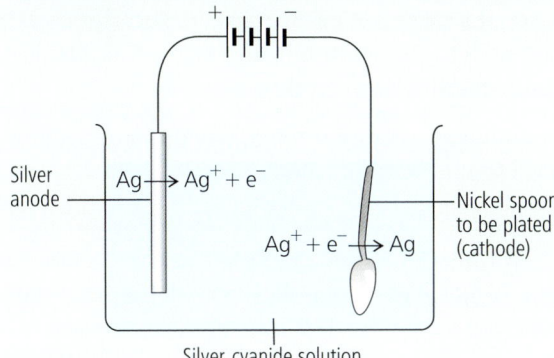

Figure 7.6.3 When an object is electroplated the metal ions formed at the anode are transferred to the cathode where they gain electrons and are deposited as a metal.

Uses of electroplating

Copper, chromium, nickel, silver and tin are the most commonly used metals for electroplating articles. Mixtures of metals can also be used for plating. There are two main reasons for electroplating: protection of metals from corrosion and improving their appearance.

Improving resistance of metals to corrosion

Some metals corrode easily. Iron and steel undergo a particular form of corrosion called rusting. Steel cans are electroplated with tin. The layer of tin protects the metal underneath from air and water so that it does not rust. If the layer of tin is scratched the metal underneath will start to corrode.

Improving the appearance of metals

Chromium plating gives a very shiny surface to objects that does not go dull. Silver plating is used for jewellery, cutlery and in electronics where it is too expensive to use solid silver. Gold plating is also used for jewellery and in specialised electronic equipment.

> **KEY POINTS**
>
> - Objects are electroplated to improve their appearance and resistance to corrosion.
> - An object can be electroplated by making the object to be plated the cathode and the plating metal the anode.
> - In electroplating, the electrolyte is a solution of an ionic compound of the plating metal.
> - When an article is electroplated, the atoms of the plating metal lose electrons at the anode and become ions. The ions of the plating metal gain electrons at the cathode and become metal atoms.

SUMMARY QUESTIONS

1 Copy and complete using the words below:

   **anode   cathode   electrolyte
   ions   solution**

   In electroplating, you make the object to be electroplated the _____ . The _____ is the plating metal. The _____ is an aqueous _____ containing _____ of the plating metal.

2 A student wants to electroplate a piece of copper with chromium. Draw a labelled diagram of the apparatus that the student could use.

3 An object is electroplated with nickel. The electrolyte contains Ni^{2+} ions. Construct the half-equations for the reactions occurring at the anode and cathode.

7.7 Extracting aluminium
Supplement

LEARNING OUTCOMES

- Describe the extraction of aluminium from purified bauxite
- Construct ionic half-equations for reactions at the anode and cathode

Figure 7.7.1 Mining bauxite.

EXAM TIP

You do not have to know the details of the purification of Al_2O_3, but you may be asked questions about purification techniques when given suitable information.

Extracting metals from the Earth

Metals such as iron and copper have been in use for thousands of years. People learned how to extract iron from its ores thousands of years ago. This required heating with carbon.

More reactive metals, such as aluminium, magnesium and sodium, cannot easily be extracted by heating with carbon.

It was not until electrolysis was discovered nearly 200 years ago that scientists could begin to work out how to get these reactive metals from their compounds. In fact, it was not until 1886 that the first small drops of molten aluminium were extracted from aluminium oxide.

Aluminium oxide from aluminium ore

Aluminium is the most abundant metal in the Earth's crust. It is found in the mineral ore **bauxite** which contains 50–65% aluminium oxide, Al_2O_3. Aluminium oxide is sometimes called alumina. The main impurities in bauxite are oxides of iron, silicon and titanium.

The first step in aluminium extraction is to purify the ore. The ore is first crushed and mixed with sodium hydroxide. The aluminium oxide reacts with the sodium hydroxide and dissolves.

$$Al_2O_3(s) + 2NaOH(aq) \longrightarrow 2NaAlO_2(aq) + H_2O(l)$$

aluminium oxide $\quad\quad\quad\quad\quad$ sodium aluminate

The impurities are insoluble in sodium hydroxide. These are filtered off. The sodium aluminate undergoes further treatment and is finally heated to make pure aluminium oxide.

Extracting aluminium from aluminium oxide

Electrolysis to produce aluminium is carried out in shallow electrolysis cells about 8 metres long and 1 metre deep. In order to carry out electrolysis, the aluminium oxide needs to be molten. Aluminium oxide melts at about 2040 °C. It is difficult to keep the electrolyte at this very high temperature for long periods of time. In addition, it is too costly because it needs so much energy – and energy is expensive. Also, aluminium oxide on its own is a poor conductor of electricity.

These problems are solved by dissolving the aluminium oxide in large amounts of molten cryolite. Cryolite, which is sodium aluminium fluoride, Na_3AlF_6, melts at about 1000 °C. Since the aluminium oxide is dissolved in the cryolite, the melting point of the electrolyte is much lower compared with pure aluminium oxide.

Dissolving the aluminium oxide in cryolite not only saves a lot of energy but also improves the electrical conductivity of the

Unit 7: Electricity and chemistry

electrolyte. Calcium fluoride, CaF_2, is often added to lower the melting point further. In most cases the melting point of the electrolyte is about 900 °C.

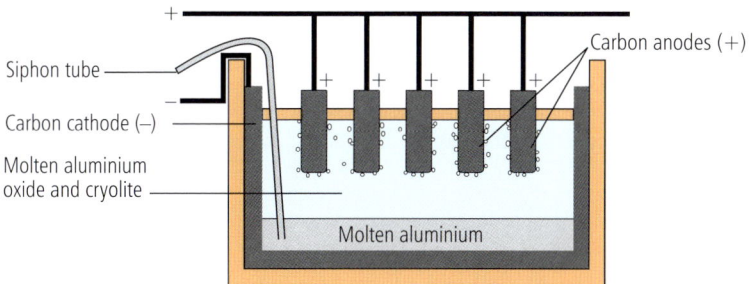

Figure 7.7.2 The electrolytic cell used in the extraction of aluminium.

Electrolysis is carried out using graphite (carbon) electrodes. The overall equation for this electrolysis is:

$$2Al_2O_3 \longrightarrow 4Al + 3O_2$$

The cathode is the carbon lining of the steel electrolysis cell. Several anodes, which can be raised or lowered, dip into the electrolyte. The very high electric current (40 000 amps) used in this electrolysis not only decomposes the aluminium oxide but also keeps the electrolyte molten.

At the cathode, aluminium ions gain electrons and are reduced to aluminium metal. The molten aluminium metal falls to the bottom of the cell. It is removed from time to time using a siphon tube:

$$Al^{3+} + 3e^- \longrightarrow Al$$

At the anode, the oxide ions lose electrons and are oxidised to oxygen:

$$2O^{2-} \longrightarrow O_2 + 4e^-$$

The oxygen formed reacts with the hot carbon anodes to form carbon dioxide gas which escapes into the air. Because the carbon anodes 'burn away' they need to be replaced from time to time.

SUMMARY QUESTIONS

1 Copy and complete using the words below:

conductivity   cryolite   energy   lower   pure

Aluminium is extracted by the electrolysis of molten aluminium oxide dissolved in _____ This mixture melts at a much _____ temperature than _____ aluminium oxide. Therefore a lot of _____ is saved and the electrical _____ of the electrolyte is improved.

2 Explain why during the electrolysis of aluminium oxide, the carbon anodes have to be replaced from time to time.

3 Balance this half-equation by adding electrons: $4Al^{3+} \longrightarrow 4Al$

EXAM TIP

You do not have to learn the diagram of the cell used to extract aluminium, but you should be able to label the different parts.

EXAM TIP

Remember that reduction (gain of electrons) always happens at the cathode and oxidation (loss of electrons) happens at the anode.

KEY POINTS

- The electrolytic cell for the extraction of aluminium has carbon anodes and a carbon cathode.

- The electrolyte in the cell is molten aluminium oxide, dissolved in molten cryolite.

- Cryolite improves the conductivity and lowers the melting point of the electrolyte, reducing energy costs.

- During the electrolysis of molten aluminium oxide, aluminium forms at the cathode and oxygen is released at the anode.

- The anodes have to be replaced regularly because they react with the oxygen formed during electrolysis, releasing carbon dioxide gas.

SUMMARY QUESTIONS

1. Classify the following as either conductors or insulators:
   a. molten sodium chloride
   b. iron
   c. plastic
   d. ceramic
   e. aqueous sodium bromide

2. Define these terms:
   a. electrolysis
   b. electrode
   c. ion
   d. insulator
   e. electrolyte

3. Match each electrolyte on the left with two of the possible products of electrolysis on the right. Each product can be used once, more than once or not at all.

molten lead chloride	chlorine
concentrated aqueous sodium chloride	hydrogen
	lead
dilute sulfuric acid	oxygen
molten aluminium oxide	aluminium

4. Draw and label the apparatus used to demonstrate the electrolysis of molten zinc chloride.

5. Predict the products and state the observations at each electrode during electrolysis of these molten salts:
   a. zinc chloride
   b. potassium bromide
   c. lead iodide

6. Copy and complete the paragraph using these words in the list below:

   anode attracted cathode deposited electrolyte electrons gain smaller tin

   Tin can be plated onto steel by electrolysis. The steel is made the _____ and the tin is made the _____. The _____ is an aqueous solution of _____(II) chloride. During electrolysis, the tin anode gets _____ in size because tin loses _____ and tin(II) ions go into solution. These ions are _____ to the steel cathode. At the cathode, tin(II) ions _____ electrons and tin is _____ on the surface of the steel.

7. a. Describe and explain the observations at the cathode and anode when a dilute aqueous solution of sodium chloride is electrolysed.
   a. Write half-equations for the reactions at (i) the cathode (ii) the anode.

Practice questions

1. Choose the correct statement about the electrolysis of molten lead(II) bromide.
   A Lead is formed at the positive electrode.
   B Hydrogen is formed at the negative electrode.
   C Bromine is formed at the negative electrode.
   D Bromine is formed at the positive electrode.
   (Paper 1)

2. Choose the correct statement about the transfer of charge during electrolysis.
   A Electrons move through the electrolyte.
   B At the cathode, positive ions accept electrons.
   C Negative ions get reduced at the anode.
   D Ions move round the external circuit from the cathode to the anode.
   (Paper 2)

3. Lead(II) bromide is electrolysed using the apparatus shown.

   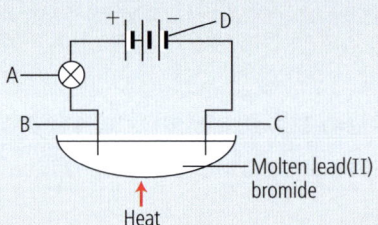

   (a) Give the letter which represents the anode. [1]
   (b) State the products formed and the observations made at
       (i) the anode and [2]
       (ii) the cathode. [2]
   (c) Complete the equation for this electrolysis:
       $PbBr_2(l) \longrightarrow$ _____ + _____ [2]
   (d) The electrolyte is molten lead(II) bromide. State the meaning of the term electrolyte. [1]
   (e) (i) Suggest a suitable substance which can be used to make the electrodes. [1]

    (ii) State two properties that a suitable electrode should have. [2]

(Paper 3)

4 The table shows the electrical conductivity of substances A–F.

compound	A	B	C	D	E	F
conductivity/ S/m	0.7	0	0	1.1	0.8	0.7

  (a) Give the letter of the substance in the table that is the best electrical conductor. [1]

  (b) Steel-cored aluminium cables are used for conducting high-voltage electricity over long distances.
    (i) Give two reasons why aluminium is used for these cables. [2]
    (ii) State the main purpose of the steel core. [1]

(Paper 3)

5 An aqueous solution of lithium iodide is electrolysed using carbon electrodes.

  (a) Explain why aqueous lithium iodide conducts electricity but solid lithium iodide does not conduct. [2]

  (b) State the names of the products formed at (i) the anode and (ii) the cathode. [2]

  (c) Write half-equations for the reactions at (i) the anode and (ii) the cathode. [2]

  (d) Explain why lithium is not deposited at the cathode. [2]

  (e) Suggest the name of another compound that will give the same products at the anode and cathode when electrolysed. [1]

(Paper 4)

6 The electrolysis of a concentrated solution of sodium bromide produces bromine at the anode and hydrogen at the cathode.

  (a) (i) Explain why sodium is not formed at the cathode. [2]
    (ii) Explain why bromine and not oxygen is formed at the anode. [2]

  (b) Write half-equations for the reactions occurring at
    (i) the cathode [1]
    (ii) the anode. [1]

  (c) Name the electrode where oxidation takes place. Explain your answer. [1]

  (d) Predict the products of electrolysis at the cathode and anode if a very dilute solution of sodium chloride is electrolysed. [2]

(Paper 4)

7 Aluminium is extracted by electrolysis. The electrolyte is a mixture of aluminium oxide and cryolite. The electrodes are made of graphite.

  (a) State the name of the ore containing aluminium oxide. [1]

  (b) Write an overall equation for this electrolysis. [1]

  (c) Write half-equations for the reactions at
    (i) the anode [1]
    (ii) the cathode. [1]

  (d) Explain the purpose of the cryolite. [2]

  (e) The graphite anodes have to be renewed regularly. Explain why giving a relevant equation. [2]

(Paper 4)

8 When very dilute aqueous magnesium iodide is electrolysed, hydrogen and iodine are formed as products.

  (a) (i) Explain why hydrogen and not magnesium is formed at the cathode. [2]
    (ii) Write the equation for the reaction at the anode. [1]

  (b) Describe the transfer of charge during this electrolysis in terms of:
    (i) the movement of electrons and ions [4]
    (ii) oxidation and reduction. [2]

(Paper 4)

8.1 Energy transfer in chemical reactions

LEARNING OUTCOMES

- Identify physical and chemical changes and understand the differences between them
- Describe the meaning of the terms exothermic and endothermic

Physical and chemical changes

A **physical change** is one in which no new substance is formed. Melting, boiling, condensing and freezing are examples of physical change. Physical changes are easily reversible. For example, we can change liquid water to steam by heating the water to its boiling point. When the steam cools, it condenses and returns to its original liquid state.

Dissolving is often thought of a physical change. We can dissolve salt (sodium chloride in water) and we can get the salt back again by evaporating the water. The salt is exactly the same substance and the same mass as we started with. Unlike melting or boiling, however, where the temperature remains constant, we do observe a temperature change when we dissolve salt in water.

Chemical changes involve the formation of new substances during a chemical reaction.

- Thermal energy (heat energy) is always transferred during a chemical change.
- One or more new substances are formed.

Thermal energy transfer and temperature

Temperature and thermal energy (heat energy) are not the same. The temperature of a substance depends on the average kinetic energy of its particles. Kinetic energy is the energy associated with movement. The higher the temperature, the greater is the average kinetic energy of the particles.

When we heat a substance, thermal energy is transferred to increase the kinetic energy of the particles of the substances being heated. The temperature of hot water does not depend on how much water there is. But the thermal energy transferred does depend on the amount of water.

Figure 8.1.1 Boiling and condensation are both physical changes. The water in the kettle boils to form steam. The steam condenses back to water droplets when it cools.

EXAM TIP

We use the term thermal energy rather than 'heat' in this syllabus. Thermal energy can be thought of as the energy associated with vibrating and moving particles. The greater the vibrations or movement of the particles, the greater is the thermal energy.

Exothermic or endothermic?

Reactions that release thermal energy to the surroundings, leading to an increase in temperature of the **surroundings**, are called **exothermic** reactions. Dissolving magnesium chloride in water is an exothermic change, as is the chemical reaction of an acid with an alkali. Many chemical reactions are exothermic. When we add zinc to copper(II) sulfate solution in a test tube, the temperature of the reaction mixture increases. The thermal energy released goes to warming up the surroundings. The surroundings include:

Unit 8: Chemical energetics

- the contents of the test tube (i.e. the reaction solution)
- the air around the test tube

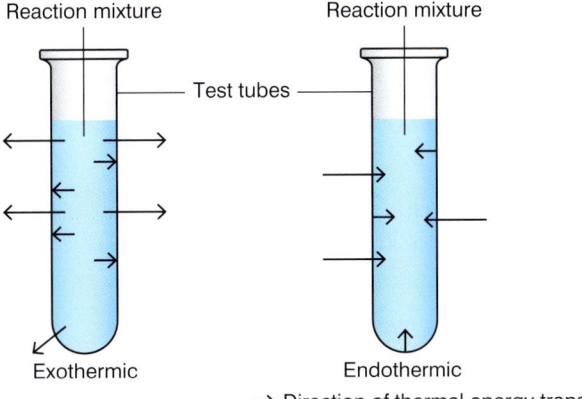

→ Direction of thermal energy transfer

Figure 8.1.2 Exothermic reactions release energy to the surroundings. Endothermic reactions absorb energy from the surroundings.

- the test tube itself
- thermometers or stirring rods dipped into the test tube.

Reactions which take in thermal energy from the surroundings are called **endothermic** reactions. We say they absorb thermal energy. The thermal energy absorbed is taken in by the reaction mixture and so lowers the temperature of the surroundings. The test tube gets cold. Examples of endothermic reactions are photosynthesis and reactions where continuous heating is needed, such as the thermal decomposition of calcium carbonate.

Supplement

Another word for the thermal energy change which occurs during a chemical reaction is **enthalpy change**. The symbol for enthalpy change is ΔH (pronounced delta H).

SUMMARY QUESTIONS

1. **a** Give one difference between physical and chemical changes.
   b Give two examples of physical change and two examples of chemical change.

2. Copy and complete using the words below:

   absorbs   decreases   exothermic   increases   reaction   surroundings

   An endothermic reaction _____ thermal energy from the _____. When this happens the temperature of the reaction mixture _____. An _____ reaction gives out thermal energy. When this happens the _____ mixture _____ in temperature.

3. State the meaning of these terms:
   a kinetic energy
   b surroundings

> **EXAM TIP**
> Remember that in **EN**dothermic reactions the thermal energy **EN**ters (is taken in). In **EX**othermic reactions the thermal energy **EX**its (goes out).

> **KEY POINTS**
> - In physical changes, no new substance is formed. Physical changes are reversible.
> - In chemical changes, one or more new substances are formed. Some chemical changes cannot be reversed, although others can be reversed.
> - An exothermic reaction releases thermal energy to the surroundings. The temperature of the surroundings increases.
> - An endothermic change absorbs thermal energy from the surroundings. The temperature of the surroundings decreases.

8.2 Reaction pathway diagrams

LEARNING OUTCOMES
- Interpret reaction pathway diagrams for exothermic and endothermic reactions
- S Define activation energy
- Draw and label reaction pathway diagrams to include activation energy

EXAM TIP

Reaction pathway diagrams are sometimes called 'energy level diagrams' and the x-axis is sometimes labelled 'progress of reaction'.

Reaction pathway diagrams

We can show exothermic and endothermic changes with the help of **reaction pathway diagrams**. These diagrams show:
- the energy of the reactants and products on the vertical axis (y-axis). We do not usually write the exact values of the energy on this axis.
- the reactants and products, with the reactants on the left, the products on the right and the horizontal axis (x-axis) labelled as 'reaction pathway'.
- the enthalpy change, shown by an arrow drawn from the energy level of the reactants up or down to the energy level of the products.

The reaction pathway diagram for an exothermic reaction shows that:
- the energy of the reactants is higher than the energy of the products.
- the arrow goes downwards to show that energy is released (given out).
- the value of the enthalpy change (ΔH) is negative because (energy of products – energy of reactants) is negative.

The reaction pathway diagram for an endothermic reaction shows that:
- the energy of the reactants is lower than the energy of the products.
- the arrow goes up to show that energy is absorbed (taken in).
- the value of the enthalpy change (ΔH) is positive because (energy of products – energy of reactants) is positive.

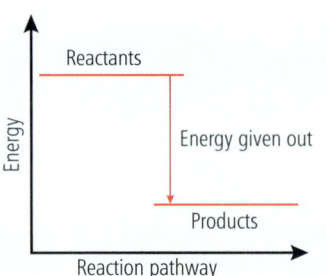

Figure 8.2.1 Exothermic reaction: the thermal energy of the reactants is greater than the thermal energy of the products.

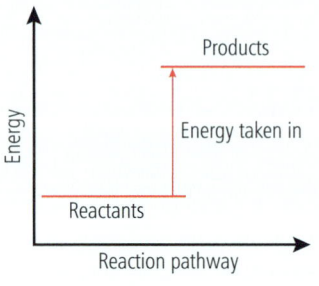

Figure 8.2.2 Endothermic reaction: the thermal energy of the reactants is lower than the thermal energy of the products.

Supplement

Activation energy, E_a

For a chemical reaction to happen, some of the bonds in the reactants must break. In order to break the bonds, energy is needed. The minimum energy that colliding particles must have in order to react is called the **activation energy**, E_a.

We can modify the reaction pathway diagrams to show the activation energy as a 'hump' between the reactants and the products. A full reaction pathway diagram also shows an arrow

Unit 8: Chemical energetics

going upwards, from the energy level of the reactats to the peak of the 'hump', to show the activation energy, E_a

Figures 8.2.4 and 8.2.5 show full reaction pathway diagrams for an exothermic reaction and an endothermic reaction.

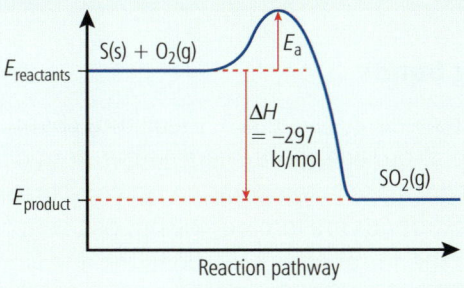

Figure 8.2.4 Full reaction pathway diagram for an exothermic reaction.

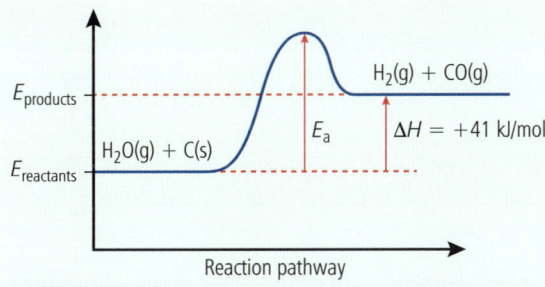

Figure 8.2.5 Full reaction pathway diagram for an endothermic reaction.

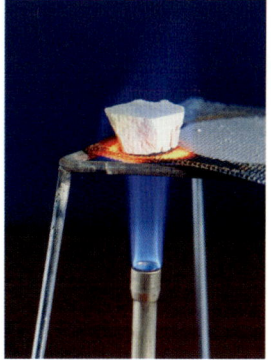

Figure 8.2.3 Calcium oxide (lime) is manufactured by heating calcium carbonate continuously. This is an endothermic reaction.

EXAM TIP

When drawing reaction pathway diagrams to include activation energy, remember that the activation energy arrow always goes upwards from the energy level of the reactants to the top of the energy 'hump'.

KEY POINTS

- In a simple reaction pathway diagram for an exothermic reaction, the reactants have more energy than the products. In a reaction pathway diagram for an endothermic reaction, the reactants have less energy than the products.

- Activation energy, E_a, is the minimum energy that colliding particles must have in order to react.

- Full reaction pathway diagrams show the symbols for the reactants and products, the enthalpy change and the activation energy.

SUMMARY QUESTIONS

1. Define activation energy.

2. Draw a reaction pathway diagram for this reaction:

   Fe + S → FeS   energy released to surroundings = 100 kJ/mol

3. Copy and complete using the words below:

   activation   break   downwards   negative   products   reactants   upwards

   The complete reaction pathway diagram for an exothermic reaction shows that the energy of the _____ is higher than the energy of the _____. The arrow showing the enthalpy change points _____. This shows that the value of the enthalpy change is _____. The arrow showing the _____ energy, E_a, always points _____ because energy is needed to _____ bonds to start the reaction.

99

8.3 Bond energy calculations
Supplement

LEARNING OUTCOMES
- Describe bond breaking as an endothermic process and bond making as an exothermic process
- Calculate the overall energy change of a reaction using bond energies and determine if the reaction is exothermic or endothermic

Figure 8.3.1 When fuels are burned the reaction is exothermic: more energy is released in bond forming than is needed in bond breaking.

EXAM TIP
Make sure you know that bond breaking is endothermic and bond forming is exothermic.

EXAM TIP
Remember that when new bonds are formed energy is transferred to increase the kinetic energy of the particles in the surroundings. So the temperature of the surroundings increases.

Making and breaking bonds

If we snap a stick, we put in energy to break it. Breaking a chemical bond is similar – we have to put energy in. Energy is transferred from the surroundings, which decrease in temperature. So bond breaking is endothermic. When new bonds are formed, the opposite happens – energy is transferred to the surroundings. So bond making is exothermic.

We can explain exothermic and endothermic reactions in terms of bond breaking and making:

- exothermic reactions: the energy taken in to break the bonds in the reactants is less than the energy given out when new bonds are made
- endothermic reactions: the energy taken in to break the bonds in the reactants is more than the energy given out when new bonds are made.

The difference between the thermal energy of the reactants and products is shown by the symbol ΔH (enthalpy change). If thermal energy is given out, ΔH is given a negative sign. If thermal energy is absorbed, ΔH is given a positive sign. For example in the reaction:

$$Zn(s) + CuSO_4(aq) \longrightarrow Cu(s) + ZnSO_4(aq) \qquad \Delta H = -212 \text{ kJ/mol}$$

Bond energies

Each type of bond has a particular amount of energy needed to break it. This is called the **bond energy**. Bond energy is the amount of energy needed to break one mole of a particular bond in one mole of gaseous molecules. The symbol for bond energy is E.

It needs 436 kJ to break the single bonds in one mole of hydrogen molecules. We can write this as:

$E(H—H) = +436$ kJ/mol

It needs 498 kJ to break the double bonds in one mole of oxygen molecules. We can write this as:

$E(O=O) = +498$ kJ/mol

Values of bond energies are always positive because they refer to bonds being broken.

Unit 8: Chemical energetics

Bond energy calculations

We can use bond energies to calculate how much energy is released or absorbed in a reaction. The 'balance sheet' method for doing this is shown in the worked example. Note that you have to take into account:

- the number of moles of each reactant and product in the stoichiometric equation
- the number of bonds of a particular type in each molecule, for example each molecule of water H_2O has two O—H bonds.

Worked example

Calculate the energy change in the reaction:

$2H_2(g) + O_2(g) \rightarrow 2H_2O(g)$

Bond energy values in kJ/mol: H—H 436; O=O 498; O—H 464

bonds broken (endothermic)		bonds formed (exothermic)	
2 H—H = 2 × 436	= 872 kJ	4 × O—H = 4 × 464	= 1856 kJ
1 O=O	= 498 kJ		
total	= 1370 kJ	total	= 1856 kJ

The calculation shows that when all the bonds in hydrogen and water are broken, the energy change is +1370 kJ. The positive sign shows that the energy change is endothermic.

When new bonds are formed, the amount of energy released is the same as the amount of energy absorbed when the same type of bond is broken, but the sign is reversed. So, when two moles of water are formed from hydrogen and oxygen atoms, the energy change is −1856 kJ. The negative sign shows that the energy change is exothermic.

For the calculation above, the overall energy change is:

+1370 − 1856 kJ = −486 kJ

So ΔH = −486 kJ

The negative sign for ΔH shows that the energy change for the reaction is exothermic.

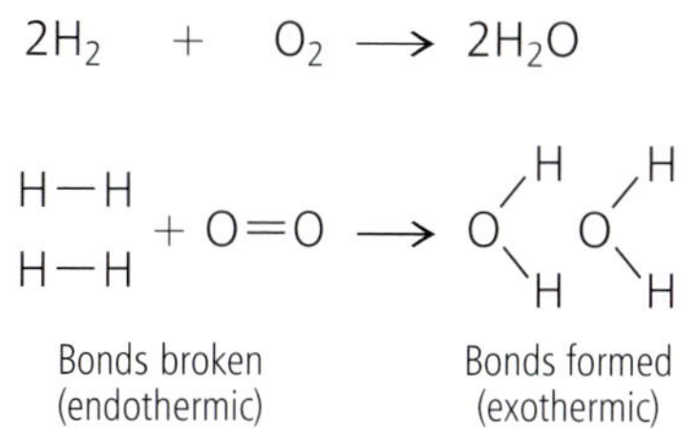

$2H_2 + O_2 \rightarrow 2H_2O$

H—H + O=O → (H₂O structure)
H—H

Bonds broken (endothermic)   Bonds formed (exothermic)

Figure 8.3.2 Bonding of the reactants and products in the equation for the formation of water.

EXAM TIP

When more complex molecules react, only some of the bonds are broken. It is a common error to think that during a reaction all the atoms in the reactants are turned into separate atoms.

KEY POINTS

- Bond breaking is endothermic and bond making is exothermic.
- The enthalpy change in a reaction is given by the symbol ΔH. For an exothermic reaction the value for ΔH is negative. For an endothermic reaction the value for ΔH is positive.
- The energy absorbed or released in a reaction can be calculated using bond energy values.

SUMMARY QUESTIONS

1. When ethane burns in oxygen to form carbon dioxide and water, energy is transferred to the surroundings. By referring to bond breaking and bond making, explain why energy is transferred to the surroundings.

2. When calcium carbonate is heated to a high temperature, the following reaction occurs:

   $CaCO_3(s) \rightarrow CaO(s) + CO_2(g)$
   ΔH = + 572 kJ/mol

   Draw a reaction pathway diagram for this reaction, including the activation energy.

3. Calculate the energy change in kJ when fluorine reacts with hydrogen to form hydrogen fluoride:

   $H_2 + F_2 \rightarrow 2HF$

   Include the correct sign for the energy change in your answer.

   Bond energy values in kJ/mol: H—H 436; F—F 158; H—F 568

8.4 Fuels and energy production

LEARNING OUTCOMES

- Name the fossil fuels coal, natural gas and petroleum
- Name methane as the main constituent of natural gas
- Give word equations for the complete combustion of carbon-containing fuels
- Describe the transfer of thermal energy by burning fuels
- State that carbon monoxide is formed from the incomplete combustion of carbon-containing fuels
- **S** Construct chemical equations for the complete combustion of carbon-containing fuels

Figure 8.4.1 Wood is a useful fuel but pollutes the atmosphere when it is burned.

The variety of fuels

A **fuel** is a substance that can be burned to transfer thermal energy to the surroundings. **Burning** a fuel is an exothermic reaction. We use a variety of fuels for heating, lighting, cooking, transport and making electricity. Many of the fuels we currently use are fossil fuels.

Petroleum (crude oil) is a complex mixture of compounds containing carbon and hydrogen.

Natural gas is largely methane.

Coal is very polluting and causes acid rain and increases global warming. Petroleum fractions and natural gas are less polluting but still contribute to global warming.

PRACTICAL

Comparing the energy transferred (see Figure 8.4.2)

1. We put a known amount of liquid fuel into the burner. When the fuel is set alight it heats up a measured volume of water in a calorimeter. The calorimeter is a copper can.
2. We compare the energy given out by different fuels, by measuring the temperature rise when burning 1 g of each fuel.
3. To make the experiment a fair test we must keep the following the same: volume of water in the calorimeter; height of flame; and distance of the tip of the flame from the calorimeter.

Combustion

Combustion (burning) is an exothermic chemical reaction between a fuel and an oxidising agent, usually atmospheric oxygen. The products formed are usually gases.

The most commonly used fuels are **hydrocarbons**. Hydrocarbons are compounds which contain only hydrogen and carbon. The simplest hydrocarbon is methane, CH_4.

Complete combustion occurs when hydrocarbons or other fuels burn in excess oxygen. Carbon dioxide and water are the products of the complete combustion of hydrocarbons:

methane + oxygen ⟶ carbon dioxide + water

propane + oxygen ⟶ carbon dioxide + water

Unit 8: Chemical energetics

Hydrogen is a non-hydrocarbon fuel that is non-polluting. Water is the only combustion product:

hydrogen + oxygen ⟶ water

Carbon undergoes complete combustion to form carbon dioxide as the only product.

carbon + oxygen ⟶ carbon dioxide

Incomplete combustion occurs when there is not enough oxygen to burn the fuel completely. Carbon monoxide and water are the main products of the incomplete combustion of hydrocarbons, although some carbon and carbon dioxide are also formed.

> **Chemical equations for combustion reactions**
>
> The chemical equations for the complete combustion of some hydrocarbon fuels are shown.
>
> $CH_4 + 2O_2 \longrightarrow CO_2 + 2H_2O$
> $C_8H_{18} + 12\frac{1}{2}O_2 \longrightarrow 8CO_2 + 9H_2O$

EXAM TIP
Remember that the burning of fuels is always an exothermic reaction.

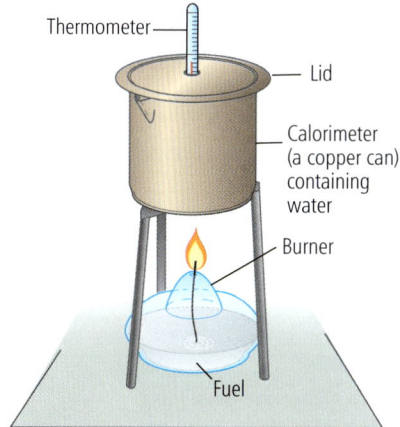

Figure 8.4.2 Comparing the energy transferred when liquid fuels are burned.

EXAM TIP
When balancing the chemical equations for the complete combustion of hydrocarbons, balance the carbon and hydrogen first. Leave the oxygen until last

KEY POINTS
- Coal, natural gas (mainly methane) and petroleum are fuels that are polluting when burned.
- When hydrocarbons are completely combusted, carbon dioxide and water are formed.
- Carbon monoxide is formed from the incomplete combustion of carbon-containing fuels.
- Chemical equations can be constructed for the complete combustion of carbon-containing fuels.

SUMMARY QUESTIONS

1. Copy and complete using the words below:

   acid   coal   exothermic   global   transferred   transport

   Fuels are used for heating, lighting and _____. The burning of fuels is an _____ reaction because thermal energy is _____ to the atmosphere. Fossil fuels such as _____ harm the environment by causing _____ rain and increased _____ warming.

2. Write word equations for:
   a the complete combustion of the hydrocarbon ethane
   b the complete combustion of carbon

3. Construct the chemical (symbol) equation for the complete combustion of butane, C_4H_{10}.

8.5 Fuel cells

LEARNING OUTCOMES

- State that a hydrogen–oxygen fuel cell uses hydrogen and oxygen to generate electricity, with water as the only product
- **S** Describe the advantages and disadvantages of using fuel cells in comparison with gasoline (petrol) engines in vehicles

Introducing fuel cells

Hydrogen is a non-polluting fuel. When it burns in oxygen, water is the only product formed. We can use this reaction to supply electrical energy continuously. We do this by reacting hydrogen and oxygen in a **fuel cell** in a reaction which does not involve combustion. No atmospheric pollutants are formed because water is the only product of the reaction.

PRACTICAL

A model fuel cell

1. We electrolyse a solution of sodium hydroxide. Hydrogen forms at the cathode and oxygen forms at the anode (see Figure 8.5.2).
2. We electrolyse the solution until both test tubes are filled with gas.
3. We then replace the power pack with a voltmeter. In the presence of the electrolyte, the hydrogen and oxygen react together. The voltmeter gives a reading. We have produced an electric current by combining hydrogen with oxygen.

Figure 8.5.1 Fuel cells have many advantages compared with a gasoline (petrol) engine.

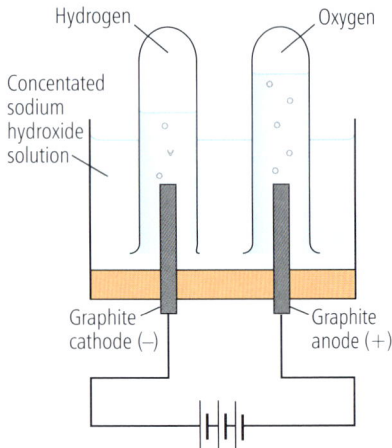

Figure 8.5.2 Electrolysing sodium hydroxide.

EXAM TIP

You do not need to remember details about the construction of a fuel cell, but you should be familiar with questions based on relevant information, including information about half-equations.

Supplement

How does a fuel cell work?

A fuel cell consists of two platinum electrodes and an electrolyte. The platinum is coated onto a porous material that allows gases to pass through it. Hydrogen gas is bubbled through the negative electrode and oxygen is bubbled through the positive electrode.

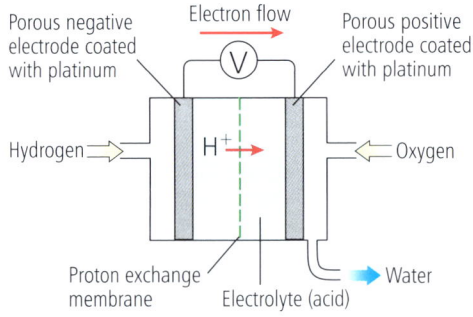

Figure 8.5.3 A hydrogen–oxygen fuel cell.

At the negative electrode, the hydrogen loses electrons and forms hydrogen ions in the electrolyte:

$$2H_2(g) \longrightarrow 4H^+(aq) + 4e^-$$

The released electrons move around the external circuit to the positive electrode. At the positive electrode, oxygen gains electrons and reacts with hydrogen ions from the acid electrolyte:

$$O_2(g) + 4H^+(aq) + 4e^- \longrightarrow 2H_2O(l)$$

The hydrogen ions removed at the positive electrode are replaced by those produced at the negative electrode. So the concentration of the electrolyte remains constant. The overall reaction is:

$$2H_2(g) + O_2(g) \longrightarrow 2H_2O(l)$$

The water is removed.

Alkaline electrolyte

Another type of fuel cell uses an alkaline electrolyte. The equations for the reactions at the electrodes are:

Negative electrode: $2H_2(g) + 4OH^-(aq) \longrightarrow 4H_2O + 4e^-$

Positive electrode: $O_2(g) + 2H_2O(l) + 4e^- \longrightarrow 4OH^-(aq)$

Advantages and disadvantages of fuel cells

Fuel cells have many advantages compared with gasoline (petrol) engines:

- water is the only product made; no CO_2 is produced (in gasoline engines the atmospheric pollutants CO_2, CO and NO_2 are formed)
- they transfer more energy per gram of fuel than gasoline
- fuel cells operate with higher efficiency than gasoline engines
- they are silent when the engine is running
- they are not as big as gasoline engines.

The disadvantages are:

- the cells are more expensive (platinum is very expensive)
- hydrogen is expensive to produce
- there are very few hydrogen filling points at garages
- the hydrogen must be stored safely as a liquid in strong containers because hydrogen is extremely flammable.

EXAM TIP

Reduction always takes place at a cathode and oxidation at an anode. So when using these terms with fuel cells, the anode is the − electrode and the cathode is the + electrode.

KEY POINTS

- A hydrogen–oxygen fuel cell used to generate electricity produces only water as a product.
- The advantages of a fuel cell compared with a gasoline (petrol) engine are that the product is non-polluting, it is more efficient and it provides more energy per gram of fuel burned.

SUMMARY QUESTIONS

1 Copy and complete using the words below:

electricity   oxygen
pollutant   water

A hydrogen–_____ fuel cell uses hydrogen and oxygen to generate _____. The only product formed is _____ which is not an atmospheric _____.

2 Write down the half-equations for the reactions taking place at each electrode in a fuel cell containing an acidic electrolyte.

3 State three advantages of a fuel cell compared with a gasoline (petrol) engine.

SUMMARY QUESTIONS

1. State whether each of the following reactions is endothermic or exothermic:
   a. burning wood
   b. a reaction where the temperature of the reaction mixture falls
   c. decomposition of calcium carbonate by heating
   d. hydrogen exploding

2. Copy and complete the paragraph using words from the list below:

   **decreases   reaction   gain   products
   reactants   surroundings**

   The _____ pathway diagram for an endothermic reaction shows that the energy of the _____ is lower than the energy of the _____. The chemicals in the reaction mixture _____ energy from the _____. The temperature of the reaction mixture _____.

3. Match the sentence starters on the left with the endings on the right.

Bond breaking is …	a source of electrical energy
Bond making is …	a hydrocarbon
Reduction takes place at …	endothermic
A fuel cell is …	exothermic
Methane is …	a cathode

Practice questions

1. Choose the phrase that best describes an endothermic reaction.
   A The reaction releases thermal energy.
   B The reaction absorbs thermal energy.
   C The temperature of the reaction mixture increases.
   D The surroundings get warmer.

   (Paper 1)

2. Choose the fuel that is **not** a fossil fuel.
   A Petroleum   B Natural gas
   C Coal   D Hydrogen

   (Paper 1)

3. Choose the correct statement about an endothermic reaction.
   A The energy absorbed to break the bonds in the reactants is greater than the energy released when new bonds formed.
   B The energy absorbed when breaking the bonds in the reactants is less than the energy released when new bonds formed.
   C The energy absorbed to break the bonds in the reactants is greater than the energy absorbed when new bonds formed.
   D The energy released when breaking the bonds in the reactants is greater than the energy absorbed when new bonds formed.

   (Paper 2)

4. When a given volume and concentration of hydrochloric acid reacts with excess potassium hydroxide, 20 kJ of energy is released.
   (a) State the name given to a chemical reaction that releases thermal energy. [1]
   (b) Describe an experiment to show that energy is released in this reaction. [3]
   (c) Copy and complete the reaction pathway diagram for this reaction. [3]

   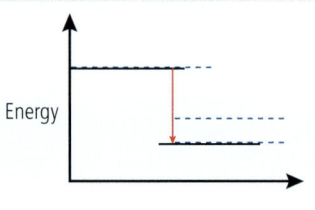

   (d) Predict the amount of thermal energy given out when the concentration of the hydrochloric acid is doubled. All other conditions are the same. [1]

   (Paper 3)

5. Equal amounts of three compounds are dissolved in the same amount of water in separate beakers and the temperature change observed. The table shows the results:

compound	initial temperature of water/°C	final temperature of solution/°C
potassium nitrate	25	18
sodium hydroxide	18	38
calcium chloride	22	40

106

(a) From the table, name one compound that: (i) releases thermal energy when it dissolves (ii) absorbs thermal energy when it dissolves. [2]

(b) State the name given to a chemical reaction that absorbs thermal energy. [1]

(c) State the name of the compound that shows the greatest temperature change when it dissolves in water. [1]

(d) When sodium hydroxide dissolves in water, suggest the temperature change when:
   (i) the amount of water is doubled. All other factors are constant. Give a reason for your answer. [2]
   (ii) the amount of sodium hydroxide is doubled. All other factors are constant. [1]

(Paper 3)

6 Natural gas contains methane, CH_4.

(a) Choose the word that best describes methane.

   carbohydrate   dioxide   hydrated hydrocarbon [1]

(b) Write a word equation for the complete combustion of methane. [2]

(c) The incomplete combustion of methane produces a different gas.
   (i) State the meaning of incomplete combustion. [1]
   (ii) State the name of the different gas formed by incomplete combustion. [1]

(d) Propane, C_3H_8, is also a hydrocarbon fuel. Copy and complete the chemical equation for the complete combustion of propane.

   $C_3H_8 + 5O_2 \rightarrow$ ____ CO_2 + ____ H_2O [1]

(Paper 3)

7 Methane burns to form carbon dioxide and water. The reaction is exothermic.

(a) State the meaning of the term exothermic. [1]

(b) Explain why this reaction is exothermic in terms of bond breaking and bond making. [2]

(c) The reaction can be represented by the equation:

$$CH_4 + 2O_2 \rightarrow CO_2 + 2H_2O$$

Draw a reaction pathway diagram for this reaction. Include the activation energy, E_a. [4]

(d) Use the bond energies below to answer the following questions.

O=O 498 kJ/mol   C=O 805 kJ/mol
C—H 413 kJ/mol   O—H 464 kJ/mol

   (i) Calculate the energy needed to break the bonds in the reactants. [2]
   (ii) Calculate the energy released when the bonds in the products are formed. [2]
   (iii) Calculate the overall energy change for the reaction. Include the correct sign for the energy change. [2]

(Paper 4)

8 Hydrogen can be used as a fuel.

(a) Explain in terms of the environment the advantage of hydrogen as a fuel compared with gasoline (petrol). [3]

(b) Hydrogen burns in oxygen to form water:

   $2H_2 + O_2 \rightarrow 2H_2O$

   Use the bond energies below to calculate the energy change for this reaction. [3]

   O=O 498 kJ/mol   H—H 436 kJ/mol
   O—H 464 kJ/mol

(c) Hydrogen and oxygen react in a fuel cell containing an electrolyte of potassium hydroxide. At the negative electrode hydrogen reacts with hydroxide ions to form water. At the positive electrode oxygen reacts with water to form hydroxide ions.
   (i) Write a half-equation for the reaction at the negative electrode. [2]
   (ii) Write a half-equation for the reaction at the positive electrode. [2]
   (iii) Explain why an electric current is produced in the external circuit. [3]

(Paper 4)

9.1 Investigating rate of reaction

LEARNING OUTCOMES

- Describe practical methods for investigating rate of reaction
- Suggest advantages and disadvantages of the experimental methods and apparatus used

Introducing rate

Some reactions, such rusting, are very slow. Other reactions are very fast. When you mix solutions of silver nitrate and potassium iodide, you get a yellow precipitate straight away.

The **rate of reaction** tells us how rapidly the products are formed from the reactants. There are many different methods to show us how fast a reaction proceeds.

Following the progress of a reaction

To find the rate of reaction we can either:

- measure how quickly the reactants are used up, or
- measure how quickly the products are formed.

To calculate the rate of reaction we have to measure something that changes with time.

Examples include the volume of gas, the mass of the reaction mixture or the amount of light passing through a solution.

Figure 9.1.1 We run at different rates. The chemical reactions in our bodies supply our muscles with glucose at the rate needed.

Following the change in mass

You can use this method for reactions that give off a gas which is allowed to escape. As the reaction takes place the mass of the reaction mixture decreases. You record the mass at intervals of time. The loss of mass is equal to the mass of gas given off. When the total mass is plotted against time you get a graph like the one shown here.

Modern balances can be attached to a data logger and computer so that you can record the loss of mass continuously.

EXAM TIP

Remember that the rate of a reaction involves two things: the change in amount of product or reactant and the time it takes for this change to happen.

EXAM TIP

Remember that change in mass is not a suitable method for investigating a reaction where hydrogen is released. Due to the low density of hydrogen gas, the change in mass is too small.

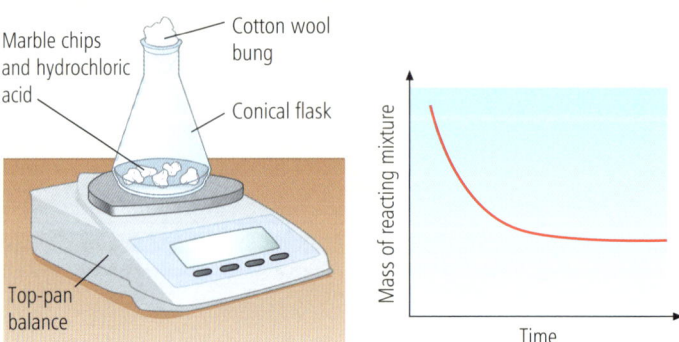

Figure 9.1.2 Following a reaction by change in mass.

Unit 9: Rates of reaction

Following a reaction by change in gas volume

Using a gas syringe to measure volume

The volume of gas is recorded at suitable time intervals. Then you can draw a graph of volume of gas given off against time.

If this method is used, it is important that there are no leaks in the apparatus so that no gas escapes.

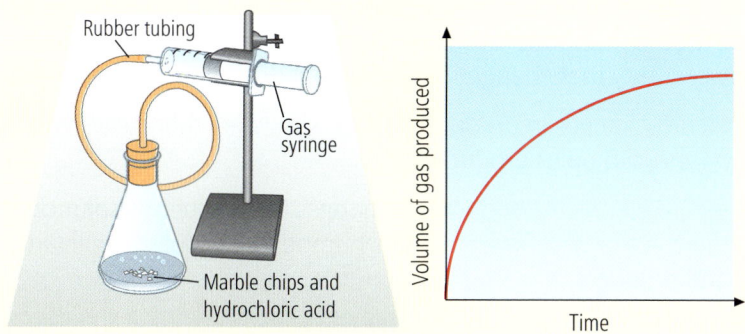

Figure 9.1.3 Following a reaction using a gas syringe.

Using an upturned measuring cylinder of water to measure volume

1 Set up the apparatus, with the measuring cylinder full of water.

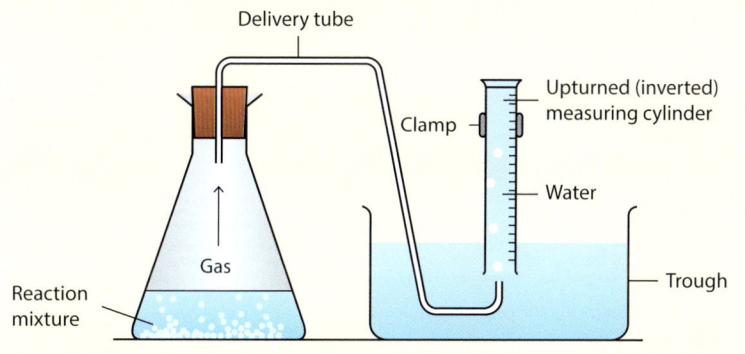

Figure 9.1.4 Following a reaction using a measuring cylinder.

2 The gas collects above the water. The pressure of the gas pushes the water downwards. The volume is recorded at time intervals.

EXAM TIP

You cannot use the measuring cylinder method for gases that are soluble in water because they will dissolve in the water.

KEY POINTS

- We can follow the progress of a chemical reaction by measuring how fast the reactants are used up or how fast the products are formed.
- We can use loss of mass of reactant with time or change in the volume of gas with time to follow the progress of a reaction.
- There are advantages and disadvantages of using some experimental methods for following the progress of a reaction where a gas is given off.

SUMMARY QUESTIONS

1 Copy and complete using the words below:

decrease   mass   mixture   rate

The _____ of reaction can be followed by measuring the _____ in mass of the reaction _____. This is equal to the _____ of gas produced.

2 Sketch a graph to show how the volume of gas given off changes with time for this reaction:

$Mg + 2HCl \longrightarrow MgCl_2 + H_2$

3 Put the following in order of increasing rate of reaction:

cement setting   firework exploding   iron rusting

4 Sulfur dioxide gas is very soluble in water. Explain why you cannot use the upturned measuring cylinder method for collecting and measuring the volume of this gas.

9.2 Evaluating experiments

LEARNING OUTCOMES

- Describe practical methods for investigating rate of reaction
- Suggest advantages and disadvantages of experimental methods and apparatus
- S Evaluate practical methods for investigating rate of reaction

A variety of methods

In Topic 9.1 we learned that change in mass or gas volume with time can be used to show the progress of a reaction. There are also many other methods for measuring rate of reaction. You can use any property that changes during a reaction. Some examples are:

- record changes in pH during a reaction where there are acids or bases used up or produced
- record electrical conductivity changes in an aqueous solution if there is a large difference in the number of ions in the reactants and products
- record differences in the light absorbed by an aqueous solution if the reactants and products are different colours.

Following the progress of a precipitation reaction

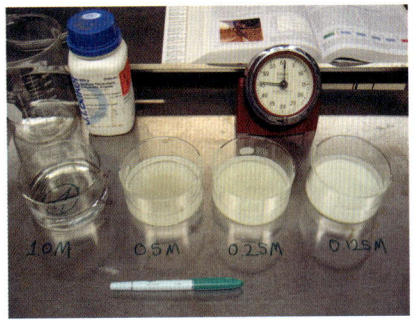

Figure 9.2.1 A precipitate of sulfur forms slowly when hydrochloric acid is added to sodium thiosulfate at room temperature.

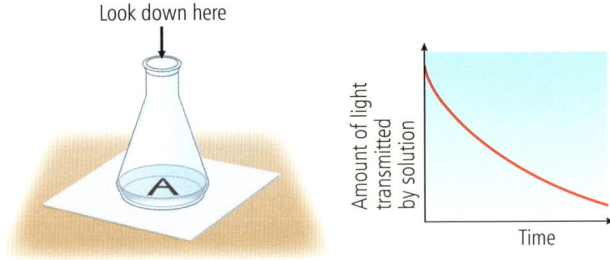

Figure 9.2.2 Following the progress of a precipitation reaction.

If a precipitate is one of the products in a reaction, the solution goes cloudy. We can follow the reaction by placing paper with a letter A on it underneath the flask containing one of the reactants. Start the reaction by adding the second reactant and recording the time taken for the letter A to 'disappear'.

We can also measure the amount of light passing through the solution using a colorimeter (or a light meter attached to a data logger and computer). This gives a graph as shown in Figure 9.2.2.

Topic 9.6 shows you how the reaction between sodium thiosulfate and hydrochloric acid can be used to investigate the effect of temperature on rate of reaction

Constants and variables

Many factors such as temperature, concentration of reagent and particle size can affect the rate of reaction. How can you compare the results of a series of experiments in a fair way? You need to keep some things constant (**control variables**) and change only the variable that you are investigating (the **independent variable**).

Unit 9: Rates of reaction

For example, you might want to find out how the concentration of hydrochloric acid affects the rate of reaction of the acid with calcium carbonate.

If so, you must keep the temperature and particle size of the calcium carbonate the same in each experiment. You change only the concentration of the hydrochloric acid (independent variable) and record the times taken (the **dependent variable**) to collect particular volumes of gas as the experiment proceeds.

> **EXAM TIP**
>
> The accuracy of measurements is how close the measured values are to the true value.

Supplement

Evaluating experiments

We need to think about the following things when evaluating how successful an experiment is.

- The **accuracy** of the instruments that you use. Using instruments that measure more accurately may improve the accuracy of your results.
- Using measuring instruments carefully will also improve the accuracy of your results.
- Repeat your measurements.
- The spread of a set of repeat readings will give you an idea of the reliability of that measurement.
- The mean value is more likely to be accurate.
- Repeat the experiments in exactly the same way.
- Is the range of data good enough? For example: you want to show the effect of temperature on reaction rate by reacting a fixed amount of magnesium with excess acid by measuring the volume of gas released. Using temperatures of 20 °C, 22 °C, 25 °C and 28 °C is less likely to give a good result than using a wider range of values such as 20 °C, 30 °C, 40 °C and 45 °C.
- Is the apparatus suitable for the experiment? The correct arrangement and use of the apparatus is important. If gases are to be dried, they must be dried using a substance that does not react with them or dissolve them.

> **KEY POINTS**
>
> - We can follow the progress of a chemical reaction by measuring changes in pH, electrical conductivity, light transmitted or the time taken for a precipitate to make a letter 'disappear'.
> - Evaluation includes ideas about (i) the reliability and accuracy of the results, and (ii) problems with the design of the experiment.

SUMMARY QUESTIONS

1. Suggest two ways of measuring the progress of this reaction:

   $2H^+(aq) + Ca(s) \longrightarrow Ca^{2+}(aq) + H_2(g)$

   (Hint: (aq) means dissolved in water, (s) means solid and (g) means gas.)

2. Copy and complete using the words below.

   constant   dependent   measure   variable

   A control variable is kept _____ to ensure that the experiment is carried out in a fair way. The _____ variable is the variable we _____ to find the effect of changing the independent _____.

3. Look at Figure 8.4.2 on p. 103 and read about the experiment. Then answer these questions.

   a List three variables in this experiment that need to be controlled.

   b A calculation showed that the experimental value of the thermal energy transferred to the water was very different from the true value. Suggest three reasons why it was different even though no errors were made in the readings.

111

9.3 Interpreting data

LEARNING OUTCOMES
- Interpret data, including graphs, from rate of reaction experiments

Figure 9.3.1 Reactions such as rusting proceed very slowly.

EXAM TIP
The rate of reaction is always greatest at the start of a reaction and then decreases as the reaction proceeds. This applies to graphs of loss of mass with time as well as graphs of increase in volume of gas with time.

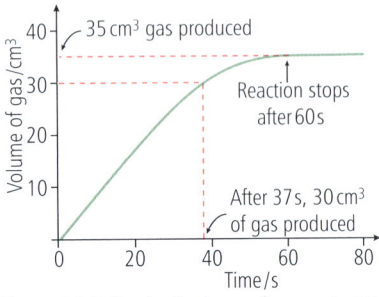

Figure 9.3.3 Graph of volume of gas against time.

Calculating rate of reaction

In a reaction such as:

$$Mg + 2HCl \longrightarrow MgCl_2 + H_2$$

the time taken for the magnesium to disappear completely can be used as a measure of the rate of reaction. The average reaction rate can be worked out by using this equation:

$$\text{average rate of reaction} = \frac{\text{amount of product formed or reactant used up}}{\text{time}}$$

A more accurate definition of rate of reaction is:

$$\text{average rate of reaction} = \frac{\text{change in concentration of reactant or product}}{\text{time}}$$

Interpreting graphs involving rate of reaction

We often need to know how the reaction rate changes as the reaction proceeds. If we look at the graph of how the volume of gas given off in a reaction changes with time, the gradient of the line at any given point (e.g. y/x in Figure 9.3.2) gives us the reaction rate at any particular time. We see that the reaction is fastest near the start but then gets slower and slower until it finally stops.

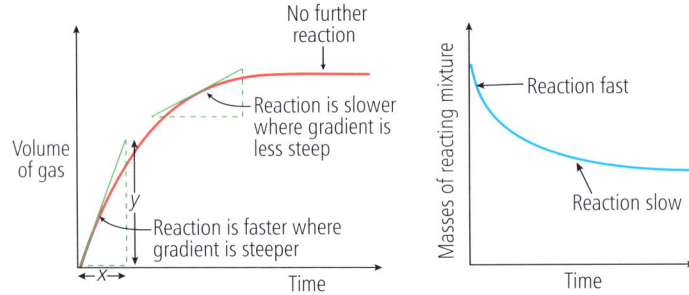

Figure 9.3.2 How volume of gas and mass of reaction mixture change with time

If we look at a graph of loss of mass of the reaction mixture against time we can see a similar pattern.

From reading the left-hand graph in Figure 9.3.2 we can also find out:

- how long it takes for a reaction to produce a given volume of gas
- the volume of gas produced in a given time.

This is shown in more detail in Figure 9.3.3.

Comparing rates using different concentrations of acids

You may be asked to draw graphs to show how the progress of a reaction changes under different conditions. We will take the reaction between magnesium and dilute hydrochloric acid as an example:

$$Mg + 2HCl \longrightarrow MgCl_2 + H_2$$

Unit 9: Rates of reaction

The reaction can be carried out using either excess magnesium or excess hydrochloric acid.

Figure 9.3.4 shows how the volume of hydrogen changes when magnesium is in excess and three different concentrations of hydrochloric acid are used. All other conditions are then kept the same. The final volume of hydrogen produced is higher as the concentration of hydrochloric acid increases. This is because there is enough magnesium to react with all the hydrochloric acid. Hydrochloric acid is the limiting reactant.

Figure 9.3.5 shows how the volume of hydrogen changes when hydrochloric acid is in excess and three different concentrations of hydrochloric acid are used. All other conditions are kept the same. The final volume of hydrogen produced is the same in each case. This is because there is enough hydrochloric acid to react with all the magnesium. The final volume of gas depends on the amount of magnesium present. Magnesium is the limiting reactant.

> **EXAM TIP**
>
> Make sure that you know which reactant is in excess in order to compare graphs of volume or mass against time.

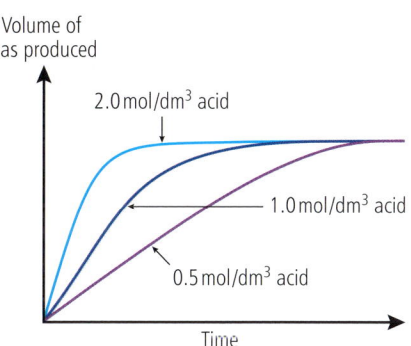

Figure 9.3.5 Volume of H_2 released with time using three concentrations of dilute HCl, with the HCl in excess.

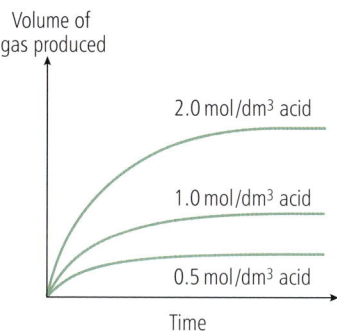

Figure 9.3.4 Volume of H_2 released with time using three concentrations of HCl, with Mg in excess.

SUMMARY QUESTIONS

1 Sketch a graph to show how the mass of the reaction mixture changes with time for the reaction:

$$MgCO_3 + 2HCl \longrightarrow MgCl_2 + CO_2 + H_2O$$

On your graph show:
   a where the reaction had just stopped
   b where the reaction is fastest
   c where the reaction is very slow

2 A student adds some pieces of calcium carbonate to some dilute acid. At the start of the reaction many bubbles are seen. After 10 minutes some calcium carbonate is still present but the bubbles have stopped. Which reactant is in excess? Explain your answer.

3 Copy and complete using the words below:

   fast   slows   stops

   When excess calcium carbonate reacts with dilute hydrochloric acid, the reaction is very _____ at first but then _____ down until it _____ completely.

> **KEY POINTS**
>
> - Average rate of reaction is calculated by dividing change in the amount of reactant or product by time.
> - As a reaction proceeds, the rate of reaction decreases as one or more of the reactants gets used up.
> - A reaction stops when one of the reactants (the limiting reactant) is completely used up.

9.4 Surfaces and reaction rate

LEARNING OUTCOMES

- Describe the effect of surface area of solids on rate of reaction
- Describe a catalyst as a substance that speeds up a reaction but remains chemically unchanged at the end of the reaction
- **S** Explain that a catalyst lowers the activation energy so that a greater proportion of collisions are successful

Surface area and rate of reaction

If you want to make a fire quickly, you are more likely to succeed if you try to light small, thin pieces of wood rather than large pieces. We see a similar effect when solids react with solutions. A large lump of marble reacts slowly with hydrochloric acid but powdered marble reacts very quickly. Why is this?

The rate of reaction depends on how frequently the particles of acid collide with the particles on the surface of the marble. For the same mass of marble, the greater the surface area, the more particles there are available to react. If we cut up the marble into smaller pieces, the surface area and the number of particles of marble which can react with the acid are both increased.

Investigating the effect of surface area

Marble is a form of calcium carbonate. You can investigate the effect of increasing the surface area by reacting the same mass of different-sized marble chips with hydrochloric acid. You can carry out the reaction either by measuring the loss in mass as carbon dioxide is released or by recording the volume of carbon dioxide given off.

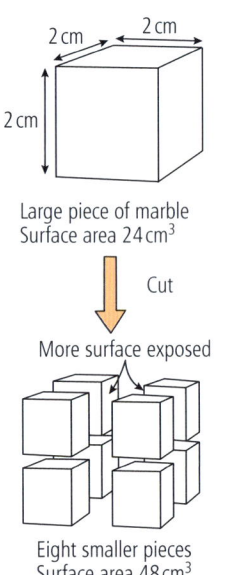

Figure 9.4.1 Cutting a cube into smaller ones increases the total surface area.

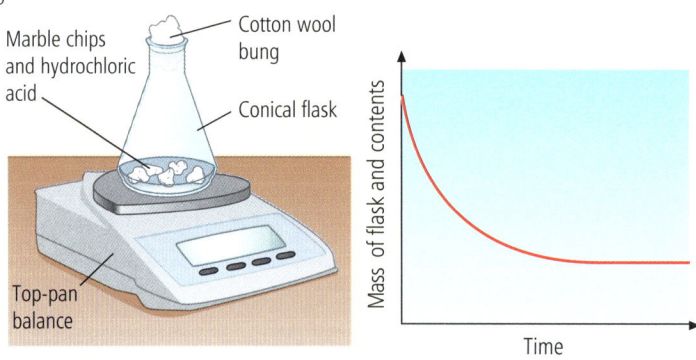

Figure 9.4.2 Investigating the effect of surface area.

EXAM TIP

It is a common misconception to think that larger particles have a larger surface area than smaller ones. Think of a large cube cut up – by cutting, you are exposing more surfaces.

Catalysts

A **catalyst** is a substance that speeds up a chemical reaction.

The catalyst is not used up in the reaction. It remains chemically unchanged at the end of the reaction. All living things contain particular types of catalysts called **enzymes**. These are specific types of protein which speed up all the chemical reactions in living things.

Unit 9: Rates of reaction

Supplement

Catalysts and collision theory

A catalyst works by lowering the activation energy, E_a, so that a greater proportion of the collisions between particles are successful (Figure 9.4.3).

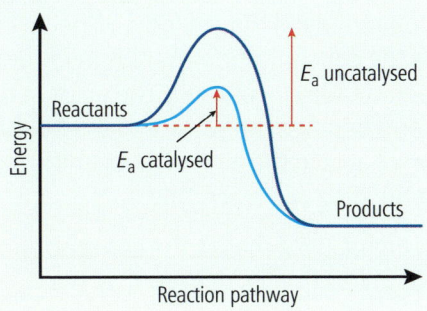

Figure 9.4.3 Reaction pathway diagram showing catalysis.

There is more information about activation energy in Topics 8.2 and 9.6.

There are two types of catalyst: (i) solid catalysts and (ii) catalysts that work in solution.

All living things contain particular types of catalysts called enzymes. These are specific types of protein which speed up all the chemical reactions in living things.

SUMMARY QUESTIONS

1 Copy and complete using the words below:

again   increases   rate   unchanged

A catalyst is a substance which _____ the _____ of a chemical reaction. The catalyst is _____ at the end of the reaction, so it can be used _____.

2 Factories that cut reactive metals often have special fans to remove metal dust. Explain why metal dust in the atmosphere can be dangerous by referring to surface area.

3 The inside of a catalytic converter in some cars has thousands of tiny beads coated with the catalyst. Explain why these beads are used rather than large lumps of catalyst.

S 4 Explain how a catalyst makes a reaction faster.

EXAM TIP

When defining a catalyst, the best definition is 'a substance that speeds up a reaction'. Phrases such as 'a substance which changes the rate of a reaction' are rather vague.

EXAM TIP

It is a common error to suggest that a catalyst does not change during a chemical reaction. It takes part in the reaction and may undergo a series of changes, but it always remains unchanged at the end of the reaction.

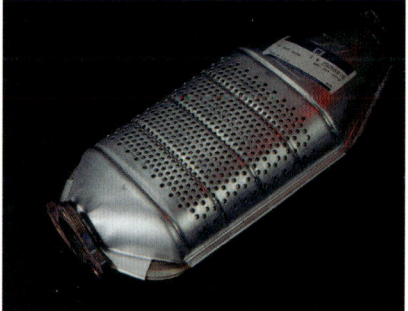

Figure 9.4.4 Catalytic converters are important in reducing pollutant gases from cars.

KEY POINTS

- Increasing the surface area of a solid reactant increases the rate of reaction.
- Smaller particles of solid have a larger surface area than larger particles with the same total volume.
- A catalyst increases the rate of a chemical reaction.
- A catalyst lowers the activation energy, E_a, of a reaction so that a greater proportion of collisions are successful.

9.5 Concentration and rate of reaction

LEARNING OUTCOMES

- Describe the effect of concentration of solutions and pressure of gases on rate of reaction
- **S** Explain using the **collision theory** how increase in concentration of solution and increase in pressure increase the rate of reaction

How concentration affects the rate of reaction

The Taj Mahal, a beautiful building in India, is being eroded away! Over recent years the concentration of acids in rainwater all over the world has been steadily increasing. The acids in the air react with the marble and damage the surface of the building. The higher the concentration of acid in the air, the quicker the exterior marble or limestone of a building will react.

You can see from Figure 9.5.2 that as the concentration of acid increases, the rate of reaction increases.

The same applies to reactions with gases. The rate of reaction of gases increases as the pressure increases. As the pressure increases, there is a greater number of particles per unit volume (greater concentration of particles).

Figure 9.5.1 This marble statue has been damaged by acid rain. The greater the concentration of acid in the rain, the greater the damage caused.

How changing the concentration of acid affects reaction rate

1. The rate of reaction of marble chips (calcium carbonate) is carried out with hydrochloric acid of different concentrations.

$$CaCO_3 + 2HCl \longrightarrow CaCl_2 + CO_2 + H_2O$$

2. We can follow the reaction by measuring the increasing volume of carbon dioxide given off as the reaction proceeds. The temperature, mass and size of the marble chips are kept constant, as well as the volume of dilute hydrochloric acid used. Just the concentration of hydrochloric acid is varied but it is always in excess.

3. For each concentration of hydrochloric acid used, we record the volume of CO_2 gas produced at time intervals.

A graph of the results is shown.

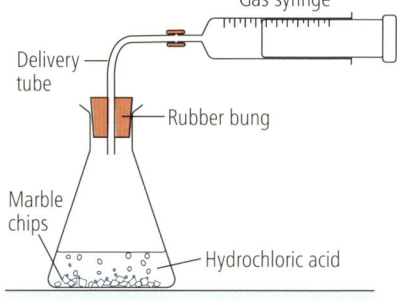

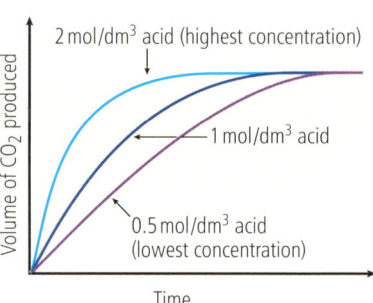

Figure 9.5.2 How changing the concentration of acid affects reaction rate.

Unit 9: Rates of reaction

Supplement
Concentration and the collision theory

A concentrated solution has more particles of solute per unit volume than a dilute solution. In the reaction between calcium carbonate and hydrochloric acid, the important solute particles are the hydrogen ions in the hydrochloric acid (see Topic 11.4).

A reaction occurs when particles collide with enough energy. The more concentrated the hydrochloric acid, the more hydrogen ions there are per unit volume to collide and react with the carbonate particles in the solid calcium carbonate. The rate of reaction depends on the number of successful collisions per second. The number of collisions per second is called the **collision frequency**. If the frequency of successful collisions is greater, the rate of reaction is faster.

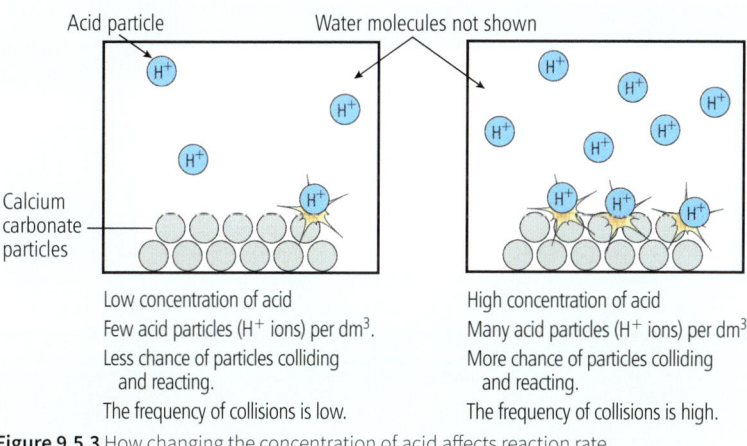

Figure 9.5.3 How changing the concentration of acid affects reaction rate.

In a reaction involving gases, increasing the pressure has a similar effect to increasing the concentration in a liquid. Increasing the pressure increases the number of particles per unit volume so that they collide with a greater frequency. So the rate of reaction increases.

> **EXAM TIP**
> When explaining the effect of concentration on reaction rate don't just refer to more collisions between the particles. It is the more frequent collisions of the particles which is important.

> **EXAM TIP**
> Remember that increasing the concentration of a reactant has no effect on the force with which the reacting particles hit each other.

> **KEY POINTS**
> - Increasing the concentration of reactants increases the rate of reaction.
> - Increasing the pressure of reacting gases increases the rate of reaction.
> - Increasing the concentration of reactants increases number of particles per unit volume and the frequency of collision of the reacting particles, and so increases reaction rate. **S**

SUMMARY QUESTIONS

1. Copy and complete using the words below:

   concentration   increases   rate

   The _____ of a chemical reaction _____ when the _____ of one or more of the reactants increases.

2. **a** When excess barium carbonate reacts with dilute hydrochloric acid, carbon dioxide gas is given off. Sketch a graph to show how the mass of the reaction mixture changes.

   b On the same set of axes, sketch the graph you would expect if you repeat the experiment using hydrochloric acid of half the concentration.

3. Copy and complete using the words below: **S**

   collide   collision   concentrated   frequently   unit

   The effect of increasing reactant concentration on the rate of reaction can be explained by _____ theory. When the reactants are more _____ there are more reactant particles present per _____ volume so they _____ more _____.

4. Use ideas from the collision theory to suggest why a reaction slows down as time goes on.

117

9.6 Temperature and rate of reaction

LEARNING OUTCOMES

- Describe the effect of temperature on the rate of reaction
- **S** Explain using the collision theory how increasing the temperature increases the rate of reaction

How temperature affects rate of reaction

We can change the rate of a chemical reaction by heating or cooling the reaction mixture. Food goes mouldy quicker when left out of the refrigerator. This is because the reactions which make food rot are faster at higher temperatures. Some surgical operations are carried out below normal room temperature to slow down the chemical reactions in the body.

Investigating the effect of temperature on reaction rate

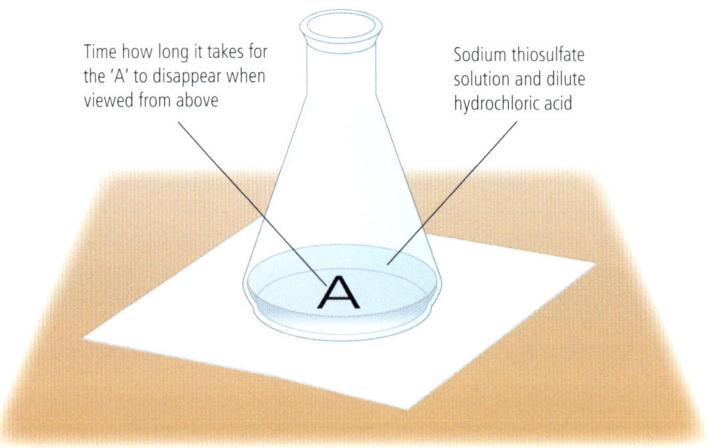

Figure 9.6.1 Reaction rate is affected by temperature.

When we react sodium thiosulfate solution with dilute hydrochloric acid, a precipitate of sulfur is formed.

1. Place the flask of sodium thiosulfate on top of the letter A on the paper.
2. Add hydrochloric acid and record the time taken for the letter A to 'disappear'.
3. Repeat the experiment at different temperatures using the same volumes of acid and thiosulfate warmed up separately before mixing them together. Do not heat above 50 °C.

You can plot a graph of the time taken for the letter A to 'disappear' against temperature. The shorter the time taken for the letter A to 'disappear', the faster the reaction is. This is because rate of reaction is proportional to amount of product formed (or reactant used) divided by time. If we plot a graph of 1/time against temperature (Figure 9.6.2), we see that for every 10 °C rise in temperature, the rate of reaction approximately doubles.

EXAM TIP

When comparing rates of reaction it is important to use words like 'faster' or 'slower' not just 'fast' or 'slow'.

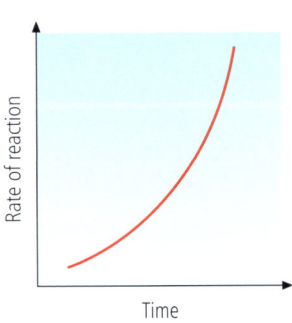

Figure 9.6.2 How rate of reaction changes with temperature.

Unit 9: Rates of reaction

Supplement
Temperature and the collision theory

There are two ways we can explain why increasing the temperature increases the rate of reaction. The second of these is the major factor. This is the answer you should give.

1. When we heat up a reaction mixture, the particles move faster. The kinetic energy of the particles increases. So the frequency of collisions is increased. This results in an increased rate of reaction.

2. As the temperature increases more particles have energy equal to or greater than the activation energy. So as the temperature increases, a greater proportion of the collisions are successful and result in a reaction taking place. The collisions are more effective at a higher temperature.

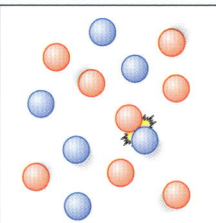 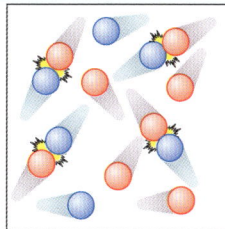

Lower temperature. Particles have less energy. They move more slowly and collide less frequently. The collisions are less effective.

Higher temperature. Particles have more energy. They move faster and collide more frequently. The collisions are more effective.

Figure 9.6.3 Increase in temperature increases reaction rate.

Figure 9.6.4 The faster you move, the more likely you are to bump into something and the bump will be harder too!

EXAM TIP
When explaining how increasing the temperature affects reaction rate, the most important points are that particles have greater kinetic energy and more particles have energy greater than the activation energy.

KEY POINTS
- The higher the temperature the greater the rate of reaction.
- The rate of reaction increases with an increase in temperature because:
  - the particles have more kinetic energy
  - more of the colliding particles have energy above the activation energy
  - there is an increase in the frequency of collisions.

SUMMARY QUESTIONS

1. Copy and complete using the words below:

   rate   slower   time

   The longer the _____ taken for a reaction to be complete, the _____ the _____ of reaction.

2. A student followed the rate of reaction of calcium carbonate with hydrochloric acid at different temperatures. At each temperature she recorded how long it took to collect 30 cm^3 of gas. Sketch a graph to show how the time changed with temperature.

3. Copy and complete the paragraph using these words:

   activation   effective   energy   faster   increases   kinetic   more

   When the temperature of a reaction mixture is increased the particles move _____ because the _____ _____ the particles increases. At a higher temperature there are also _____ particles with energy greater than the _____ energy. So the collisions are more _____ and the rate of reaction _____.

119

SUMMARY QUESTIONS

1 Match the type of reaction on the left with the appropriate method for measuring reaction rate on the right:

a purple solution changes to a colourless solution	measure the volume of gas produced
hydrogen is released during a reaction	measure change in electrical conductivity
two solutions react slowly to form a precipitate	measure the light transmitted through a solution
there are more ions in solution in the reactants than in the products	see how long it takes for a letter 'A' under a flask to disappear

2 Write definitions of the following:
   a rate of reaction
   b catalyst
   c gradient

3 Sketch a graph to show how the mass of solid changes over time in the reaction:
   $CaCO_3(s) \longrightarrow CaO(s) + CO_2(g)$

4 Complete the following phrases:
   a Increasing the surface area of a solid _____ the rate of reaction.
   b As a reaction proceeds the rate of reaction _____.

5 What effect does each of the following have on the rate of reaction?
   a diluting the reaction mixture
   b using large lumps of solid rather than small ones

6 Use your knowledge of the collision theory to describe and explain how increasing the pressure affects the rate of the following reaction:
   $2N_2O(g) \longrightarrow 2N_2(g) + O_2(g)$

7 Write definitions of the following:
   a limiting reactant
   b activation energy
   c collision frequency

8 Use the collision theory to suggest why food cooks more quickly when the temperature is higher.

Practice questions

1 Which one of the following statements about catalysts is true? [1]
   A They are always non-metals.
   B They do not take part in chemical reactions.
   C They have no effect on rate of reaction.
   D Their mass remains unchanged at the end of the reaction.

(Paper 1)

2 The table shows the volume of oxygen given off when hydrogen peroxide decomposes at 40 °C in the presence of a catalyst.

Time/s	0	5	10	20	30	40	50	60
Volume of oxygen/cm³	0	22	34	48	56	59	60	60

(a) Plot a graph of the results with time on the x-axis and volume of oxygen on the y-axis. [3]
(b) On the same axes sketch a curve for the same reaction but carried out at a temperature of 50 °C. [2]
(c) What results would you expect if the catalyst was not present? Explain your answer. [2]

(Paper 3)

3 A student compared how well different compounds catalysed the reaction between zinc and hydrochloric acid. The results are shown in the table.

Compound	Time taken for all the zinc to react/s
No catalyst	500
Copper(II) sulfate	150
Copper(II) chloride	175
Manganese(IV) oxide	390
Sodium chloride	500
Sodium sulfate	500

(a) What do you understand by the term catalyst? [1]
(b) Which is the best catalyst for this reaction? [1]
(c) What things must you keep constant in these experiments if it is to be a fair test?

(d) Which compounds in the table are not catalysts? [1]

(Paper 3)

4 The graph shows how the volume of carbon dioxide given off changes when hydrochloric acid of three different concentrations reacts with large pieces of calcium carbonate.

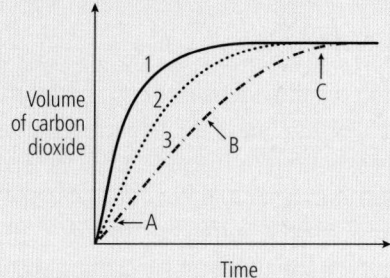

(a) Which line (1, 2 or 3) is the graph for the most concentrated acid? [1]

(b) In line 3, state and explain where the rate of reaction is fastest (at A, B or C). [2]

(c) Draw a diagram of the apparatus you can use to obtain these results. [3]

(d) Suggest two ways of increasing the rate of this reaction other than by increasing the concentration of acid. [2]

(Paper 3)

5 Magnesium ribbon reacts with hydrochloric acid. The equation is:

$Mg + 2HCl \longrightarrow MgCl_2 + H_2$

(a) Describe and explain two methods that can be used to follow the progress of this reaction. [4]

(b) For one of these methods describe how you can calculate the rate of reaction. [4]

(c) Using the kinetic particle theory, explain how and why the rate of reaction changes when:
  (i) the concentration of hydrochloric acid is increased [3]
  (ii) the temperature of the reaction mixture is lowered [3]
  (iii) powdered magnesium is used rather than magnesium ribbon. [3]

(Paper 4)

6 A student investigated the reaction between 4.0 g calcium carbonate and 30 cm³ of 2.0 mol/dm³ hydrochloric acid.

$CaCO_3 + 2HCl \longrightarrow CaCl_2 + CO_2 + H_2O$

(a) Draw a sketch graph to show how the loss in mass of the reaction mixture changes with time. [2]

(b) Describe how the rate of reaction changes with time by referring to the change of gradient of the graph. [2]

(c) Use the idea about colliding particles to explain why the rate of reaction decreases with time. [3]

(d) (i) Calculate the number of moles of calcium carbonate and hydrochloric acid at the start of the reaction. [2]
   (ii) Was the hydrochloric acid or the calcium carbonate the limiting reactant? Explain your answer. [2]

(e) In another experiment the student investigated how the rate of this reaction varies with temperature. The results are shown in the table.

Temperature/°C	20	30	40	50
Time taken for a piece of calcium carbonate to dissolve/s	64	32	16	8

  (i) Draw a suitable graph to display these results. [3]
  (ii) From your graph, predict how many seconds it would take the piece of calcium carbonate to dissolve at 60 °C. [1]

(Paper 4)

7 Zinc reacts with dilute sulfuric acid:
$Zn + H_2SO_4 \longrightarrow ZnSO_4 + H_2$

(a) Explain using the collision theory why large pieces of zinc react more slowly than the same mass of zinc powder when the same concentration of acid is used. [2]

(b) Explain using the collision theory why 50 cm³ of 1.0 mol/dm³ sulfuric acid reacts with 6.5 g of zinc powder more slowly than 50 cm³ of 1.5 mol/dm³ sulfuric acid. [3]

(c) Show by calculation that sulfuric acid is the limiting reactant. [2]

(d) Write an ionic equation for this reaction, including state symbols. [2]

(Paper 4)

10.1 Reversible reactions

LEARNING OUTCOMES

- Recognise that some chemical reactions are reversible
- Describe the reversible reactions of anhydrous and hydrated salts
- **S** State the meaning of equilibrium in terms of rates of forward and backward reactions and concentrations of reactants and products

Heating hydrated salts

When you heat crystals of blue copper(II) sulfate they decompose (break down) to form a white powder.

$$CuSO_4 \cdot 5H_2O(s) \xrightarrow{heat} CuSO_4(s) + 5H_2O(l)$$
hydrated copper(II) sulfate (blue) → anhydrous copper(II) sulfate (white)

The blue copper(II) sulfate crystals have water as part of their structure. We say that the copper(II) sulfate is hydrated. We call the water in the salt crystals, water of crystallisation. When the blue crystals are heated the water of crystallisation is lost. We are left with white anhydrous copper(II) sulfate. Anhydrous means 'without water'. We can reverse the reaction by adding water back to the white copper(II) sulfate:

$$CuSO_4(s) + 5H_2O(l) \longrightarrow CuSO_4 \cdot 5H_2O(s)$$

Because this reaction is reversible, we can write both reactions in the same equation. The sign $\rightleftharpoons$ is the symbol for a reversible change. It is used for reactions which we call **reversible reactions**.

$$CuSO_4 \cdot 5H_2O(s) \underset{\text{backward (reverse) reaction}}{\overset{\text{forward reaction}}{\rightleftharpoons}} CuSO_4(s) + 5H_2O(l)$$

Another reversible reaction used to test for water is:

$$CoCl_2 \cdot 6H_2O(s) \underset{\text{add water}}{\overset{\text{heat}}{\rightleftharpoons}} CoCl_2(s) + 6H_2O(l)$$
hydrated cobalt(II) chloride (red/pink) → anhydrous cobalt(II) chloride (blue)

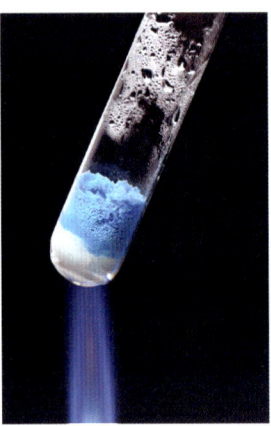

Figure 10.1.1 Blue copper(II) sulfate crystals turn white when heated.

EXAM TIP

Make sure that you know the meanings of the terms anhydrous, hydrated and water of crystallisation.

EXAM TIP

Remember that the concentrations are constant at equilibrium because the rate of the forward reaction equals the rate of the backward reaction.

Supplement

Equilibrium

When we change blue copper(II) sulfate to white copper(II) sulfate the two reactions are separate. We don't heat and add water at the same time. In some reactions, however, both the forward and reverse reactions are going on at the same time. We call these **equilibrium** reactions. Equilibrium reactions have particular features:

- at equilibrium the rate of the forward reaction is equal to the rate of the backward reaction
- at equilibrium the concentration of the reactants and products does not change

Unit 10: Chemical reactions

- the reactants or products must not escape from the reaction mixture. We call this a **closed system**.

- CaCO$_3$ (s)
- CaO (s)
- × CO$_2$ (s)

 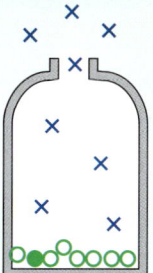

CaCO$_3$(s) $\rightleftharpoons$ CaO(s) + CO$_2$(g)   CaCO$_3$(s) $\rightarrow$ CaO(s) + CO$_2$(g)

This is a closed system. The calcium carbonate is decomposing to calcium oxide but the carbon dioxide is not lost.

This is an open system. The calcium carbonate is decomposing to calcium oxide and the carbon dioxide is lost.

Figure 10.1.2 Comparison of a closed system and an open system.

The equilibrium can be approached from either direction. We can start with only the reactants or only the products. When we heat hydrogen and iodine in a sealed tube:

$$H_2(g) + I_2(g) \rightleftharpoons 2HI(g)$$

we get specific concentrations of H$_2$(g), I$_2$(g) and HI(g) at equilibrium. These concentrations are the same whether we start from hydrogen and iodine or from hydrogen iodide alone.

○ Hydrogen   ● Iodine   ☐ Hydrogen iodide

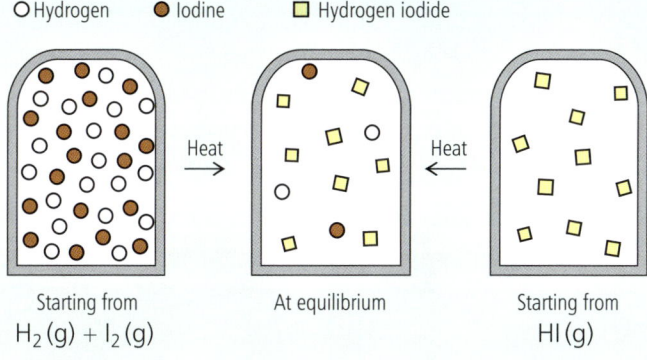

Starting from H$_2$(g) + I$_2$(g)   At equilibrium   Starting from HI(g)

Figure 10.1.3 At equilibrium H$_2$(g), I$_2$(g) and HI(g) are at fixed concentrations whether we start from reactants or products.

The **position of equilibrium** tells us how far the reaction goes in favour of reactants or products. If the concentration of products is greater than the concentration of the reactants in a reaction mixture at equilibrium, we say that the position of equilibrium is to the right. It favours the products. If the concentration of reactants is greater than the concentration of the products, we say that the position of equilibrium is to the left. It favours the reactants.

KEY POINTS

- In a reversible reaction, shown by $\rightleftharpoons$, the products can react to form the original reactants again.
- An equilibrium reaction can take place only in a closed system.
- At equilibrium, the concentrations of reactants and products are no longer changing.
- At equilibrium, the rate of the forward reaction equals the rate of the backward reaction.

SUMMARY QUESTIONS

1 Copy and complete using the words below:

   **anhydrous   chloride
   crystallisation   pink
   reversible   white**

   When we heat hydrated copper(II) sulfate it loses its water of _____. Its colour changes from blue to _____. When we add water to _____ cobalt(II) _____ it turns _____. These are _____ reactions.

2 State three features of an equilibrium reaction.

3 State the meaning of the terms:

   a position of equilibrium
   b closed system

123

10.2 Shifting the equilibrium
Supplement

LEARNING OUTCOMES

- Predict and explain how equilibrium reactions are affected by changing the concentration, pressure or temperature
- Recognise that catalysts do not affect the position of equilibrium

Changing the concentration of reactants or products or changing the temperature or pressure has an effect on an equilibrium reaction. The reaction tries to oppose the changes that you make. Catalysts do not affect the position of equilibrium. They just speed up the forward and reverse reactions equally.

Changing the concentration

When the concentration of a reactant is increased, the equilibrium moves to the right. Adding more reactants unbalances the equilibrium. So the equilibrium moves to the right to form more products until the equilibrium is restored.

This reaction is in equilibrium. The concentration of reactants is twice that of the products.

More reactants have been added. The equilibrium has been disturbed.

The equilibrium shifts to the right. Reactants are changed into products until the equilibrium is restored. The relative concentrations of reactants and products are the same as before.

Figure 10.2.1 Equilibrium is a balancing act!

When the concentration of a product is increased, the equilibrium moves to the left. Products are changed to reactants until equilibrium is restored.

PRACTICAL

Changing the direction of a reaction

You pass chlorine gas through a tube containing the brown liquid iodine monochloride. As more chlorine is passed through, the brown liquid turns into yellow crystals. When you tip out the chlorine gas, the yellow crystals turn back into the brown liquid.

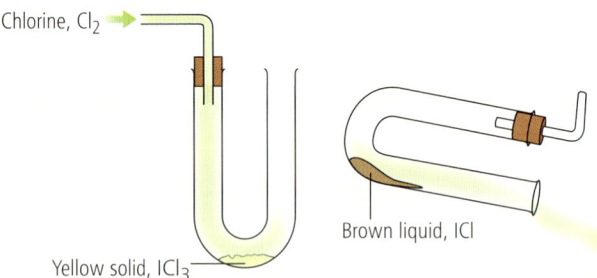

Figure 10.2.2 The equilibrium reaction $Cl_2(g) + ICl(l) \rightleftharpoons ICl_3(s)$.

By changing the concentration of chlorine we have made the reaction go backwards or forwards.

$$Cl_2(g) + ICl(l) \rightleftharpoons ICl_3(s)$$

iodine monochloride (brown liquid)   iodine trichloride (yellow crystals)

A high concentration of chlorine pushes the reaction from left to right. The forward reaction is favoured. Removing the chlorine favours the reverse reaction.

Unit 10: Chemical reactions

Changing the pressure

Change in pressure only affects reactions where there is a gas in the chemical equation and there are different numbers of moles of gas on each side of the equation. Increasing the pressure moves the equilibrium to the side with the smaller volume of gas (fewer moles of gas). So in the reaction:

$$2SO_2(g) + O_2(g) \rightleftharpoons 2SO_3(g)$$

there are three volumes (three moles) of gas on the left and only two on the right. So increase in pressure moves the position of equilibrium further to the right. More SO_2 and O_2 combine to form SO_3. This happens because increasing the pressure squashes the molecules closer together – you are increasing their concentration. The reaction mixture tries to overcome this by moving the equilibrium to the right so that the overall number of molecules is reduced. Decreasing the pressure has the opposite effect. It pushes the reaction to the left. When there are equal volumes or moles of gas on both sides of the chemical equation, increasing the pressure has no effect.

Changing the temperature

If a reaction is exothermic in the forward reaction, it will be endothermic in the backward (reverse) reaction. When we produce ammonia from nitrogen and hydrogen the forward reaction is exothermic, so the reverse reaction is endothermic.

$$N_2(g) + 3H_2(g) \underset{endothermic}{\overset{exothermic}{\rightleftharpoons}} 2NH_3(g)$$

The table shows the effect of temperature on the percentage yield of ammonia at 300 atmospheres pressure.

temperature/°C	350	400	450	500
percentage yield of ammonia	65	49	37	20

You can see that the yield of ammonia gets lower as the temperature increases. For an exothermic forward reaction, when we increase the temperature the equilibrium shifts in favour of the backward endothermic reaction. Shifting the equilibrium to the left cools down the reaction mixture, opposing the increase in temperature.

In an endothermic forward reaction the opposite happens. Increasing the temperature moves the equilibrium further to the right, in the direction of absorbing more thermal heat.

EXAM TIP
Remember that if the equilibrium conditions are changed the reaction always tries to act in the direction that opposes the change.

SUMMARY QUESTIONS

1. Copy and complete using the words below:

   **concentration   disturbed
   equilibrium   products
   right**

   In an _____ reaction, when you increase the _____ of a reactant, the reaction moves to the _____. The equilibrium is _____ so more of the reactants are changed to _____. This continues until equilibrium is restored.

2. Describe and explain how each of the following changes affects this equilibrium reaction:

   $$CO(g) + 2H_2(g) \rightleftharpoons CH_3OH(g)$$

   The forward reaction is endothermic.

   a increasing the pressure
   b decreasing the concentration of $CO(g)$
   c decreasing the temperature

KEY POINTS

In an equilibrium reaction:
- increasing the concentration of a reactant moves the reaction in the direction of the products until the equilibrium is restored
- increasing the pressure moves the reaction in the direction of the lower number of gas molecules in the chemical equation
- for an endothermic forward reaction, increasing the temperature moves the reaction further to the right in the direction of the products. For an exothermic forward reaction, increasing the temperature moves the reaction further to the left.

10.3 Redox reactions

LEARNING OUTCOMES

- Define oxidation and reduction in terms of oxygen gain and loss
- Redox reactions involve simultaneous oxidation and reduction
- Define oxidation and reduction in terms of loss and gain of electrons

Oxidation and reduction

When we burn magnesium in oxygen, magnesium oxide is formed:

$$2Mg(s) + O_2(g) \longrightarrow 2MgO(s)$$

We say that magnesium has been oxidised – it has gained oxygen. **Oxidation** is the gain of oxygen. When we remove oxygen from a compound we say that the compound has been reduced.

PRACTICAL

The reduction of copper(II) oxide

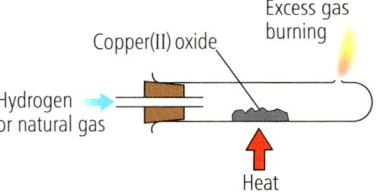

Figure 10.3.2 Reducing copper(II) oxide.

We pass hydrogen or natural gas over heated copper(II) oxide. The black copper(II) oxide changes to pink copper. We have removed the oxygen from the copper(II) oxide. So the copper(II) oxide has been reduced by the hydrogen.

Reduction is the loss of oxygen from a compound.

When you look at the reaction between copper(II) oxide and hydrogen, you can see that both oxidation and reduction have taken place at the same time. We say they are simultaneous.

Figure 10.3.1 Redox reactions are involved when the fuel in this rocket burns.

$$CuO(s) + H_2(g) \rightarrow Cu(s) + H_2O(l)$$

(reduction / oxidation)

copper oxide | hydrogen | copper | water

The copper(II) oxide has been reduced because it has lost oxygen. The hydrogen has been oxidised because it has gained oxygen. Because the reduction and oxidation have taken place simultaneously (at the same time). We call this a **redox** reaction.

EXAM TIP

When identifying a substance that is reduced, a common error is to write 'oxygen is removed from copper' (an element) instead of oxygen is removed from copper oxide' (a compound).

Remember OIL RIG – Oxidation Is Loss (of electrons), Reduction Is Gain (of electrons).

Where hydrogen takes part in a reaction, we can also define reduction as the addition of hydrogen to a compound and oxidation as the removal of hydrogen from a compound. This is often used in organic reactions, for example:

$$C_2H_4 + H_2 \rightarrow C_2H_6$$

ethene | hydrogen | ethane

In this reaction we say that the hydrogen has reduced ethene to ethane.

Unit 10: Chemical reactions

Oxidation numbers

We often put a Roman number after the name of an element in a compound. This number is called the **oxidation number**. It refers to a particular element in the compound. For example:

iron(II) chloride   iron(III) chloride   copper(I) oxide   copper(II) oxide

You can see that there are two types of iron chloride. So we have to find a way of telling these apart. Oxidation numbers help us to do this. Oxidation numbers tell us about:

- the type of ion present in the compound. For example, copper(I) oxide has Cu^+ ions and copper(II) oxide has Cu^{2+} ions
- how oxidised an element in a compound is. For example, the formula for manganese(II) oxide is MnO but the formula for manganese(IV) oxide is MnO_2. This shows us that manganese in manganese(IV) oxide is more oxidised than in manganese(II) oxide. You can also tell this by the fact that it has more oxygen.

Oxidation numbers can also be used for non-metallic elements. For example, $KClO_3$ is potassium chlorate(V) and $KClO_4$ is potassium chlorate(VII). The chlorine in potassium chlorate(VII) is more oxidised because it is associated with more oxygen.

Other useful examples are copper(II) sulfate and potassium manganate(VII).

Supplement

Electron transfer in redox reactions

We saw in Topic 7.4 that we can write half-equations for electrode reactions. We do this by balancing the equations by adding one or more electrons to either side of the equation. We can now give a wider definition of oxidation and reduction that does not involve oxygen or hydrogen:

Oxidation is loss of electrons. Reduction is gain of electrons.

So when we electrolyse molten sodium chloride we can split up the overall equation into two parts to show what happens to each ion:

overall equation: $2NaCl(l) \longrightarrow 2Na(l) + Cl_2(g)$

sodium: $2Na^+ + 2e^- \longrightarrow 2Na$

The sodium ions have been reduced because they have gained electrons.

chloride: $2Cl^- \longrightarrow Cl_2 + 2e^-$

The chloride ions have been oxidised because they have lost electrons.

KEY POINTS

- Oxidation is the gain of oxygen by a substance. Reduction is the loss of oxygen from a compound.
- Redox reactions involve both oxidation and reduction at the same time.
- Oxidation numbers give information about the degree of oxidation or reduction of compounds.
- Oxidation is loss of electrons. Reduction is gain of electrons.

SUMMARY QUESTIONS

1. Copy and complete using the words below:

   loses   oxidised   reduced   sulfur

   When sulfur burns in oxygen _____ dioxide is formed. The sulfur is _____ because it gains oxygen. When copper(II) oxide reacts with hydrogen, the copper(II) oxide is _____ to copper. This is because the copper(II) oxide _____ oxygen.

2. Identify the substance which is oxidised and the substance which is reduced in each of these equations:

   a $2Na + O_2 \longrightarrow Na_2O_2$
   b $PbO + C \longrightarrow Pb + CO$
   c $CuO + CO \longrightarrow Cu + CO_2$

3. The equation for the electrolysis of molten calcium chloride is:

   $CaCl_2(l) \longrightarrow Ca(l) + Cl_2(g)$

   Write two half-equations for this reaction. State which shows an oxidation and which shows a reduction.

10.4 More about redox reactions

Supplement

LEARNING OUTCOMES

- Define oxidation and reduction in terms of loss and gain of electrons
- Define redox reactions in terms of change in oxidation numbers

Figure 10.4.1 These sunglasses darken in the sunlight because of redox reactions.

EXAM TIP

When referring to a particular species make sure that you use the correct terms. For example, it is a common error to refer to copper ions as just 'copper'. It's even better if you write copper(II) ions or Cu^{2+} ions!

EXAM TIP

When balancing with electrons make sure that the charges balance. This is best done by adding the correct number of electrons to one or other side of the half-equation.

Oxidation, reduction and electron transfer

We can identify many redox reactions by the fact that we can write half-equations for them.

For example, when zinc metal reacts with copper(II) sulfate solution the equation is:

$$Zn(s) + CuSO_4(aq) \longrightarrow ZnSO_4(aq) + Cu(s)$$

We can divide this into two half-equations to show what happens to each **species**. (A species is a word we use when we want to refer generally to atoms, ions, molecules or electrons.)

$$Zn(s) \longrightarrow Zn^{2+}(aq) + 2e^-$$

The zinc atoms have lost electrons. They have been oxidised to zinc ions.

$$Cu^{2+}(aq) + 2e^- \longrightarrow Cu$$

The copper ions have gained electrons. They have been reduced to copper atoms.

Here is an example of electron transfer in redox reaction where the ionic equation has been given.

$$Mg + 2Fe^{3+} \longrightarrow Mg^{2+} + 2Fe^{2+}$$

oxidation: $Mg \longrightarrow Mg^{2+} + 2e^-$

Magnesium atoms have lost electrons. Magnesium atoms have been oxidised.

reduction: $2Fe^{3+} + 2e^- \longrightarrow 2Fe^{2+}$ OR $Fe^{3+} + e^- \longrightarrow Fe^{2+}$

Iron(III) ions have gained electrons. Iron(III) ions have been reduced.

Oxidation numbers

We can use **oxidation numbers** (ox. no.) to follow what happens in a redox reaction.

Oxidation numbers are sometimes called oxidation numbers. We use a set of rules (oxidation number rules) to find the oxidation number of an element in a compound.

1. An element that is uncombined has an ox. no. of zero. For example, Zn = 0, S = 0.
2. The ox. no. of a monatomic ion (the ion that comes from a single atom) is the same as the charge on the ion. So sodium ion = +1, aluminium ion = +3, oxide = –2, chloride = –1. You will notice that these are the same as the charge on the ion.

Unit 10: Chemical reactions

3 The total ox. no. of all the elements in a compound is zero. For example, in $MgCl_2$:

  ox. no. of Mg = +2
  ox. no. of 2Cl⁻ = 2 × −1 = −2

4 For a molecular ion, e.g. NO_3^- or SO_4^{2-}, the total ox. no. equals the charge on the ion:

  ox. no. of N and 3O in NO_3^- = +5 + (3 × −2) = −1
  ox. no. of S and 4O in SO_4^{2-} = +6 + (4 × −2) = −2

Using oxidation number rules

For elements that form more than one ion, the oxidation number can be worked out from the formula of the oxide. Elements in **covalent compounds** can also be given oxidation numbers but you sometimes have to use additional rules.

Example

Deduce the ox. no. of iron in Fe_2O_3

Using rule 3: 2Fe + 3O = 0

Using rule 2: ox. no. of O^{2-} = −2

So 2Fe + 3 × (−2) = 0   So 2Fe + −6 = 0

So each Fe is +3 to balance the −6.

Oxidation numbers in redox reactions

An increase in oxidation number of an element is oxidation. A decrease in oxidation number of an element is reduction.

In the reaction:

$$Zn + Cu^{2+} \rightarrow Zn^{2+} + Cu$$
oxidation numbers: 0   +2   +2   0

(reduction: $Cu^{2+} \rightarrow Cu$; oxidation: $Zn \rightarrow Zn^{2+}$)

Zinc has increased its ox. no. from 0 to +2 so it is oxidised to zinc ions. Copper has decreased its ox.no. from +2 to 0 so copper(II) ions are reduced to copper atoms.

KEY POINTS

- Oxidation is loss of electrons and reduction is gain of electrons.
- An increase in oxidation number of a species is oxidation. A decrease in oxidation number of a species is reduction.

EXAM TIP

Oxygen atoms in compounds always have an oxidation number of −2 (except in peroxides where it is −1).

SUMMARY QUESTIONS

1 Copy and complete using the words below:

**decrease   increase
number   oxidation
redox   species**

Changes in oxidation _____ can be used to follow what happens in a _____ reaction. Oxidation is an _____ in the _____ number of a species. Reduction is a _____ in the oxidation number of a _____.

2 Deduce the oxidation number of:

  a Al in $AlCl_3$
  b K in KCl
  c Fe in $FeBr_2$
  d O in MgO
  e S in SO_3

3 Which species have been oxidised and which have been reduced in the reaction:

  $Mg + FeO \longrightarrow MgO + Fe$

  Explain your answer in terms of electron transfer and ox. no. changes.

10.5 Oxidising agents and reducing agents
Supplement

LEARNING OUTCOMES

- Identify redox reactions by colour changes using potassium manganate(VII) or potassium iodide
- Define the terms oxidising agent and reducing agent
- Identify oxidising agents and reducing agents in redox equations

Identifying oxidising agents and reducing agents

An **oxidising agent** is a substance which oxidises another substance, and is itself reduced, in a redox reaction.

A **reducing agent** is a substance which reduces another substance, and is itself oxidised, in a redox reaction.

Example 1: the reaction between copper(II) oxide and hydrogen

$$CuO(s) + H_2(g) \longrightarrow Cu(s) + H_2O(l)$$

- The hydrogen is the reducing agent because it has removed the oxygen from the copper(II) oxide. Hydrogen has become oxidised.
- The copper(II) oxide is the oxidising agent because it gives its oxygen to hydrogen to form water. Copper(II) has been reduced.

Example 2: the reaction between magnesium and silver ions, Ag^+

In terms of ox. no. changes:

$$Mg(s) + 2Ag^+(aq) \longrightarrow Mg^{2+}(aq) + 2Ag(s)$$

ox. no.　　　0　　　　+1　　　　　　+2　　　　　0

- Magnesium is the reducing agent because it decreases the ox. no. of each Ag^+ ion from +1 to 0. The magnesium atoms are oxidised to magnesium ions.
- The Ag^+ ion is the oxidising agent because it increases the ox. no. of magnesium from 0 to +2. The silver ions are reduced to silver atoms.

Colour changes in redox reactions

We can use particular elements or compounds as oxidising agents or reducing agents.

Test for reducing agents

Aqueous potassium manganate(VII), $KMnO_4$, in acidic solution, is a good oxidising agent. When it oxidises a substance its colour changes from purple to colourless. This can be used as a test for reducing agents when the reducing agent is in excess.

When you add a solution of potassium manganate(VII) drop by drop to a solution containing iron(II) ions, the solution changes

Figure 10.5.1 When acidified potassium manganate(VII) is added to iron(II) ions, the potassium manganate(VII) decolourises and a very pale yellow colour of iron(III) ions becomes visible.

EXAM TIP

Remember that reducing agents lose electrons to another species and oxidising agents gain electrons from another species.

Unit 10: Chemical reactions

colour from light green to yellow.

The iron(II) ions are oxidised by the potassium manganate(VII). A yellow solution containing iron(III) ions has been formed. The purple potassium manganate(VII), which contains MnO_4^- ions, becomes colourless when it reacts. So you only see the colour of the iron(III) ions. When you add excess potassium manganate(VII) you see a purple colour of the manganate(VII) ions in the beaker.

The equation for this reaction is shown. The equation is complex but shows you the main points about the reaction:

$$\underset{\substack{\text{Purple}\\\text{ox. no. +7}}}{MnO_4^-} + \underset{\substack{\text{reducing agent}\\+2}}{5Fe^{2+}} + 8H^+ \longrightarrow \underset{\substack{\text{colourless}\\+2}}{Mn^{2+}} + \underset{+3}{5Fe^{3+}} + 4H_2O$$

(5e⁻ transferred; Fe^{2+} oxidized)

Test for oxidising agents

Aqueous potassium iodide in acidic solution is a good reducing agent. When it reduces a substance its colour changes from colourless to brown. This is due to the oxidation of colourless iodide ions, I^-, to brown aqueous iodine, I_2. An example of this test for oxidising agents is the reduction of hydrogen peroxide to water:

$$\underset{\substack{\text{colourless}\\\text{ox. no. -1}}}{2I^-} + \underset{\substack{\text{oxidising}\\\text{agent }H_2O_2}}{H_2O_2} + 2H^+ \longrightarrow \underset{\substack{\text{brown}\\0}}{I_2} + \underset{-2}{2H_2O}$$

(2e– transferred; oxygen atoms reduced)

SUMMARY QUESTIONS

1. Identify the oxidising agent in each of these equations:
   a. $2Al + Fe_2O_3 \longrightarrow Al_2O_3 + 2Fe$
   b. $CH_4 + 2O_2 \longrightarrow CO_2 + 2H_2O$
   c. $Zn^{2+} + Mg \longrightarrow Zn + Mg^{2+}$

2. Identify the reducing agent in each of these equations:
   a. $Fe_2O_3 + 3CO \longrightarrow 2Fe + 3CO_2$
   b. $2I^- + Br_2 \longrightarrow I_2 + 2Br^-$
   c. $ZnCl_2 + 2Li \longrightarrow Zn + 2LiCl$

3. Excess potassium iodide solution is added to a solution of potassium manganate(VII) until the potassium iodide is in excess.

   State the colour change you would observe in this reaction and identify the oxidising agent and reducing agent.

EXAM TIP

Make sure that you don't confuse oxidising agents and reducing agents with the compounds that are being oxidised or reduced. An oxidising agent oxidises another substance. When it does this the oxidising agent itself is reduced.

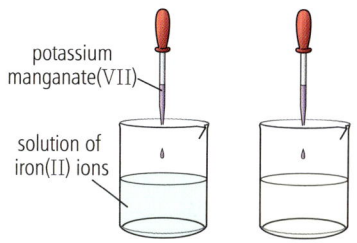

Figure 10.5.2 A test for reducing agents using potassium manganate(VII).

EXAM TIP

You do not need to worry about the equations for the test for reducing agents. You just have to know that potassium manganate(VII) turns colourless in the presence of excess reducing agent.

KEY POINTS

- An oxidising agent is a substance which oxidises another substance and is itself reduced, in a redox reaction.
- A reducing agent is a substance which reduces another substance and is itself oxidised in a redox reaction.
- The colour changes of potassium manganate(VII) and potassium iodide can be used to test for oxidising and reducing agents.

SUMMARY QUESTIONS

1 Match the words on the left with the phrases on the right.

oxidation	a reaction that can go in the forward or backward direction
reversible reaction	loss of hydrogen or addition of oxygen
reduction	without water
anhydrous	removal of oxygen or addition of hydrogen

2 State whether the underlined elements or compounds are oxidised or reduced when they react:
   a $\underline{PbO} + C \longrightarrow Pb + CO$
   b $\underline{S} + O_2 \longrightarrow SO_2$
   c $CuO + \underline{H_2} \longrightarrow Cu + H_2O$
   d $\underline{CO_2} + C \longrightarrow 2CO$

3 a State the name of a substance used to test for a reducing agent and state the colour change.
   b State the name of a substance used to test for an oxidising agent and state the colour change.

4 State three characteristics of an equilibrium reaction.

5 How do each of the following affect this endothermic reaction?
   $CaCO_3(s) \rightleftharpoons CaO(s) + CO_2(g)$
   a increasing the concentration of carbon dioxide
   b decreasing the temperature
   c allowing the carbon dioxide to escape
   d increasing the pressure

6 Describe the colour change when hydrated cobalt chloride is heated and suggest what happens to cause this colour change.

7 For each of these reactions, state whether oxidation or reduction is taking place:
   a $Cu^{2+} + 2e^- \longrightarrow Cu$
   b $Fe^{2+} \longrightarrow Fe^{3+} + e^-$
   c $2H^+ + 2e^- \longrightarrow H_2$

8 Which is the oxidising agent in each of these equations?
   a $Cl_2 + 2KBr \longrightarrow 2KCl + Br_2$
   b $Zn + CuSO_4 \longrightarrow ZnSO_4 + Cu$
   c $5Fe^{2+} + MnO_4^- + 8H^+ \longrightarrow 5Fe^{3+} + Mn^{2+} + 4H_2O$

Practice questions

1 Blue copper(II) sulfate has the formula $CuSO_4 \cdot 5H_2O$. Which phrase best describes blue copper(II) sulfate?
   A Hydrated salt
   B Anhydrous salt
   C Crystalline element
   D Dehydrated salt

(Paper 1)

2 Which one of the following statements about this reaction is correct?
$N_2(g) + 3H_2(g) \rightleftharpoons 2NH_3(g)$ ΔH −92.4 kJ/mol
   A Increasing the temperature shifts the equilibrium to the left.
   B Removing the ammonia shifts the equilibrium to the left.
   C Decreasing the pressure shifts the equilibrium to the right.
   D Decreasing the concentration of hydrogen shifts the equilibrium to the right.

(Paper 2)

3 When blue copper(II) sulfate is heated it turns white.
$CuSO_4 \cdot 5H_2O \longrightarrow CuSO_4 + 5H_2O$
   (a) (i) How can you reverse this reaction? [1]
   (ii) Write an equation for the reverse reaction. [1]
   (b) Complete the following sentences using some of the words from the list.
   alcohol   anhydrous   crystallisation   hydrated   red   water
   When _____ copper(II) sulfate is heated it loses its water of _____ and turns white. White copper(II) sulfate is called _____ copper(II) sulfate. [3]
   (c) Aqueous copper(II) sulfate is reduced by zinc.
   (i) Copy and complete the equation for this reaction.
   $CuSO_4 + Zn \longrightarrow$ _____ $+ ZnSO_4$ [1]
   (ii) State the meaning of the term reduction. [1]

(d) State which compound gets reduced in this equation. Explain your answer.
$Fe_2O_3 + 3CO \longrightarrow 2Fe + 3CO_2$ [2]
(Paper 3)

4 A mixture of hydrogen and iodine is put into a closed tube and heated at 400 °C.
$H_2(g) + I_2(g) \rightleftharpoons 2HI(g)$

(a) At first the rate of the forward reaction decreases with time. Suggest a reason for this. [2]

(b) Explain why the rate of the backward reaction increases with time. [2]

(c) After a time equilibrium is reached. How do the rates of the forward and the backward reaction compare at equilibrium? [1]

(d) State and explain the effect of increasing the concentration of iodine on this equilibrium. [2]

(e) Increasing the pressure does not have any effect on this equilibrium. Explain why not. [1]

(f) The reaction is exothermic. Predict the effect of increasing the temperature on this equilibrium. Explain your answer. [2]

(Paper 4)

5 Aqueous chlorine reacts with an aqueous solution of sodium iodide:
$Cl_2(aq) + 2NaI(aq) \longrightarrow 2NaCl(aq) + I_2(aq)$

(a) Deduce the oxidation numbers of
  (i) Cl_2 [1]
  (ii) I in NaI [1]
  (iii) Identify the oxidising agent in this reaction. Explain your answer. [2]

(b) Write an ionic equation for this reaction. [2]

(c) Use your ionic equation to help you write two half-equations for this reaction. [2]

(d) Chlorine reacts with iodine to form a brown liquid called iodine monochloride, ICl. In the presence of excess chlorine, iodine monochloride reacts further to form yellow crystals of iodine trichloride, ICl_3.
$Cl_2(g) + ICl(l) \rightleftharpoons ICl_3(s)$
  (i) What is the meaning of the symbol $\rightleftharpoons$? [1]

(ii) Describe and explain what you would observe when the chlorine gas supply is turned off. [3]

(iii) Describe and explain the effect of decreasing the pressure on this reaction. [2]

(Paper 4)

6 Oxygen combines with nitrogen.
$N_2(g) + O_2(g) \rightleftharpoons 2NO(g)$

(a) Explain why this is a redox reaction and identify the reducing agent. [3]

(b) The reaction of nitrogen with oxygen is endothermic. Describe and explain how an increase in temperature affects this equilibrium. [2]

(c) Use these oxidation numbers to answer these questions: S in H_2SO_4 is +6, O in compounds is −2, H in compounds is +1.
S reacts with concentrated nitric acid;
$S + 6HNO_3 \longrightarrow H_2SO_4 + 6NO_2 + 2H_2O$
  (i) Explain how you know that nitric acid is an oxidising agent in this reaction. [3]
  (ii) Deduce the oxidation number of nitrogen in nitric acid. [1]
  (iii) Describe this redox reaction in terms of electron transfer. [2]

(Paper 4)

7 (a) State two characteristics of an equilibrium reaction. [2]

(b) Hypochlorous acid, HClO, and hydrochloric acid are formed when chlorine reacts with water in a closed container.
$Cl_2(g) + H_2O(l) \rightleftharpoons HClO(aq) + HCl(aq)$
  (i) Describe and explain what happens to the position of equilibrium when the concentration of chlorine is decreased. [2]
  (ii) Describe and explain the effect of increasing the pressure on this equilibrium. [2]

(c) Deduce the oxidation number of Cl in
  (i) HClO (ii) HCl. [2]

(Paper 4)

11.1 How acidic?

LEARNING OUTCOMES

- State that aqueous solutions of acids contain H⁺ ions and aqueous solutions of alkalis contain OH⁻ ions
- Describe acidity and alkalinity in terms of the pH scale
- Describe the use of universal indicator to find pH

Acids and alkalis

Many **acids** exist naturally in plants and foods. Citric acid is found in oranges and lemons, ethanoic acid is found in vinegar, and methanoic acid is present in nettles. The common laboratory acids are:

hydrochloric acid HCl   sulfuric acid H_2SO_4 nitric acid HNO_3

All acids form hydrogen ions, H^+, when dissolved in water. It is the hydrogen ions that make a solution acidic.

Alkalis are the opposite of acids in the way they react. Common laboratory alkalis are:

sodium hydroxide NaOH   calcium hydroxide $Ca(OH)_2$
ammonia NH_3

Alkalis form hydroxide ions, OH^-, when they dissolve in water.

Substances that are neither acidic nor alkaline are **neutral**. Pure water is neutral.

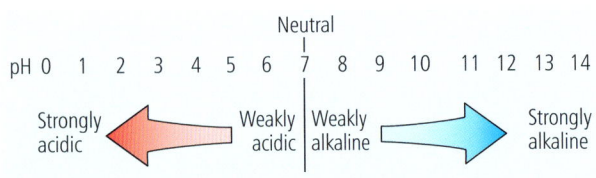

Figure 11.1.1 The pH scale.

The pH scale

We use the **pH scale** to show us if a solution is acidic, alkaline or neutral. We can also use this scale to find out exactly how acidic or alkaline a substance is.

The pH scale runs from 0 to 14. Acids have a pH below 7. Alkalis have a pH above 7. The lower the pH, the more acidic the solution is. A solution with a lower pH has a higher concentration of hydrogen ions. The higher the pH, the more alkaline the solution is. A higher pH means a higher concentration of hydroxide ions. We can measure pH using universal **indicator** paper or a pH electrode connected to a pH meter.

EXAM TIP

Remember that a lower acidity gives a higher pH and a higher acidity gives a lower pH.

PRACTICAL

Comparing the pH of household products

1. Put the pH electrode into a solution of a substance such as vinegar, lemon juice, washing powder or soap.
2. Record the reading on the pH meter.

You can also use a pH sensor connected to a data logger and computer to see how the pH changes as you add an acid to an alkali in the beaker.

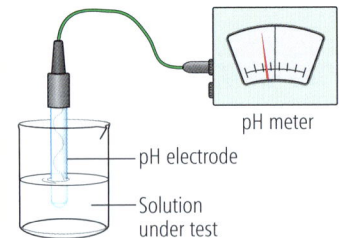

Figure 11.1.2 Using a pH electrode.

Using universal indicator

An acid–alkali indicator is a chemical which changes colour when we add an acid or an alkali. Universal indicator is a mixture of indicators which shows a range of colours depending on the pH.

Unit 11: Acids and bases

You can use universal indicator solution or universal indicator paper to measure the pH. When you take a drop of solution under test and place it on universal indicator paper, the paper turns a particular colour. The colour of the indicator paper is then matched against a colour chart showing the pH for different colours. A typical colour chart is shown in Figure 11.1.3.

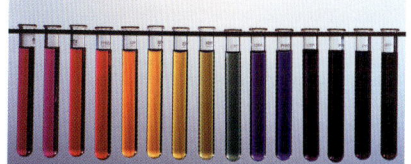

Figure 11.1.4 The colour of universal indicator changes with pH.

EXAM TIP

When using universal indicator to test the pH of a solution, it is often better to add a few drops of solution to universal indicator paper rather than to add universal indicator to the solution. In this way your solution is not contaminated with universal indicator.

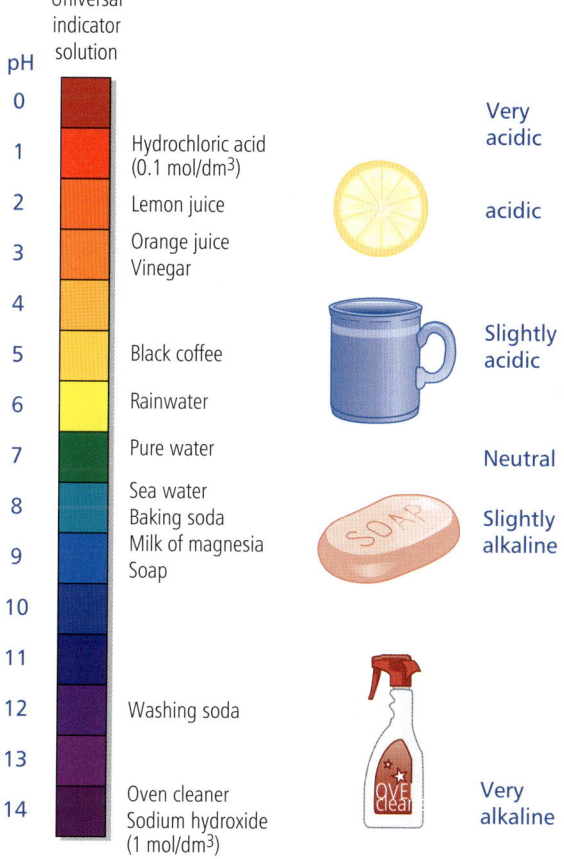

Figure 11.1.3 A colour chart for universal indicator showing the pH values for some common substances.

KEY POINTS

- In aqueous solutions, acids contain H^+ ions and alkalis contain OH^- ions.
- The pH scale is used to show the acidity or alkalinity of a solution.
- Solutions with a pH below 7 are acidic. Solutions with a pH above 7 are alkaline.
- Universal indicator, or a pH electrode and meter, can be used to find the pH of a solution.

SUMMARY QUESTIONS

1. Copy and complete using these words:

   acidic   alkaline   high   neither   neutral   scale   seven   universal

   The pH _____ shows how acidic or _____ a solution is. Strongly _____ solutions have a low pH. Strongly alkaline solutions have a _____ pH. A solution that is _____ acidic nor alkaline is called a _____ solution. It has a pH of _____. The pH of a solution can be found using _____ indicator or a pH meter.

2. Classify these pH values as acidic, alkaline or neutral and give the colour with universal indicator:

   a pH 5   b pH 13   c pH 7   d pH 8

3. Describe how to use universal indicator paper to distinguish between a solution of sodium hydroxide and a solution of ethanoic acid.

135

11.2 Properties of acids

LEARNING OUTCOMES

- Describe how acids react with metals, metal oxides, metal hydroxides and carbonates
- Describe acids in terms of their effect on litmus

The litmus test

Litmus is an indicator that has two colours, red and blue. If a solution is acidic it turns blue litmus red. If a solution is alkaline it turns red litmus blue. You can use litmus as a solution or as litmus test paper. If you use it as a test paper, it should be damp.

The simplest definition of an acid is that an acid is a substance that dissolves in water (aq) to form hydrogen ions.

$$HCl(g) + aq \longrightarrow H^+(aq) + Cl^-(aq)$$
hydrogen chloride   water   hydrochloric acid

Chemical properties of acids

Reaction of acids with metals

Many metals react with dilute acids to form a salt and hydrogen. A **salt** is a compound formed when a metal or an ammonium group (NH_4) replaces hydrogen in an acid. Some examples of the salts formed are:

- chlorides: for example magnesium chloride, $MgCl_2$ (from hydrochloric acid, HCl)
- sulfates: for example zinc sulfate, $ZnSO_4$ (from sulfuric acid, H_2SO_4)
- nitrates: for example lithium nitrate, $LiNO_3$ (from nitric acid, HNO_3).

Note that nitric acid does not give off hydrogen with metals unless it is very dilute. Unreactive metals, such as silver and copper, do not react with dilute acids.

The equation for the reaction of magnesium with hydrochloric acid is:

$$Mg(s) + 2HCl(aq) \longrightarrow MgCl_2(aq) + H_2(g)$$
magnesium + hydrochloric acid $\longrightarrow$ magnesium chloride + hydrogen

Figure 11.2.1 Acids are found in many foods that we eat. They also help us start our cars! That's because some car batteries contain sulfuric acid.

EXAM TIP

Remember the pattern:
<u>m</u>etal + <u>a</u>cid → <u>s</u>alt + <u>h</u>ydrogen (spells 'mash').

EXAM TIP

The reaction of an acid with a metal is a redox reaction (see Topic 10.4). The metal atoms are oxidised to metal ions and the H^+ ions are reduced to hydrogen molecules.

PRACTICAL

The reaction of magnesium with hydrochloric acid

You add magnesium to dilute hydrochloric acid. You collect the hydrogen gas over water. When the tube is full of gas, you can test the gas to see if it is hydrogen. (Hydrogen gives a 'pop' when a lighted splint is put in the mouth of the test tube.)

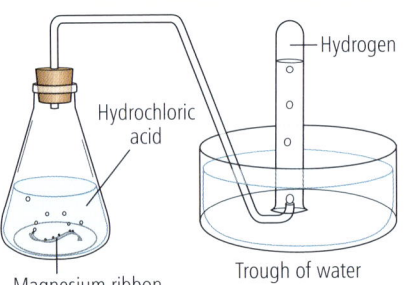

Figure 11.2.2 The reaction of magnesium with hydrochloric acid.

Unit 11: Acids and bases

Reaction of acids with metal oxides

Acids react with metal oxides to form a salt and water.

$$\text{acid} + \text{metal oxide} \longrightarrow \text{salt} + \text{water}$$

Example:

$$\underset{\text{copper(II) oxide}}{CuO(s)} + \underset{\text{sulfuric acid}}{H_2SO_4(aq)} \longrightarrow \underset{\text{copper(II) sulfate}}{CuSO_4(aq)} + \underset{\text{water}}{H_2O(l)}$$

Reaction of acids with metal hydroxides or aqueous ammonia

Acids react with many metal hydroxides to form a salt and water.

Example:

$$\underset{\text{sodium hydroxide}}{NaOH(aq)} + \underset{\text{nitric acid}}{HNO_3(aq)} \longrightarrow \underset{\text{sodium nitrate}}{NaNO_3(aq)} + \underset{\text{water}}{H_2O(l)}$$

Aqueous ammonia, $NH_3(aq)$, reacts in a similar way to metal hydroxides. But water does not appear in the equation because we do not usually write aqueous ammonia to show the hydroxide ions in solution. For example:

$$\underset{\text{aqueous ammonia}}{NH_3(aq)} + \underset{\text{hydrochloric acid}}{HCl(aq)} \longrightarrow \underset{\text{ammonium chloride}}{NH_4Cl(aq)}$$

These are all examples of **neutralisation** reactions. When the acid reacts completely with the alkali we say it has neutralised the alkali.

Reaction of acids with carbonates

Carbonates react with acids to form a salt, water and carbon dioxide. The carbon dioxide bubbles off as a gas.

$$\text{acid} + \text{metal carbonate} \longrightarrow \text{salt} + \text{water} + \text{carbon dioxide}$$

Example:

$$\underset{\text{copper(II) carbonate}}{CuCO_3(s)} + \underset{\text{sulfuric acid}}{H_2SO_4(aq)} \longrightarrow \underset{\text{copper(II) sulfate}}{CuSO_4(aq)} + \underset{\text{water}}{H_2O(l)} + \underset{\text{carbon dioxide}}{CO_2(g)}$$

Hydrogencarbonates react in a similar way. For example:

$$\underset{\text{sodium hydrogencarbonate}}{NaHCO_3(s)} + \underset{\text{hydrochloric acid}}{HCl(aq)} \longrightarrow \underset{\text{sodium chloride}}{NaCl(aq)} + \underset{\text{water}}{H_2O(l)} + \underset{\text{carbon dioxide}}{CO_2(g)}$$

KEY POINTS

- Many metals react with acids to form a salt and hydrogen.
- Acids react with metal oxides and hydroxides to form a salt and water.
- Acids react with carbonates to form a salt, water and carbon dioxide.

EXAM TIP

Don't forget that water (not hydrogen) is formed when acids react with oxides, hydroxides and carbonates, and that carbonates also form carbon dioxide.

SUMMARY QUESTIONS

1 Copy and complete using these words:

**dissolves   hydrogen
hydroxides   oxides
salt   water**

An acid is a substance that _____ in water to form _____ ions. Many acids react with metals to form a metal _____ and hydrogen. When acids react with metal _____ or _____ a salt and _____ are formed.

2 Name the salts formed when:

   a sulfuric acid reacts with zinc

   b calcium oxide reacts with hydrochloric acid

   c magnesium carbonate reacts with nitric acid

3 Write word equations for the reaction of:

   a magnesium with (very dilute) nitric acid

   b copper(II) carbonate with hydrochloric acid

   c sodium hydroxide with sulfuric acid

11.3 Bases

LEARNING OUTCOMES

- State that metal oxides and hydroxides are bases
- Describe alkalis as soluble bases
- Describe the characteristic properties of bases as reactions with acids and with ammonium salts and effect on litmus
- Recognise that a neutralisation reaction occurs between an acid and a base
- Describe the neutralisation of an acid with an alkali as:

  $H^+(aq) + OH^-(aq) \rightarrow H_2O(l)$

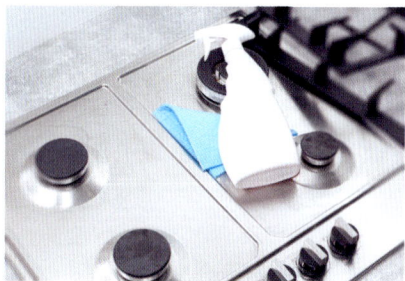

Figure 11.3.1 Alkalis are present in many cleaning products.

EXAM TIP

We do not usually write aqueous ammonia by showing the hydroxide ions in solution. We use a simpler way, where Example 2 is:
$NH_3(aq) + HCl(aq) \rightarrow NH_4Cl(aq)$.

What are bases?

Many substances we use in the home are **bases**. Dishwasher tablets, washing powder and cleaning liquids all contain bases.

Bases are oxides or hydroxides of metals. Some bases such as iron(III) oxide and copper(II) oxide are insoluble water. Others such as sodium hydroxide are soluble in water.

An alkali is a base which is soluble in water. Alkalis dissolve in water to form hydroxide ions. Ammonia is a base even though it does not contain a metal. This is because ammonia dissolves in water to produce OH^- ions (hydroxide ions):

$$NH_3(g) + H_2O(l) \rightleftharpoons NH_4^+(aq) + OH^-(aq)$$
ammonia + water ⇌ aqueous ammonia

Alkalis turns damp red litmus paper blue. So aqueous ammonia and aqueous sodium hydroxide are both alkalis.

Reactions of bases with acids

Bases (metal oxides and hydroxides) react with acids to form a salt and water.

Example 1:

$$CaO(s) + 2HCl(aq) \rightarrow CaCl_2(aq) + H_2O(l)$$
calcium oxide (base) + hydrochloric acid (acid) → calcium chloride (salt) + water

Example 2:

$$NH_4^+(aq) + OH^-(aq) + HCl \rightarrow NH_4Cl(aq) + H_2O(l)$$
aqueous ammonia (base) + hydrochloric acid (acid) → ammonium chloride (salt) + water

Neutralisation

When a base reacts with an acid to form a salt we call the reaction a neutralisation reaction.

An example of a neutralisation reaction to form a sulfate is:

$$H_2SO_4(aq) + 2NaOH(aq) \rightarrow Na_2SO_4(aq) + 2H_2O(l)$$
sulfuric acid + sodium hydroxide → sodium sulfate + water

The neutralisation of any acid with any alkali can be described by the equation

$$H^+(aq) + OH^-(aq) \rightarrow H_2O(l)$$

For many neutralisation reactions we can tell when the acid has just neutralised the alkali: A blue litmus paper dipped into the solution will turn red when the acid is just in excess.

Unit 11: Acids and bases

Reaction of alkalis with ammonium salts

When warmed with an alkali, ammonium salts decompose to form a metal salt, ammonia and water.

$$KOH(aq) + NH_4Cl(aq) \xrightarrow{heat} KCl(aq) + NH_3(g) + H_2O(l)$$
potassium hydroxide + ammonium chloride → potassium chloride + ammonia + water

PRACTICAL

Reacting ammonium sulfate with an alkali

1 When you warm a solution of an ammonium salt with sodium hydroxide solution a gas is given off.

2 You test this gas with damp red litmus paper. The litmus paper turns blue. This shows that an alkaline gas is present. If you are heating an ammonium salt, the gas will be ammonia.

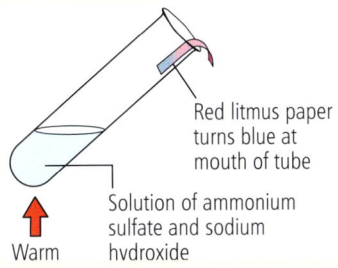

Figure 11.3.2 Ammonia gas turns damp red litmus paper blue.

EXAM TIP

It is important to understand the naming of salts. A salt made from hydrochloric acid is a chlor<u>ide</u>, and a salt made from sulfuric acid is a sulf<u>ate</u>.

EXAM TIP

You must be able to tell the difference between the ammonia molecule, NH_3, and the ammon<u>ium</u> ion, NH_4^+, in ammon<u>ium</u> salts (e.g. ammonium chlor<u>ide</u>).

Neutralising excess acidity in the soil

Soil can become acidic due to acid rain and the use of excess fertiliser. Many crop plants grow better if the soil is neutral. If soil acidity drops below pH 5.5, many plants will not grow well.

We can remove excess acidity from the soil by adding calcium oxide (lime). This neutralises the acid.

$$CaO(s) + 2H^+(aq) \rightarrow Ca^{2+}(aq) + H_2O(l)$$
calcium oxide + acid → calcium ions + water

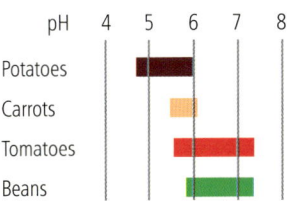

Figure 11.3.3 Crops plants grow best in these pH ranges.

SUMMARY QUESTIONS

1 Copy and complete using these words:

acids   ammonia   hydroxide   salt   soluble

Alkalis are _____ bases. Examples of alkalis are sodium _____ and _____. Alkalis react with _____ to form a _____ and water.

2 Write word equations for the reactions of:
 a ammonium sulfate with potassium hydroxide
 b calcium oxide with hydrochloric acid
 c magnesium hydroxide with nitric acid

3 Copy and complete these equations:
 a $H_2SO_4 + \_\_KOH \rightarrow K_2SO_4 + \_\_\_\_\_$
 b $Ca(OH)_2 + \_\_HCl \rightarrow \_\_\_\_\_ + \_\_\_\_\_H_2O$

KEY POINTS

- Metal oxides and hydroxides are bases.
- An alkali is a soluble base.
- Bases react with acids to form a salt and water.
- Neutralisation can be described as $H^+(aq) + OH^-(aq) \rightarrow H_2O(l)$
- Ammonia gas is released when ammonium salts are heated with an alkali.

139

11.4 More about acids and bases
Supplement

LEARNING OUTCOMES
- Define acids and bases in terms of proton transfer
- Explain the difference between strong and weak acids and bases
- Write ionic equations for acid–base reactions

More about neutralisation

When we write ionic equations we cancel out the spectator ions. So for the reaction of sodium hydroxide with hydrochloric acid:

Separating into ions then cancelling the spectator ions, we have:

$$\cancel{Na^+(aq)} + OH^-(aq) + H^+(aq) + \cancel{Cl^-(aq)} \longrightarrow \cancel{Na^+(aq)} + \cancel{Cl^-(aq)} + H_2O(l)$$

This leaves us with a very simple ionic equation:

$$OH^-(aq) + H^+(aq) \longrightarrow H_2O(l)$$

If you carry out the same process with any combination of acid and alkali, you get the same ionic equation.

Acids, bases and protons

A hydrogen ion is formed by the removal of the single electron from a hydrogen atom. Nearly all hydrogen atoms have a single proton and no neutrons in their nuclei. So a hydrogen ion is nothing more than a proton. We can define acids and bases more generally by seeing what happens to the hydrogen ions (protons) when acids and bases react.

An acid is a **proton donor** – it gives a proton to a base.

$$\underset{\text{acid}}{H^+(aq)} + \underset{\text{base}}{OH^-(aq)} \longrightarrow \underset{\text{water}}{H_2O(l)}$$

(proton transfer)

A base is a **proton acceptor** – it removes protons from an acid. You can see this in the equation above – the hydroxide ion is acting as a base as it has accepted a proton from the acid to form water.

We can see how this works in a more general case, the reaction of an ammonium salt with sodium hydroxide.

$$\underset{\text{ammonium chloride}}{NH_4Cl(aq)} + \underset{\text{sodium hydroxide}}{NaOH(aq)} \longrightarrow \underset{\text{ammonia}}{NH_3(g)} + \underset{\text{sodium chloride}}{NaCl(aq)} + \underset{\text{water}}{H_2O(l)}$$

The ionic equation is:

$$NH_4^+(aq) + OH^-(aq) \longrightarrow NH_3(g) + H_2O(l)$$

In this equation there are no hydrogen ions but there is a transfer of protons. The base, OH^-, has accepted a proton from the ammonium ion, NH_4^+. So the ammonium ion must be acting as an acid.

EXAM TIP

It is incorrect to use the words 'strong' and 'weak' when referring to the concentration of acids or alkalis. Use 'concentrated' or 'dilute'. Strong and weak refer to the degree of dissociation of the acid or base, *not* the concentration.

Unit 11: Acids and bases

$$NH_4^+(aq) + OH^-(aq) \longrightarrow NH_3(g) + H_2O(l)$$

proton transfer from acid to base

Strong and weak acids

The laboratory acids hydrochloric, sulfuric and nitric acids are all **strong acids**. A strong acid is completely ionised when it dissolves in water. There are no molecules of the acid present. We say that a strong acid is completely **dissociated** (broken up into its ions) in aqueous solution.

$$HCl(g) + aq \longrightarrow H^+(aq) + Cl^-(aq)$$

Weak acids are generally organic acids, such as ethanoic acid. A weak acid is only partially dissociated in water. There are lots of acid molecules present but very few ions.

$$CH_3COOH(l) + aq \rightleftharpoons CH_3COO^-(aq) + H^+(aq)$$

ethanoic acid (lots of molecules) — ethanoate ion (very few ions)

We can tell if an acid is strong or weak by measuring its pH and rate of reaction with metals or metal carbonates.

- A strong acid has a lower pH than a weak acid of the same concentration. This is because there is a greater concentration of hydrogen ions in the strong acid compared with the weak acid.

- A strong acid reacts faster with a metal, metal oxide or metal carbonate than a weak acid of the same concentration. This is also because there is a greater concentration of hydrogen ions in the strong acid compared with the weak acid.

Strong and weak bases

All hydroxides of Group I metals are strong bases. Strong bases are completely dissociated in water.

For example, sodium hydroxide:

$$NaOH(s) + aq \longrightarrow Na^+(aq) + OH^-(aq)$$

Ammonia is a weak base because it is only partially dissociated in water. There are very few ammonium ions.

$$NH_3(g) + H_2O(l) \rightleftharpoons NH_4^+(aq) + OH^-(aq)$$

Weak bases have a less alkaline pH (lower pH) than strong bases of the same concentration.

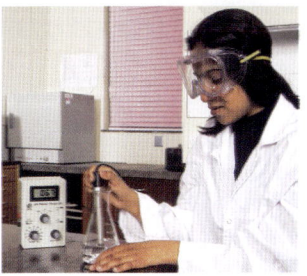

Figure 11.4.1 Measuring the pH of a solution.

EXAM TIP

Notice that the dissociation of a weak acid or base is written as an equilibrium reaction using the reversible reaction sign.

KEY POINTS

- An acid is a proton donor and a base is a proton acceptor.
- Strong acids and bases are completely dissociated in aqueous solution.
- Weak acids and bases are partially dissociated in aqueous solution.

SUMMARY QUESTIONS

1 Copy and complete using these words:

base dissolves hydrogen ions proton

When an acid _____ in water, hydrogen _____ are formed. A _____ ion is a proton. An acid is a _____ donor. It gives its protons to a _____.

2 State the meaning of the term weak acid.

3 Describe one way to tell the difference between a strong and a weak acid of the same concentration.

11.5 Acid–base titrations and indicators

LEARNING OUTCOMES

- Describe how to do an acid–base titration
- Describe how to identify the end point of a titration
- Describe the effect of acid and alkalis on litmus, thymolphthalein and methyl orange

Doing a titration

In Topic 6.8 we learned how to find the concentration of an alkali needed to completely react with an acid using a procedure called an acid–base titration. We use the following sequence to carry out a titration:

1. Measure a known volume of alkali into the conical flask using a volumetric pipette which has first been washed with a little of the alkali you are using (Figure 11.5.1).
2. Add a few drops of acid–base indicator solution to the alkali in the flask.
3. Fill a clean burette with acid after washing the burette with a little of the acid you are using.
4. Record the burette reading.
5. Open the burette tap and let the acid flow into the flask. Keep swirling the flask gently to make sure that the acid and alkali mix and react (Figure 11.5.2).
6. Keep adding the acid slowly until the indicator changes colour. This is the **end point**.
7. Record the reading on the burette. The final reading minus the initial reading is called the titre. The first time you do this gives you the rough titre or 'range-finder' titre.
8. Repeat this process at least three times. You can add the acid rapidly until you are a few cm³ from the end point. Then add the acid drop by drop so that you can get an accurate titre. Ideally, the titres will be within 0.1 cm³ of each other.
9. If you are doing calculations to find the concentration of the alkali in the flask, you take the average of the accurate titres. You can ignore any titres that appear to be inconsistent.

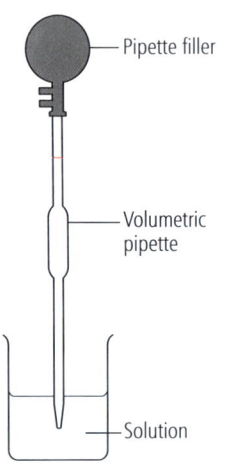

Figure 11.5.1 Filling a volumetric pipette.

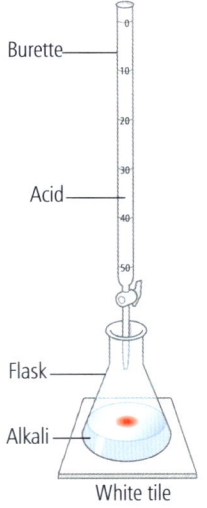

Figure 11.5.2 Titration apparatus.

Indicators

We use an indicator to find when the acid has just reacted with all the alkali. We call this the end point of the titration. At the end point the indicator changes colour. The indicator we choose depends on whether we use a strong or weak acid or alkali:

- For a strong acid such as hydrochloric acid and **strong alkali** such as sodium hydroxide, we can use any indicator. Litmus is suitable. This goes from blue to red when there is excess acid.

Unit 11: Acids and bases

- If we titrate a **weak alkali**, such as ammonia, with a strong acid we use methyl orange indicator. Methyl orange goes from yellow to red when there is excess acid.

- If we titrate a weak acid, for example ethanoic acid, with a strong base we use thymolphthalein indicator. Thymolphthalein goes from blue to colourless when there is excess acid.

See Topic 2.1 for more about the use of the pipette and burette, and Topic 6.8 for how to do the calculation from titration results

pH changes during a titration

Figure 11.5.4 shows how the pH changes during a titration as we add hydrochloric acid to aqueous ammonia. We see that the pH changes suddenly at the end point. For an indicator to be suitable for this titration we have to choose an indicator that changes colour during this sudden pH change. Slightly different curves are produced for different acids and alkalis.

The table shows the pH values at which three different indicators change colour.

indicator	pH range and colour
litmus	red at pH 5 to blue at pH 8
methyl orange	red at pH 3.2 to yellow at pH 4.4
thymolphthalein	colourless at pH 8.3 to blue at pH 10.6

From this table you can see that you cannot use thymolphthalein for the titration shown in Figure 11.5.4 because it does not change colour at the point where there is a sudden change in pH.

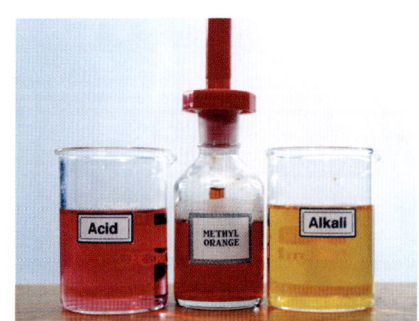

Figure 11.5.3 Methyl orange is red in hydrochloric acid but yellow in sodium hydroxide an alkali.

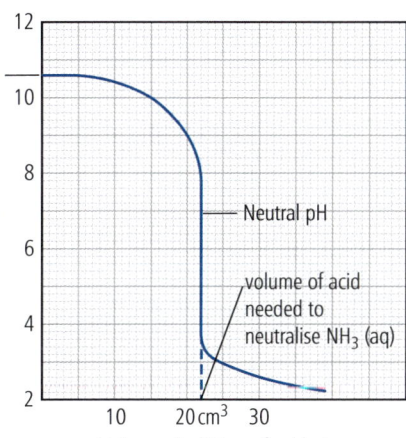

Figure 11.5.4 At the end point of a titration of NH_3(aq) with HCl(aq), the pH changes suddenly.

EXAM TIP

You do not have to understand the shape of the pH titration curve in Figure 11.5.4, but you may be asked to interpret data from titration curves such as the pH at the start.

SUMMARY QUESTIONS

1 Copy and complete using these words:

   burette   colour   end   indicator   titration   volumetric

   We can find the concentration of alkali needed to completely react with an acid using a procedure called an acid–base _____. An accurate volume of aqueous alkali is placed in a flask using a _____ pipette. A few drops of _____ are then added to the flask. Acid is added from a _____ until the indicator changes _____. This shows the _____ point.

2 Give the correct colour changes:

   a methyl orange is _____ at pH 2 and _____ at pH 8
   b thymolphthalein is _____ at pH 4 and _____ at pH 12
   c litmus is _____ at pH 10 and _____ at pH 4

KEY POINTS

- An acid–base titration involves the correct use of a volumetric pipette and burette.

- The end point of a titration is reached when a suitable indicator changes colour.

- Methyl orange is red in acid and yellow in alkali, and thymolphthalein is colourless in acid and blue in alkali.

11.6 Oxides

LEARNING OUTCOMES

- Classify oxides as acidic or basic related to their non-metallic or metallic character
- **S** Classify zinc oxide and aluminium oxide as amphoteric

Figure 11.6.1 Lime (CaO), a basic oxide, is spread on the soil to decrease its acidity.

EXAM TIP

Make sure you know that most metal oxides are basic oxides and most non-metal oxides are acidic oxides.

Four types of oxide

Oxides are compounds of metals or non-metals with oxygen. There are four types of oxide: basic, acidic, amphoteric and neutral. We tell the difference between the four types by their typical chemical reactions.

Basic oxides

Most metal oxides are **basic oxides**. Examples are calcium oxide, CaO, and copper(II) oxide, CuO. Many basic oxides are formed by the direct combination of a metal with oxygen. Basic oxides react with acids to form a salt and water. For example:

$$CaO(s) + 2HCl(aq) \longrightarrow CaCl_2(aq) + H_2O(l)$$
calcium oxide + hydrochloric acid → calcium chloride + water

$$CuO(s) + H_2SO_4(aq) \longrightarrow CuSO_4(aq) + H_2O(l)$$
copper(II) oxide + sulfuric acid → copper(II) sulfate + water

Basic oxides do not react with alkalis.

Many basic oxides do not react with water. But those from Group I and many from Group II in the Periodic Table react to form a metal hydroxide. An alkaline solution is formed which turns red litmus blue.

$$BaO(s) + H_2O(l) \longrightarrow Ba(OH)_2(aq)$$
barium oxide + water → barium hydroxide

Acidic oxides

Most non-metal oxides are acidic oxides. Examples are sulfur dioxide, SO_2, and carbon dioxide, CO_2. Many are formed by direct reaction with oxygen.

$$S(s) + O_2(g) \longrightarrow SO_2(g)$$
sulfur → sulfur dioxide

Acidic oxides react with alkalis to form a salt and water. For example:

$$CO_2(g) + 2NaOH(aq) \longrightarrow Na_2CO_3(aq) + H_2O(l)$$
carbon dioxide + sodium hydroxide → sodium carbonate + water

Some acidic oxides react with bases such as metal oxides when heated. For example:

$$SiO_2 + CaO \xrightarrow{heat} CaSiO_3$$
silicon(IV) oxide + calcium oxide → calcium silicate

Many acidic oxides react with water to form acidic solutions. For example:

$$SO_2(g) + H_2O(l) \longrightarrow H_2SO_3(aq)$$
sulfur dioxide + water → sulfurous acid

Unit 11: Acids and bases

Supplement

Neutral oxides

Neutral oxides do not react with acids or bases. Examples of neutral oxides are nitrogen(I) oxide, N_2O, nitrogen(II) oxide, NO, and carbon monoxide, CO. Most are oxides of particular non-metals that have lower oxidation numbers. For example:

- carbon monoxide is a neutral oxide but carbon dioxide, CO_2, is an acidic oxide
- nitrogen(I) oxide is a neutral oxide but nitrogen dioxide (nitrogen(IV) oxide), NO_2, is an acidic oxide.

Amphoteric oxides

The word amphoteric means 'both of them'. **Amphoteric oxides** have both acidic and basic properties. The oxides of aluminium and zinc are examples. They form salts when they react with acids. They also react with alkalis to form complex salts.

Examples:

$$ZnO(s) + 2HNO_3(aq) \longrightarrow Zn(NO_3)_2(aq) + H_2O(l)$$
zinc oxide + nitric acid → zinc nitrate + water

$$ZnO(s) + 2NaOH(aq) \longrightarrow Na_2ZnO_2(aq) + H_2O(l)$$
zinc oxide + sodium hydroxide → sodium zincate + water

$$Al_2O_3(s) + 6HCl(aq) \longrightarrow 2AlCl_3(aq) + 3H_2O(l)$$
aluminium oxide + hydrochloric acid → aluminium chloride + water

$$Al_2O_3(s) + 2NaOH(aq) \longrightarrow 2NaAlO_2(aq) + H_2O(l)$$
aluminium oxide + sodium hydroxide → sodium aluminate + water

The zincates and aluminates have the ending -ate to show that their ions are compound ions containing oxygen. Zincate ions are ZnO_2^{2-} and aluminate ions are AlO_2^-. You may also see these ions written $Zn(OH)_4^{2-}$ and $Al(OH)_4^-$. Notice that sodium zincate and aluminate are soluble in water, like all other sodium salts.

> **EXAM TIP**
>
> You do not need to know about neutral oxides for your exam but it is useful to distinguish them from acidic oxides with the same types of atoms.

> **EXAM TIP**
>
> You will not be expected to remember the equations for the reactions of ZnO or Al_2O_3 with NaOH, but you may be expected to balance these equations given relevant formulae.

> **KEY POINTS**
>
> - The oxides of most metals are basic oxides. The oxides of most non-metals are acidic oxides.
> - The oxides of aluminium and zinc are amphoteric. They react with both acids and alkalis.

SUMMARY QUESTIONS

1. Copy and complete using these words:

   acidic   alkalis   litmus   metals   react   salt

   Most oxides of non-metals are _____ oxides. Acidic oxides react with _____ to form a _____ and water. Some acidic oxides _____ with water to form solutions of acids. These solutions turn blue _____ red. Most oxides of _____ are basic oxides.

2. How can you show by experiment that calcium oxide is a basic oxide?

3. Copy and complete these equations:
   a $ZnO(s) + 2HCl(aq) \longrightarrow$ _____(aq) + _____(l)
   b   _____(s) + _____ KOH(aq) $\longrightarrow$ _____ $KAlO_2$(aq) + H_2O(l)

SUMMARY QUESTIONS

1. Match the reactants on the left with the products on the right.

acid + carbonate	salt + water
acid + hydroxide	salt + ammonia + water
acid + metal	salt + hydrogen
ammonium salt + alkali	salt + water + carbon dioxide

2. Copy and complete these word equations:

   a zinc + hydrochloric acid → _____ + _____

   b sulfuric acid + magnesium oxide → _____ + _____

   c _____ + _____ → calcium nitrate + carbon dioxide + water

   d ammonium sulfate + _____ → potassium sulfate + ammonia + water

3. a What is meant by the term acid–base indicator?

   b State the colour of: (i) thymolphthalein in acidic solution and (ii) methyl orange at pH 3.

4. Match the substances on the left with the pH in the centre and the acidity on the right.

vinegar	pH 0	strongly acidic
dishwasher powder	pH 12	weakly acidic
soap	pH 7	weakly alkaline
distilled water	pH 4.5	neutral
concentrated hydrochloric acid	pH 7.5	strongly alkaline

5. Classify these oxides as acidic or basic:

   a copper(II) oxide

   b magnesium oxide

   c sulfur dioxide

   d phosphorus(V) oxide

6. Write balanced equations for the reaction of:

   a sodium carbonate with dilute nitric acid

   b magnesium with dilute hydrochloric acid

Practice questions

1. A solution has a pH of 9. The solution is best described as:

   A strongly alkaline
   B weakly acidic
   C weakly alkaline
   D strongly acidic.

   (Paper 1)

2. The equation between ammonium ions and hydroxide ions is shown by the equation:

   $NH_4^+(aq) + OH^-(aq) \rightarrow NH_3(g) + H_2O(l)$

   Which one of these statements about this reaction is true?

   A Ammonium ions are acting as a base.
   B A proton is transferred from the hydroxide ions to the ammonium ions.
   C Ammonium atoms are acting as proton acceptors.
   D Hydroxide ions are acting as proton acceptors.

   (Paper 2)

3. Magnesium sulfate can be made by neutralising magnesium oxide with an acid.

   (a) What do you understand by the term acid? [1]

   (b) Name the acid used to make magnesium sulfate. [1]

   (c) Copy and complete the equation for this reaction:

   MgO + _____ → $MgSO_4$ + _____ [2]

   (d) What type of oxide is magnesium oxide? Give a reason for your answer. [2]

   (e) (i) Suggest one other reaction for making magnesium sulfate other than using MgO. [1]

   (ii) Write a word equation for this reaction. [2]

   (Paper 3)

4. Ammonia is a gas that is very soluble in water.

   (a) From the following list choose the most likely pH of an aqueous solution of ammonia.

   pH 2   pH 5   pH 7   pH 10 [1]

   (b) What effect does aqueous ammonia have on red litmus paper? [1]

   (c) Ammonia is released when ammonium sulfate is warmed with sodium hydroxide solution. Write a word equation for this reaction. [2]

(d) Ammonia reacts with hydrochloric acid to form a salt.
 (i) Copy and complete the equation for this reaction by writing the formulae of an acid and an alkali.
 _____ + _____ → NH_4Cl [2]
 (ii) Name the salt formed in this reaction. [1]

(e) State the colour change when hydrochloric acid is added in excess to an aqueous solution of methyl orange. [1]

(Paper 3)

5 Hydrochloric acid is a strong acid.
(a) (i) What colour change occurs when excess hydrochloric is added to thymolphthalein in alkaline solution? [2]
 (ii) Describe how universal indicator is used to find the pH of hydrochloric acid. [2]
(b) Hydrochloric acid reacts with magnesium.
 (i) Write a word equation for this reaction. [2]
 (ii) Name the salt formed in this reaction. [1]
(c) Hydrochloric acid also reacts with magnesium carbonate. Copy and complete the symbol equation for this reaction.
 $MgCO_3$ + ___ HCl → $MgCl_2$ + ___ + ___ [3]
(d) When hydrochloric acid reacts with sodium sulfite, sulfur dioxide is produced. What type of oxide is sulfur dioxide? [1]
(e) Describe how the pH of the mixture changes as hydrochloric acid is added to a solution of potassium hydroxide until hydrochloric acid is in excess. [2]

(Paper 3)

6 Ethanoic acid is a weak acid.
(a) Suggest how you can use universal indicator paper to show that ethanoic acid is a weak acid. [3]
(b) Describe one other method you can use to show that ethanoic acid is a weak acid. Give two observations you can make. [4]
(c) Write a balanced equation to show the reaction of ethanoic acid, CH_3COOH, with magnesium. [2]

(d) Ethanoic acid reacts with potassium hydroxide:
 $CH_3COOH + KOH → CH_3COOK + H_2O$
 Use ideas about proton transfer to explain how this equation shows that ethanoic acid is an acid and potassium hydroxide is a base. [2]
(e) Potassium hydroxide is a strong base. What is the meaning of the term strong base? [1]

(Paper 4)

7 Calcium oxide reacts with hydrochloric acid to form calcium chloride and water.
(a) Write a symbol equation for this reaction. [2]
(b) Calcium oxide is a base. Define a base. [1]
(c) Write an equation for the dissociation of ethanoic acid in water. [2]
(d) Write an ionic equation to show how the oxide ion reacts with one type of ion in hydrochloric acid to form water. [2]
(e) Zinc oxide reacts with both acids and alkalis.
 (i) State the name given to an oxide that reacts with both acids and alkalis. [1]
 (ii) Copy and complete the equation for the reaction of zinc oxide with sodium hydroxide.
 $ZnO(s) +$ _____ $(aq) →$ $Na_2ZnO_2(aq) +$ _____ (l) [2]

(Paper 4)

8 Ammonia is a weak base.
(a) What is the meaning of the term *weak* in the term weak base? [2]
(b) Write a balanced equation to show the reaction of ammonia with sulfuric acid. [2]
(c) When ammonium salts are warmed with hydroxide ions, ammonia is released.
 $NH_4^+ + OH^- → NH_3 + H_2O$
 Explain how ammonium ions are acting as an acid in this reaction. [2]
(d) When ammonia reacts with water an equilibrium mixture of reactants and products is formed.
 $NH_3 + H_2O \rightleftharpoons NH_4^+ + OH^-$
 Describe each of the reactants and products as either acids or bases. Give reasons for your answers. [4]

(Paper 4)

12.1 Making salts from metals, bases or carbonates

LEARNING OUTCOMES

- Describe the preparation, separation and purification of soluble salts from metals, insoluble bases or insoluble carbonates

Figure 12.1.1 Salts come in many different colours and crystal forms.

EXAM TIP

You should know that at the point of crystallisation a saturated solution is formed.

EXAM TIP

You know when the reaction is complete because there is no more bubbling or fizzing.

EXAM TIP

When you make a salt using excess metal or insoluble oxide (or carbonate), don't forget that you must filter off the excess solid immediately after the reaction.

How do we make salts?

A salt is a compound formed when a metal or an ammonium group replaces hydrogen in an acid. Many salts are soluble in water although a few, such as lead chloride, are insoluble. There are four ways of making salts:

- reacting a metal with an acid
- reacting an insoluble base with an acid
- neutralising an alkali with an acid by the titration method
- by precipitation.

Not all these methods are suitable for making a particular salt. So we have to choose the method that best fits the type of salt we want to make. We also have to choose a suitable method of purifying the salt.

Soluble salts from metals

We can make soluble salts by reacting acids with metals above hydrogen in the reactivity series. So we can make salts of magnesium, zinc, aluminium and iron in this way. For example, to make zinc sulfate we carry out the following reaction:

$$Zn(s) + H_2SO_4(aq) \longrightarrow ZnSO_4(aq) + H_2(g)$$
$$\text{zinc} + \text{sulfuric acid} \longrightarrow \text{zinc sulfate} + \text{hydrogen}$$

We cannot use this type of reaction for making salts of copper, lead and silver which are too close to or below hydrogen in the reactivity series. Also it is not a good idea to prepare salts of very reactive metals, such as sodium and potassium, using this method. The reaction of these metals with the acid is too violent. A titration method is more suitable for salts of these very reactive metals.

This is the procedure:

1. Add the metal to the acid in a flask so that the metal is in excess. The acid is the limiting reactant.
2. Warm the flask gently to complete the reaction.
3. Filter off the excess metal. The filtrate is a solution of the metal salt.
4. Put the filtrate into an evaporating basin and evaporate the water until the crystallisation point is reached. Then you leave the salt to crystallise at room temperature.
5. Filter off the crystals and wash them with a tiny amount of cold solvent so they don't dissolve.
6. Dry the crystals between sheets of filter paper.

Unit 12: Making and identifying salts

Soluble salts from insoluble bases and insoluble carbonates

We can make soluble salts of many metals by reacting an insoluble base or an insoluble carbonate with an acid. We use this method for making salts of metals that are low in the reactivity series.

Example: making copper(II) sulfate from copper(II) oxide and sulfuric acid

$$CuO(s) + H_2SO_4(aq) \rightarrow CuSO_4(aq) + H_2O(l)$$

copper(II) oxide + sulfuric acid → copper(II) sulfate + water

The method for making salts from insoluble oxides or carbonates is exactly the same as that given in points 1 to 6 on page 148 for making a salt from excess metal and an acid.

PRACTICAL

Making copper(II) sulfate from copper(II) oxide

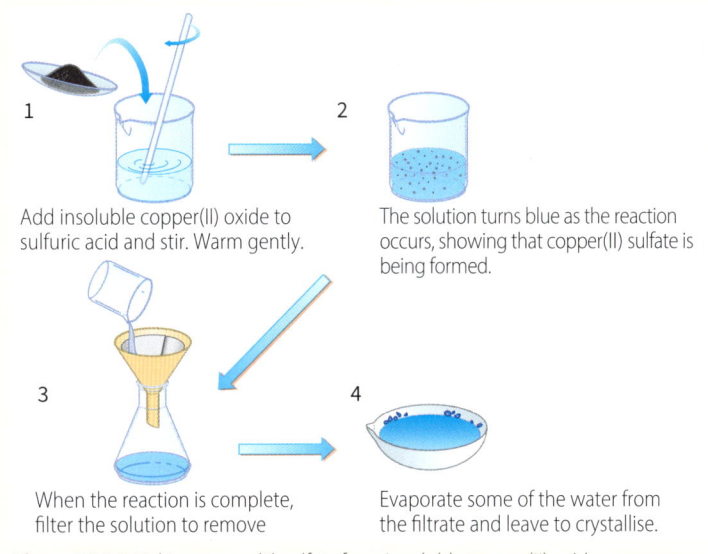

1. Add insoluble copper(II) oxide to sulfuric acid and stir. Warm gently.
2. The solution turns blue as the reaction occurs, showing that copper(II) sulfate is being formed.
3. When the reaction is complete, filter the solution to remove
4. Evaporate some of the water from the filtrate and leave to crystallise.

Figure 12.1.2 Making copper(II) sulfate from insoluble copper(II) oxide.

EXAM TIP

When washing a salt, it is best to put the crystals in a filter paper in a filter funnel. Use a solvent that does not dissolve the salt but evaporates when you leave the crystals to dry. Suitable solvents are ethanol or propanone.

EXAM TIP

When making salts from metal oxides or metal carbonates, remember that the metal oxide or carbonate is always in excess.

KEY POINTS

- Soluble salts of a metal above hydrogen in the reactivity series are made by the reaction of excess metal with an acid.
- Soluble salts can be made by the reaction of excess insoluble oxide or carbonate with an acid.

SUMMARY QUESTIONS

1 Copy and complete using the words below:

filter   filtrate   insoluble   limiting   neutralised   oxide

You can make a soluble metal salt by reaction of an _____ metal oxide with an acid. During this reaction the acid is _____ by the metal _____. The acid is the _____ reactant. You _____ off the excess metal oxide. The _____ is a solution of the metal salt.

2 Write word equations for these reactions for making salts:

   a $Fe_2O_3(s) + 6HCl(aq) \rightarrow 2FeCl_3(aq) + 3H_2O(l)$

   b $Mg(s) + H_2SO_4(aq) \rightarrow MgSO_4(aq) + H_2(g)$

3 Write instructions for making the salt, zinc chloride, from zinc oxide.

12.2 Making salts by titration

LEARNING OUTCOMES

- Describe the preparation, separation and purification of soluble salts by a titration method

Why do we use the titration method to make a soluble salt?

When we make a soluble salt using a metal, an insoluble oxide or an insoluble carbonate, we know when the reaction is complete because we can see excess solid in the reaction mixture. In the case of metals and insoluble carbonates, the effervescence or fizzing also stops.

However, if we want to make salts of Group I metals from their hydroxides, we cannot use this method because these hydroxides are very soluble in water. So we cannot see when they are in excess. If the hydroxide remains in excess, we will get impure crystals that are a mixture of the salt and the excess hydroxide.

So we have to use a titration method to make sure that the acid has exactly neutralised the alkali. We use this method to make salts from the hydroxides of Group I elements or to make ammonium salts from aqueous ammonia.

Example 1: making sodium sulfate from sodium hydroxide

$$2NaOH(aq) + H_2SO_4(aq) \longrightarrow Na_2SO_4(aq) + 2H_2O(l)$$

sodium hydroxide + sulfuric acid → sodium sulfate + water

Example 2: making ammonium chloride from aqueous ammonia

$$NH_3(aq) + HCl(aq) \longrightarrow NH_4Cl(aq)$$

aqueous ammonia + hydrochloric acid → ammonium chloride

Figure 12.2.1 Sodium sulfate crystals, $Na_2SO_4.10H_2O$, have a lot of water of crystallisation bound up in their structure.

The titration method for making salts

We first carry out a titration using an acid–base indicator to find the correct volumes of solution to mix. Then we carry out the titration again without the indicator to prepare a sample of the salt uncontaminated by indicator.

This is the procedure:

1. Put a known volume of hydroxide or aqueous ammonia in the flask and add a suitable acid–base indicator.
2. Add acid from the burette until the indicator just changes colour. Record the volume of acid added.
3. Repeat the titration exactly as before but without the indicator until the volume of acid you recorded in step **2** has been added.

EXAM TIP

You need to learn the full procedure for making salts by titration. Remember that the acid usually goes in the burette and the alkali in the flask, and that you repeat the titration without the indicator.

Unit 12: Making and identifying salts

4 Crystallise the contents of the flask and dry the crystals as in Steps **4** to **6** in Topic 12.1 (on page 148).

Note that as an alternative to Step **3** you can shake the contents of the flask with activated carbon to remove the colour of the indicator and then filter the solution.

> **EXAM TIP**
>
> When writing about crystallisation it is NOT sufficient to state 'heat to dryness'. By doing this you will lose the water of crystallisation. You will make a powder, not crystals.

PRACTICAL

Making sodium sulfate using the titration method

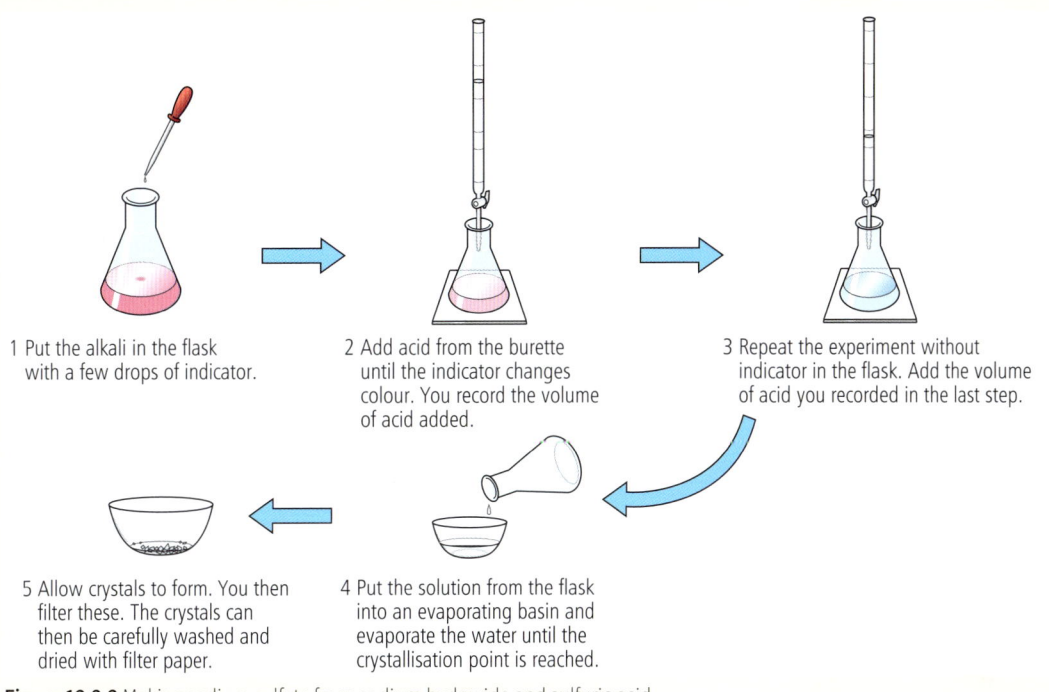

1 Put the alkali in the flask with a few drops of indicator.

2 Add acid from the burette until the indicator changes colour. You record the volume of acid added.

3 Repeat the experiment without indicator in the flask. Add the volume of acid you recorded in the last step.

4 Put the solution from the flask into an evaporating basin and evaporate the water until the crystallisation point is reached.

5 Allow crystals to form. You then filter these. The crystals can then be carefully washed and dried with filter paper.

Figure 12.2.2 Making sodium sulfate from sodium hydroxide and sulfuric acid.

SUMMARY QUESTIONS

1 Copy and complete using the words below:

alkali   burette   indicator   repeated   soluble

A titration method is used to make a _____ salt from an acid and an _____. The acid is added to the alkali using a _____ until the _____ in the flask changes colour. The process is then _____ without using the indicator.

2 Name an acid and alkali you can use to make these salts:
 a ammonium chloride
 b sodium chloride
 c ammonium nitrate

3 Name the salts formed from these acids and alkalis:
 a sodium hydroxide + nitric acid
 b ammonia + sulfuric acid
 c lithium hydroxide + hydrochloric acid

KEY POINTS

- A titration is used to make a soluble salt from an acid and an alkali.
- Salts made by the titration method include salts of Group I elements and ammonium salts.
- When making a salt using the titration method, the titration is first carried out using an indicator and then repeated without an indicator.

12.3 Making salts by precipitation

LEARNING OUTCOMES
- Describe the general solubility rules for salts
- **S** Describe the preparation of insoluble salts by precipitation

EXAM TIP

Remember to use the correct chemical terms. Sodium oxide, *reacts* with water to form sodium hydroxide but sodium chloride *dissolves* in water. It is still Na^+ and Cl^- ions.

Figure 12.3.1 Water treatment plants use iron sulfate or aluminium sulfate to precipitate some unwanted compounds in the impure water.

Soluble or insoluble?

The salts that we have discussed in Topics 12.1 and 12.2 are soluble in water. Other salts are insoluble. We make these salts by mixing two soluble compounds. The solid obtained when solutions of two soluble compounds are mixed is called a **precipitate**.

It is useful to know which salts are insoluble in order to understand the results of qualitative tests for specific ions.

Some **solubility** rules are shown in the table:

soluble compounds	insoluble compounds
all salts of Group I elements	
all nitrates	
all ammonium salts	
most chlorides, bromides and iodides	chlorides, bromides and iodides of silver and lead
most sulfates	sulfates of calcium, barium and lead
Group I hydroxides and carbonates and ammonium carbonate (calcium hydroxide is slightly soluble)	most hydroxides and carbonates

Most metal oxides are insoluble but oxides of the elements in Group I react with water.

Making salts by precipitation

If we are going to make salts by precipitation, we need to know which compounds are soluble in water and which are insoluble. We can use the solubility rules to help us.

If we want to make an insoluble salt, for example lead chloride, we:

1. identify the ions present in the insoluble salt: lead ions and chloride ions
2. use the solubility rules to choose soluble compounds including these ions: for example lead nitrate for the lead ions and sodium chloride for the chloride ions
3. add one solution to the other
4. filter off the precipitate then wash with distilled water and dry the solid.

What happens in a precipitation reaction?

We can explain why a solid precipitates by looking at the reaction between lead nitrate and sodium chloride as an example.

$$Pb(NO_3)_2(aq) + 2NaCl(aq) \longrightarrow PbCl_2(s) + 2NaNO_3(aq)$$

Unit 12: Making and identifying salts

PRACTICAL

Making an insoluble salt

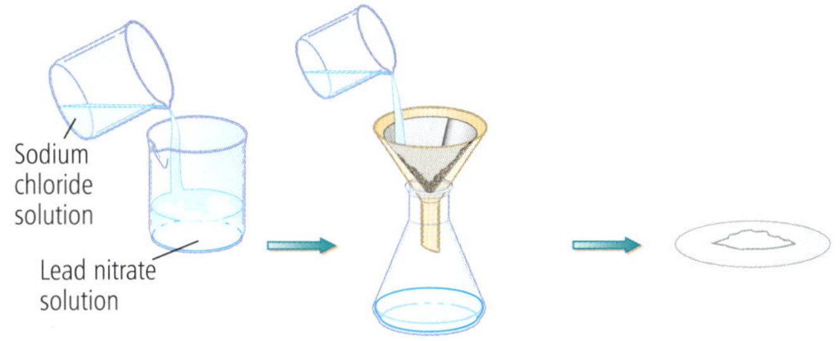

1. We add sodium chloride solution to lead nitrate solution and stir
2. The precipitate of lead chloride that forms is filtered off from the solution
3. The precipitate is washed with distilled water and dried

12.3.2 Making lead chloride.

Lead chloride, the insoluble salt, is precipitated when we mix the solutions. The lead ions in solution have a greater attraction for the chloride ions than the water molecules that keep them in solution. So the lead ions and chloride ions come together in large numbers and form a three-dimensional ionic lattice. The sodium ions and nitrate ions remain in solution. They are called the spectator ions.

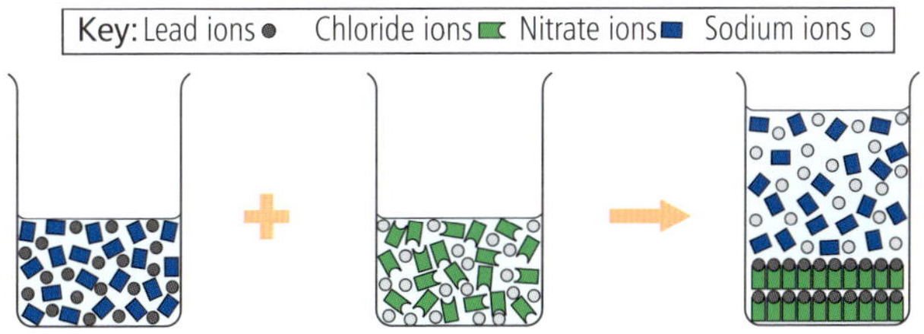

Key: Lead ions ● Chloride ions ▬ Nitrate ions ■ Sodium ions ○

Lead ions and nitrate ions are free to move in solution.

Sodium ions and chloride ions are free to move in solution.

When mixed, lead chloride forms an ionic lattice and sodium and nitrate ions remain in solution.

Figure 12.3.3 How a precipitate forms. The water molecules are not shown. They are represented by the light blue background.

EXAM TIP

Make sure that you know what types of compound are soluble or insoluble. Without this knowledge you will not be able to select precipitation as the correct method to make a particular salt.

KEY POINTS

- All salts of Group I elements, ammonium salts and nitrates are soluble in water.
- Most chlorides, bromides and iodides are soluble. Those of lead and silver are insoluble.
- Insoluble salts are made by mixing solutions of two soluble salts.

SUMMARY QUESTIONS

1. Which of these compounds are insoluble in water?
   - **a** potassium bromide
   - **b** silver bromide
   - **c** sodium hydroxide
   - **d** barium sulfate
   - **e** lead iodide

2. Copy and complete using the words below:

   attracted   insoluble   ion   lattice   precipitation   spectator

   We can make an _____ salt by mixing two solutions of soluble salts. This type of reaction is called _____. When the solutions react the two types of _____ that form the precipitate are _____ to each other and form a _____ structure. The ions that do not take part in the reaction are called _____ ions.

3. Write symbol equations for the reactions below, including state symbols:
   - **a** aqueous lead nitrate reacting with aqueous potassium iodide
   - **b** aqueous iron(II) chloride, $FeCl_2(aq)$, reacting with aqueous sodium hydroxide

12.4 What's that gas?

LEARNING OUTCOMES

- Describe tests for the gases ammonia, carbon dioxide, chlorine, hydrogen, oxygen and sulfur dioxide

Collecting gases

Many chemical reactions produce a gas as one of the products. Before we can identify a particular gas we have to collect it. The way we do this depends on:

- the density of the gas – is it heavier or lighter than air?
- the solubility of the gas in water – is it soluble or insoluble?

We can use a gas syringe to collect any gas. But it is easier to identify a gas if we collect it in a test tube. There are three ways we can collect a gas in a test tube:

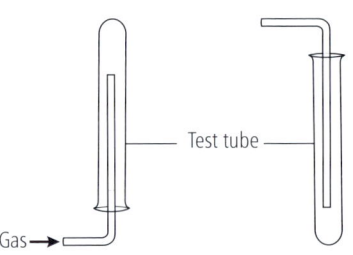
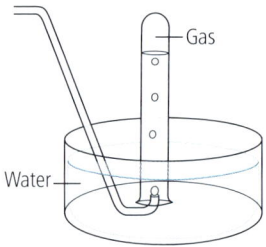

1. Downward displacement is used for gases lighter than air. For example: hydrogen, ammonia
2. Upward displacement is used for gases heavier than air. For example: carbon dioxide, chlorine, hydrogen chloride
3. Downward displacement of water is used for gases which are insoluble or slightly soluble in water. For example: hydrogen, oxygen

Figure 12.4.1 Methods for collecting gases.

After collecting the gas in the test tube you put a bung on the tube so that the gas does not escape before you identify it.

Identifying hydrogen

You put a lighted splint at the mouth of the test tube. If the gas is hydrogen it burns with a squeaky 'pop' sound. The hydrogen is reacting with oxygen in the air to cause a small explosion when a flame or spark is present.

Identifying oxygen

You put a glowing splint into the test tube. If the gas is oxygen the splint will relight. The splint is made of wood and wood is a fuel. Fuels burn better in oxygen than in air – there is no nitrogen to dilute the oxygen. So the splint will burn much better in pure oxygen – so much so that the glowing splint will relight.

The litmus test for ammonia

We can tell if a gas is acidic or alkaline by holding a piece of damp litmus paper at the mouth of the test tube. If the gas is alkaline it will turn damp red litmus paper blue. The gas is almost certainly ammonia if there is a strong sharp smell as well. If the gas given off in a reaction is acidic it will turn damp blue litmus paper red.

EXAM TIP

A common mistake is to confuse the tests for hydrogen and oxygen. It may help you to remember that 'lighted' (splint) has an 'h' in it for hydrogen and a 'glowing' (splint) has an 'o' in it for oxygen.

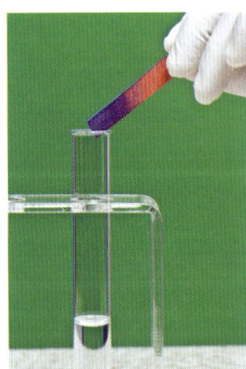

Figure 12.4.2 The litmus turns blue because an alkaline gas is given off. The gas is probably ammonia.

Unit 12: Making and identifying salts

Identifying carbon dioxide

If we think that a gas given off in a reaction is carbon dioxide, we bubble it through limewater. If carbon dioxide is present, the limewater turns milky or cloudy. A simpler way to test for carbon dioxide is to simply put a drop of limewater on the end of a flattened glass rod and hold it above the reaction mixture. But take care that the drop does not fall off!

Limewater is a solution of calcium hydroxide. This solution is colourless. But when you bubble carbon dioxide through it, a fine white precipitate of calcium carbonate is formed:

$$Ca(OH)_2(aq) + CO_2(g) \longrightarrow CaCO_3(s) + H_2O(l)$$

calcium hydroxide + carbon dioxide $\longrightarrow$ calcium carbonate + water

Carbon dioxide is an acidic oxide. So it reacts with a base to form a salt and water. If you bubble the carbon dioxide through the limewater for too long the limewater goes colourless again. This is because the calcium carbonate dissolves to form soluble calcium hydrogencarbonate.

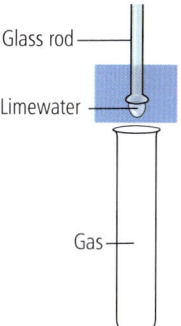

Figure 12.4.3 Testing for carbon dioxide.

Identifying chlorine

Chlorine is a poisonous green gas. So if you think chlorine is going to be released you should carry out the test in a fume cupboard. You put damp litmus paper or universal indicator paper at the mouth of the test tube. The indicator paper turns white – it is bleached.

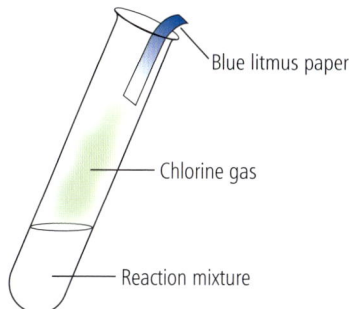

Figure 12.4.4 Chlorine bleaches litmus paper.

Identifying sulfur dioxide

Sulfur dioxide is a colourless gas with a strong acidic smell. It is poisonous, so tests for this gas should be carried out in a fume cupboard. We use an acidified solution of potassium manganate(VII), which is purple in colour, to test for sulfur dioxide. When sulfur dioxide is bubbled through acidified aqueous potassium manganate(VII), the solution turns from purple to colourless.

EXAM TIP

Make sure you know the difference between the test for chlorine gas (a molecule) and chloride ions (see Topic 12.5).

KEY POINTS

- Oxygen relights a glowing splint.
- Hydrogen gives a squeaky pop with a lighted splint.
- Chlorine bleaches damp litmus paper.
- Carbon dioxide turns limewater milky.
- Ammonia turns red litmus paper blue.
- Sulfur dioxide turns acidified potassium manganate(VII) solution colourless.

SUMMARY QUESTIONS

1 Copy and complete using the words below:

collecting   displacement   heavier   insoluble   water

There are several methods of _____ a gas. If a gas is _____ than air you collect it by upward _____ of air. If a gas is _____ in water you can collect it over _____.

2 Describe the tests for **a** ammonia **b** chlorine.

3 Describe the differences between the test for oxygen and the test for hydrogen.

155

12.5 Testing for cations

LEARNING OUTCOMES

- Describe tests for these ions in aqueous solution: Al^{3+}, Ca^{2+}, Cr^{3+}, Cu^{2+}, Fe^{2+}, Fe^{3+}, NH_4^+ and Zn^{2+}.
- Describe flame tests for Li^+, Na^+, K^+, Ca^{2+}, Ba^{2+} and Cu^{2+}.

Tests using aqueous sodium hydroxide and aqueous ammonia

Many compounds look similar in the laboratory. For example, if you have a white powder it could be sodium chloride or magnesium sulfate. Even coloured compounds may appear colourless when they are in aqueous solutions at low concentrations. For example, iron(II) sulfate has light green crystals but when dissolved in water it appears colourless unless you make a very concentrated solution.

We can identify an aqueous cation (positive ion) in a compound by adding aqueous sodium hydroxide or aqueous ammonia.

If you have a solid that you want to identify, it is best to dissolve it in a little water first and use this aqueous solution for the test. The procedure for identifying an unknown cation is:

1. Put a small amount of the solution you want to identify into a test tube.
2. Add a few drops of aqueous sodium hydroxide.
3. Observe the colour of any precipitate formed.
4. Add excess aqueous sodium hydroxide and shake the test tube.
5. Record whether or not the precipitate dissolves, and any colour changes.

This procedure can be repeated using aqueous ammonia instead of sodium hydroxide.

The table shows the results if particular cations are present.

If the alkalis added are not in excess the precipitates formed are metal hydroxides. The equations for all these reactions are similar. For copper(II) ions the equation is:

$$Cu^{2+}(aq) + 2OH^-(aq) \longrightarrow Cu(OH)_2(s)$$

copper(II) ions + hydroxide ions ⟶ copper(II) hydroxide

Figure 12.5.1 A light blue precipitate forms when we add sodium hydroxide solution to a solution containing Cu^{2+} ions.

EXAM TIP

When testing for metal ions using NaOH(aq) or NH_3(aq), make sure that you identify (i) if there is a precipitate, (ii) the colour of the precipitate, and (iii) what happens when you add excess NaOH(aq) or NH_3(aq).

metal cation	result with aqueous sodium hydroxide	result with aqueous ammonia
aluminium, Al^{3+}	white precipitate soluble in excess (colourless solution)	white precipitate insoluble in excess
calcium, Ca^{2+}	white precipitate insoluble in excess	no precipitate or very slight white precipitate
copper(II), Cu^{2+}	light blue precipitate insoluble in excess	light blue precipitate soluble in excess (dark blue solution)
chromium(III), Cr^{3+}	grey-green precipitate soluble in excess (green solution)	grey-green precipitate insoluble in excess partly dissolves on standing to form violet solution
iron(II), Fe^{2+}	green precipitate insoluble in excess (turning yellow on the surface)	green precipitate insoluble in excess (turning yellow on the surface)

Unit 12: Making and identifying salts

iron(III), Fe^{3+}	red-brown precipitate insoluble in excess	red-brown precipitate insoluble in excess
zinc, Zn^{2+}	white precipitate soluble in excess (colourless solution)	white precipitate soluble in excess (colourless solution)

We can also test for ammonium ions using sodium hydroxide solution. When a solution containing ammonium ions is heated with sodium hydroxide solution, ammonia gas is given off. This turns red litmus blue.

$$NH_4^+(aq) + OH^-(aq) \longrightarrow NH_3(g) + H_2O(l)$$
ammonium ions + hydroxide ions ⟶ ammonia + water

Flame tests

A flame test can be used to identify some cations, especially those in compounds containing elements from Groups I and II. The procedure is:

1. Clean a platinum or nichrome wire by dipping it in concentrated hydrochloric acid.
2. Place a sample of the compound on the end of the wire.
3. Hold the wire on the edge of a non-luminous (blue) Bunsen flame.
4. Note any change in the colour of the flame.

The typical flame test colours for some metal ions are shown in the table.

Metal ion	Flame colour
lithium, Li$^+$	red
sodium, Na$^+$	yellow
potassium, K$^+$	lilac
calcium, Ca^{2+}	brick-red (orange-red)
barium, Ba^{2+}	apple-green (pale green)
copper(II), Cu^{2+}	blue-green

EXAM TIP
The precipitates formed when Al^{3+} and Zn^{2+} ions react with NaOH(aq) will dissolve in excess sodium hydroxide because they form soluble aluminates and zincates (Topic 11.6).

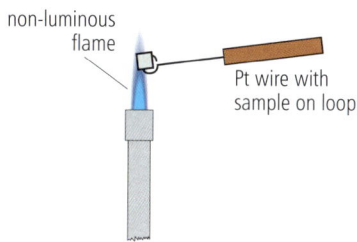

Figure 12.5.2 Doing a flame test.

KEY POINTS
- We can use aqueous sodium hydroxide and aqueous ammonia to identify aqueous cations.
- When sodium hydroxide is heated with a solution containing ammonium ions, ammonia gas is produced.
- Aluminium hydroxide and zinc hydroxide precipitates dissolve in excess sodium hydroxide but only zinc ions dissolve in excess aqueous ammonia.
- Some metal cations can be identified by their typical colours in a flame test.

SUMMARY QUESTIONS

1. Copy and complete using the words below:

   ammonia   cations   colour   hydroxide   precipitate   white   zinc

   A solution containing metal _____ can be identified using aqueous sodium _____ or aqueous _____. A _____ is formed. This often has a distinctive _____. Zinc and aluminium ions both form _____ precipitates with aqueous ammonia but only the precipitate from _____ dissolves in excess ammonia.

2. State the colour of the precipitate when solutions containing these ions react with a few drops of sodium hydroxide:
   a iron(III)   b zinc   c copper(II)

3. How you can distinguish between a solution containing zinc ions and a solution containing calcium ions?

12.6 Testing for anions

LEARNING OUTCOMES

- Describe the tests for halide ions
- Describe the tests for carbonates, nitrates, sulfates and sulfites

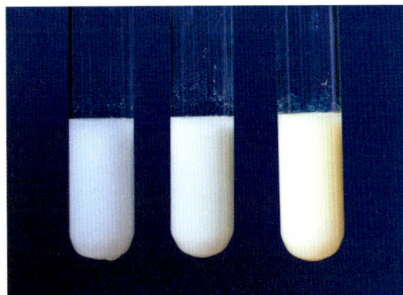

Figure 12.6.1 Chloride, bromide and iodide ions (called halide ions) give different colour precipitates with silver nitrate.

EXAM TIP

Remember that you add nitric acid and aqueous silver nitrate in the test for halide ions. If you add hydrochloric acid you will be adding chloride ions!

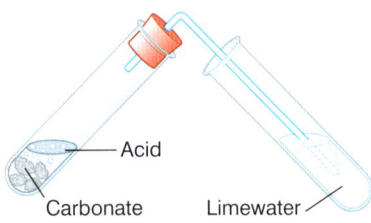

Figure 12.6.2 Testing for carbon dioxide.

Qualitative tests

Negative ions are called anions. If we want to find the type of anion present in an unknown compound we have to use a variety of tests.

When we have completed all our tests for both cations and anions we can identify the unknown compound. We can also carry out further tests to confirm our conclusions. The whole process of finding out what elements are present in a compound is called **qualitative analysis**.

Identifying chlorides, bromides and iodides

Halides are formed when a Group VII atom gains an electron to form a negatively charged ion. Halides can be identified using aqueous silver nitrate:

To a small volume of the halide solution in a test tube:

1. Add an equal volume of dilute nitric acid.
2. Add a few drops of aqueous silver nitrate.
3. Observe the colour of the precipitate.

Chlorides give a white precipitate.

Bromides give a cream precipitate.

Iodides give a pale yellow precipitate.

The precipitates are the silver halides, for example:

$$Ag^+(aq) + Br^-(aq) \longrightarrow AgBr(s)$$

silver ions + bromide ions → silver bromide

In the presence of sunlight, the silver chloride precipitate goes greyish-purple very quickly. The silver bromide goes greyish-purple slowly.

Identifying carbonate ions

We add dilute acid to the unknown compound. The unknown compound can be either a solid or a solution. If a carbonate is present we will see effervescence (bubbles of gas). We test to see if the gas given off is carbon dioxide using limewater.

Identifying nitrates

The identification of nitrates makes use of the test for ammonia. The procedure is:

1. Put an aqueous solution of the unknown compound into a test tube.
2. Add aqueous sodium hydroxide, then aluminium foil and warm gently.

Unit 12: Making and identifying salts

3 Test the gas given off with a piece of damp red litmus paper placed at the mouth of the test tube. If ammonia gas is given off, the litmus paper will turn blue. So the compound is likely to be a nitrate.

Identifying sulfates

Barium chloride or barium nitrate solution is used to test for sulfates.

The procedure is:

1 Put an aqueous solution of the unknown compound into a test tube.
2 Add an equal volume of dilute hydrochloric acid and then add an aqueous solution of a soluble barium salt. This can be barium chloride or barium nitrate.
3 If a white precipitate is formed the compound is a sulfate.

The ionic equation for this reaction is:

$$Ba^{2+}(aq) + SO_4^{2-}(aq) \longrightarrow BaSO_4(s)$$

barium ions   sulfate ions   barium sulfate (white precipitate)

Identifying sulfites

Most sulfites are insoluble in water. The exceptions are sulfites of the Group I elements and ammonium sulfite. Sulfites contain the ion SO_3^{2-}. So the formula for sodium sulfite is Na_2SO_3 and the formula for calcium sulfite is $CaSO_3$.

To test for the presence of a sulfite ion, we add dilute hydrochloric acid and warm gently. If sulfur dioxide is given off, we know that a sulfite is present. Sulfur dioxide turns acidified potassium manganate(VII) solution from purple to colourless. The test is most easily done by placing a piece of filter paper soaked in acidified potassium manganate(VII) above the test tube containing the acid and sulfite.

EXAM TIP

It is a common error to suggest that the test for nitrates involves only sodium hydroxide or only aluminium. Remember that both Al <u>and</u> NaOH are needed.

KEY POINTS

- Carbonates can be identified by adding a dilute acid then testing the gas produced with limewater.
- Halides are identified by the colour of the precipitate obtained when silver nitrate is added.
- Nitrates are identified by adding sodium hydroxide and aluminium foil, warming gently, then testing for ammonia gas.
- Sulfates are identified by adding hydrochloric acid and then barium chloride. A white precipitate shows the presence of a sulfate.
- Sulfites are identified by adding dilute hydrochloric acid, warming and testing for sulfur dioxide gas.

SUMMARY QUESTIONS

1 Name the anions present in the following precipitates:
   a Nitric acid is added to a solution followed by aqueous silver nitrate. A pale yellow precipitate is observed.
   b Hydrochloric acid is added to a solution followed by aqueous barium chloride. A white precipitate is observed.
2 State the tests to identify these anions and give the result of a positive test:
   a nitrate   b carbonate   c bromide
3 A student wants to identify the compound in solution Y. To one sample of Y she adds nitric acid followed by aqueous silver nitrate. A pale yellow precipitate is formed. To another sample of Y she adds a solution of sodium hydroxide. A blue precipitate is formed. Identify the compound in solution Y.

SUMMARY QUESTIONS

1 Match the ions or molecules on the left with the correct test reagents on the right.

iron(II) ions	silver nitrate
iodide ions	acidified barium chloride
sulfate ions	warm with aluminium powder and sodium hydroxide
carbon dioxide	sodium hydroxide
nitrate ions	limewater

2 Copy and complete using words from the list:

crystallise   evaporating   excess filtered   filtrate   salt   sulfuric   water

Zinc sulfate is a _____ that can be made by reacting _____ zinc with _____ acid. The excess zinc is _____ off. The _____ (zinc sulfate solution) is then put into an _____ basin. Some of the _____ is boiled off and the solution is left to _____.

3 Both hydrogen and oxygen can be identified by a test involving a wooden splint.
   a State the differences between the tests for these two gases.
   b State the differences in the results obtained from these tests.

4 Draw a flow chart to show how to make dry crystals of sodium chloride from hydrochloric acid and aqueous sodium hydroxide.

5 Describe three different methods of making the soluble salt iron(II) chloride. For each method describe (i) suitable reactants (ii) the reactant, if any, that is in excess.

6 Which of the following compounds are soluble in water and which are insoluble?
   a silver chloride
   b sodium bromide
   c calcium carbonate
   d sodium carbonate
   e lead sulfate
   f magnesium chloride
   g lead nitrate
   h iron(II) hydroxide

S 7 Suggest the best method for making each of the following salts. Choose from:
   (i) titration
   (ii) adding an insoluble metal or metal compound to an acid
   (iii) precipitation

   a lead bromide from lead nitrate and potassium bromide
   b potassium nitrate from nitric acid and potassium hydroxide
   c iron(III) hydroxide from iron(III) chloride and sodium hydroxide
   d copper(II) sulfate from copper(II) carbonate and sulfuric acid

Practice questions

1 Sodium hydroxide is added to solution M. A red-brown precipitate is formed. Solution M contains:
   A iron(II) ions
   B iron(III) ions
   C copper(II) ions
   D zinc ions.

(Paper 1)

2 Which one of the following ions gives an apple-green colour in the flame test?
   A copper
   B lithium
   C potassium
   D barium

(Paper 1)

3 Which one of these methods is used to prepare the salt ammonium chloride from aqueous ammonia?
   A Precipitation reaction
   B Adding ammonia to an insoluble oxide
   C Adding aqueous ammonia to a metal
   D Titration

(Paper 2)

4 Some instructions for making crystals of magnesium sulfate are given below.

   • Heat excess magnesium oxide with sulfuric acid.
   • Filter off the excess magnesium oxide.
   • Put the filtrate into an evaporating basin.

(a) Complete the instructions to produce pure dry crystals of magnesium sulfate. [4]

(b) Suggest why excess magnesium oxide was used. [1]

(c) Describe a test for sulfate ions and describe the observations if sulfate ions are present. [2]

(Paper 3)

5 Colourless crystals of M dissolve in water to form a solution with a pH of 3.5. M has the molecular formula $C_2H_4O_2$.

(a) Describe the effect of a solution of M on blue litmus paper. [1]

(b) When a solution of M is added to solid L, a colourless gas is given off. Describe how you can show that this gas is carbon dioxide. [2]

(c) The solution formed contains calcium ions. Describe how you can show that the solution contains calcium ions. [4]

(d) Identify solids L and M. Give reasons for your answer. [2]

(e) When warmed with hydrochloric acid compound T produces a gas which decolourises potassium manganate(VII). A sample of compound T gives a yellow colour in the flame test. Identify compound T. [2]

(Paper 3)

6 A student wants to make the soluble salt potassium chloride from potassium hydroxide using a titration method.

(a) Suggest a suitable substance that the student could add to potassium hydroxide to make potassium chloride by this method. [1]

(b) Draw a labelled diagram of the apparatus required to carry out this titration. [3]

(c) Describe how to carry out a titration. [3]

(d) Describe how to make colourless crystals of potassium chloride using this method. [3]

(e) Describe how you can show that the crystals contain chloride ions. [3]

(Paper 3)

7 Lead iodide, PbI_2, and barium sulfate, $BaSO_4$, are insoluble salts.

(a) Suggest two compounds in aqueous solution that you can use to make lead iodide. [2]

(b) Write an ionic equation for this reaction. Include state symbols. [3]

(c) Describe the method used to make dry lead iodide crystals. [4]

(d) Describe a test for iodide ions and state the result. [2]

(e) An equation for the formation of barium sulfate is shown.

$Ba(NO_3)_2(aq) + Na_2SO_4(aq) \rightarrow BaSO_4(s) + 2NaNO_3(aq)$

(i) Write the ionic equation for this reaction. [2]

(ii) Describe the test for nitrate ions. [3]

(Paper 4)

8 Most iron(II) and iron(III) salts are soluble in water.

(a) (i) Very dilute solutions of iron(II) chloride and iron(III) chloride can be difficult to distinguish by colour. Explain how you can distinguish solutions of these compounds. Give full details of the tests you can carry out as well as the expected results. [3]

(ii) Explain why addition of ammonia to iron(II) and iron(III) salts gives the same results as in part (i). [2]

(b) Crystals of iron(II) chloride gradually change from light green to yellowish brown when left in the air.

(i) Suggest why this colour change happens. [2]

(ii) Write an ionic half-equation for this reaction. [1]

(c) (i) Describe how you can make the soluble salt iron(II) sulfate from iron. [5]

(ii) Write an equation for the reaction in part (i). Include state symbols. [2]

(Paper 4)

13.1 The Periodic Table

LEARNING OUTCOMES

- Describe the arrangement of the elements in the Periodic Table
- Describe the change from metallic to non-metallic character across a period
- Describe the relationship between the group number and charge on the ions of an element
- Explain similarity in chemical properties in the same group in terms of electronic configuration
- **S** Explain the change from metallic to non-metallic character across a period

Arrangement of elements in the Periodic Table

The elements in the Periodic Table are arranged in order of increasing proton number (atomic number). In the Periodic Table the elements are arranged so that elements with similar properties fall under each other in vertical columns. We call these vertical columns groups.

We call the horizontal rows periods. Metals are on the left-hand side of the Periodic Table and non-metals are on the right. As we go across a period, the metallic character of the elements decreases.

Trends down the groups

Elements in the same group have similar chemical properties. This is due to the fact that elements in the same group have the same number of electrons in the outer shell of their atoms. All Group I elements have one electron in their outer shell, all Group II elements have two electrons in their outer shell, and so on for the rest of the groups.

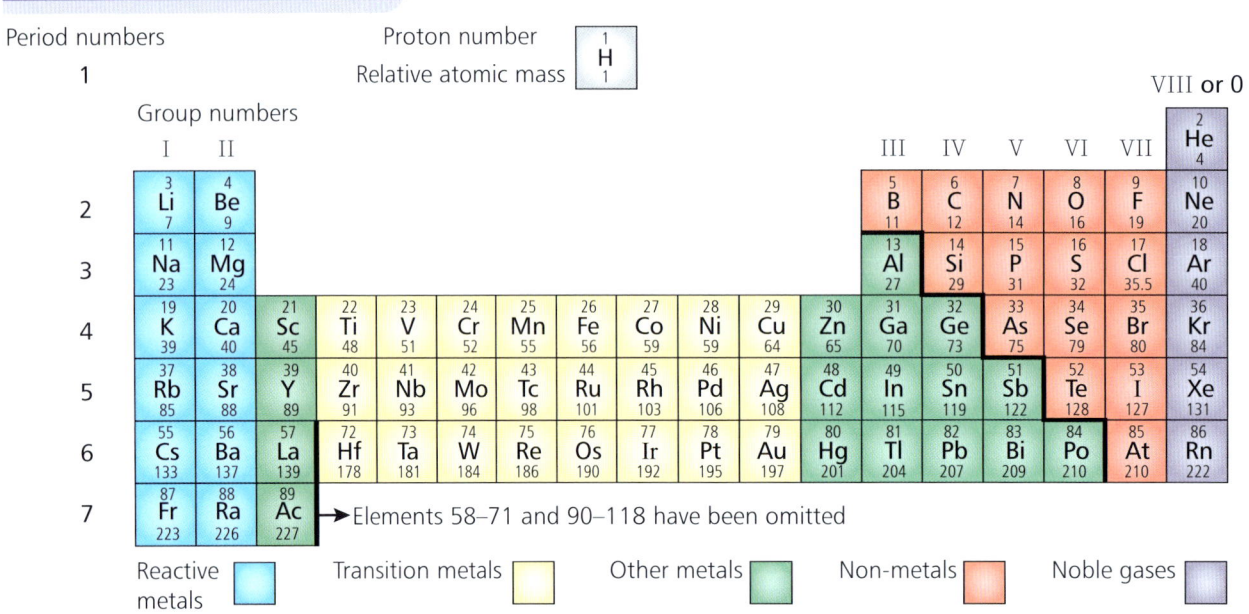

Figure 13.1.1 The Periodic Table.

EXAM TIP

You have to know where the metals and non-metals appear in the Periodic Table. You do not have to remember exactly the dividing line between metals and non-metals.

Within each group we can identify trends in physical and chemical properties down the group. In Group I the elements get more reactive going down the group but in Group VII they get less reactive going down the group.

Unit 13: The Periodic Table

Outer electrons, ions and the Periodic Table

Groups I to III are metals (except boron). Their atoms form ions by losing electrons.

Example: $Mg \rightarrow Mg^{2+} + 2e^-$

Groups IV and V have non-metals at the top and metals at the bottom (see Fig 13.1.1).

Groups VI and VII are mainly non-metals. The atoms of Groups VI and VII form ions by gaining electrons.

Example: $O + 2e^- \rightarrow O^{2-}$

The table shows how the electronic configuration and charge on the ions change across a typical period, e.g. Period 3.

Group	I	II	III	IV	V	VI	VII	VIII
electronic configuration of atom	2,8,1	2,8,2	2,8,3	2,8,4	2,8,5	2,8,6	2,8,7	2,8,8
electronic configuration of ion	2,8	2,8	2,8	rarely forms ions	2,8,8	2,8,8	2,8,8	no ions
charge on the ion	1+	2+	3+		3–	2–	1–	

Figure 13.1.2 Dmitry Mendeleev made the first Periodic Table that is familiar to us.

EXAM TIP

For Groups I, II and III, the ionic charge is positive and the same as their group number. For Groups V to VII the ionic charge is negative and is 8 minus their group number.

Supplement

As we go across a period, the metallic character of the elements decreases. This is because it becomes more difficult to form stable positive ions and therefore more difficult to form a metallic lattice as we go across a period.

On the right-hand side of the Periodic Table, the atoms of the elements accept electrons from other non-metal atoms to form covalent bonds.

KEY POINTS

- Atoms of elements in the same group often have similar properties because they have the same number of electrons in their outer shells.
- Metallic character decreases across a period.
- Metallic character decreases across a period because it becomes more difficult to form positive ions.

SUMMARY QUESTIONS

1. Copy and complete using the words below:

   chemical   eight   electrons   groups   proton   outer

   In the Periodic Table the elements are arranged in order of increasing _____ number. Across Periods 2 to 6 the number of outer shell _____ increases to a maximum of _____. Elements in many _____ have similar _____ properties because they have the same number of electrons in their _____ shell.

2. Describe how:

   a metallic character   b ionic charge

   varies across a period.

163

13.2 Group I metals

LEARNING OUTCOMES

- Describe the Group I elements (the alkali metals) as relatively soft metals
- Describe the trends in the properties of the Group I metals
- Predict the properties of other Group I elements given data

The alkali metals

The Group I metals are called the **alkali metals**. They are a family of metals with similar chemical properties. They are very reactive because they all have one electron in their outer shell which is easily removed when they react. They are unusual metals because they are soft and have relatively low melting points. The alkali metals are stored under oil to stop them reacting with oxygen in the air. When cut, they show a silvery surface that oxidises very quickly.

Physical properties of the alkali metals

The table shows some physical properties of lithium, sodium and potassium.

metal	electronic configuration	density/g per cm³	melting point /°C	boiling point /°C	hardness
lithium	2,1	0.53	181	1342	fairly soft
sodium	2,8,1	0.97	98	883	soft
potassium	2,8,8,1	0.86	63	760	very soft

There are several trends down the group:

- the melting points and boiling points decrease down the group
- the metals get softer down the group
- there is a general increase in density down the group. At first sight it seems that the densities show no trend. But if we include the other Group I elements we can see that there is one:

density in g/cm³: Li 0.53; Na 0.97; K 0.86; Rb 1.53; Cs 1.88

It is sodium and potassium that upset the pattern.

We can use these trends to predict the physical properties of other alkali metals. For example, we can predict that the melting point of rubidium will be lower than the melting point of potassium by about 20–30°C. This gives a melting point for rubidium of about 33–43°C. Its actual melting point is 39°C.

Figure 13.2.1 The alkali metals.

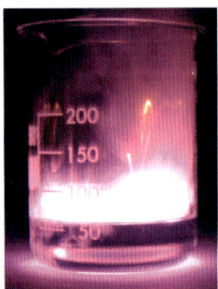

Figure 13.2.2 Alkali metals are stored in oil.

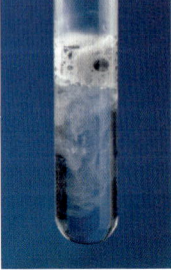

Figure 13.2.3 Potassium (on the left) reacts more rapidly with water than lithium (on the right).

The reaction of the alkali metals with water

PRACTICAL

Reacting alkali metals with water

Small pieces of lithium, sodium and potassium are dropped into a large trough of water one by one. You observe what happens. When each reaction has finished a few drops of universal indicator are added to the trough.

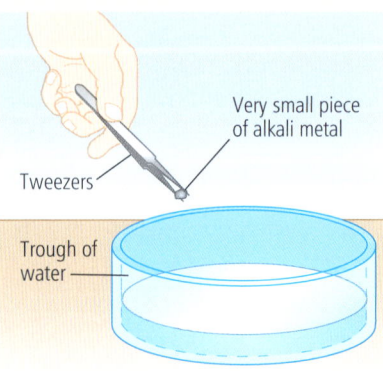

Figure 13.2.4 Adding an alkali metal to water.

Unit 13: The Periodic Table

When we add the alkali metals to water, bubbles are formed and we hear a fizzing sound. We can make other observations too. The results are shown in the table:

lithium	fizzes slowly, a few bubbles	disappears slowly	moves slowly on the surface	remains solid, no flames
sodium	fizzes quickly, many bubbles	disappears quickly	moves quickly on the surface	melts into a liquid ball, no flames
potassium	fizzes violently, even more bubbles	disappears very quickly	moves very quickly on the surface	melts into a liquid ball, violet flame

You can see that the reactions get more vigorous down the group. So you can predict that the reaction of rubidium with water will be very violent, bursting into flames very quickly, with lots of bubbles. It may even explode!

The reactions of the alkali metals with water are very similar – it's just the reactivity that is very different. The bubbles and fizzing are caused by hydrogen gas which is released in the reaction. We observe a flame with potassium because the reaction is violent enough to make the hydrogen catch fire. Sparks are often seen as well.

When we add universal indicator to the trough after the reactions, the solution turns purple. This shows that an alkali has been formed. Alkalis have OH^- ions which have come from the metal hydroxides formed. That's why the Group I metals are called alkali metals. We can write similar equations for each metal reacting with water. For example:

$$2Li(s) + 2H_2O(l) \rightarrow 2LiOH(aq) + H_2(g)$$
lithium + water → lithium hydroxide + hydrogen

$$2K(s) + 2H_2O(l) \rightarrow 2KOH(aq) + H_2(g)$$
potassium + water → potassium hydroxide + hydrogen

EXAM TIP

When you compare the physical properties of Group I metals, remember that they have 'similar properties' but not 'the same properties'. The properties change slightly down the group.

EXAM TIP

When you describe your observations, concentrate on what you can see, hear and smell or if there is a temperature change.

KEY POINTS

- The Group I elements show a trend in physical properties such as hardness, melting points and density going down the group.
- Group I elements react with water to produce hydrogen and a solution of the alkali metal hydroxide.
- The chemical reactivity of the alkali metals increases going down the group.

SUMMARY QUESTIONS

1 Copy and complete using the words below:

alkali   fire   fizzes   hydrogen   hydroxide   increases   rapidly   surface

The Group I metals are also called _____ metals. They react readily with water to produce _____ gas and an alkaline solution of a metal _____. Their reactivity _____ going down the group. Lithium _____ slowly on the _____ of the water but potassium fizzes _____ and the hydrogen produced catches _____.

2 The boiling points of some Group I metals are Li 1342°C, Na 883°C, K 760°C. Predict the boiling point of rubidium.

3 Copy and complete this symbol equation:
$2Na + \underline{\quad} H_2O \rightarrow \underline{\quad} + H_2$

13.3 Group VII elements

LEARNING OUTCOMES

- Describe the elements in Group VII (the halogens) as diatomic non-metals
- Describe the appearance of the halogens and the trend in their density and reactivity
- Describe the reactions of halogens with other halide ions
- Predict the properties of other Group VII elements given data

Figure 13.3.1 The Group VII elements.

Figure 13.3.2 Chlorine is a gas, bromine is a liquid and iodine is a solid at room temperature and pressure.

The halogens

The Group VII elements are called the **halogens**. They are non-metals that have low melting and boiling points. They all exist as diatomic molecules. This means that they have two atoms in each molecule. We shall concentrate on three of the halogens: chlorine, bromine and iodine.

Trends in physical properties

The halogens show trends in their physical properties:

halogen	electronic configuration	melting point/°C	boiling point/°C	state at r.t.p.	colour
chlorine	2,8,7	−101	−35	gas	yellow-green
bromine	2,8,18,7	−7	+59	liquid	red-brown
iodine	2,8,18,18,7	+114	+184	solid	grey-black

- The melting and boiling points of the halogens increase down the group. This is the opposite trend to the Group I metals.
- the state of the halogens at room temperature changes from gas to liquid to solid down the group.
- The density of the halogens increases down the group.

You can predict the properties of other halogens by observing the trends going down the group. For example, fluorine will have a lower boiling point than chlorine.

Trends in chemical reactivity

Reactions with Group I elements

The salts formed when metals react with halogens are called halides. Chlorine, bromine and iodine all react with sodium to form halides. For example:

$$2Na(s) + Cl_2(g) \xrightarrow{heat} 2NaCl(s)$$

sodium + chlorine → sodium chloride

The reactivity of the halogens with sodium decreases going down the group. This reactivity trend is opposite to the order of reactivity of the Group I elements.

Reactions of halogens with halide ions

When an aqueous solution of chlorine reacts with aqueous potassium bromide, potassium chloride and bromine are formed:

$$Cl_2(aq) + 2KBr(aq) \rightarrow 2KCl(aq) + Br_2(aq)$$

chlorine + potassium bromide → potassium chloride + bromine

Unit 13: The Periodic Table

We call this a **displacement** reaction because one type of atom has replaced another. In this case chlorine has replaced the bromine in the potassium bromide. We say that chlorine has displaced the bromine. We can use displacement reactions to show that the trend in reactivity of the halogens decreases down the group.

PRACTICAL

Reacting halogens with halides

In this experiment we use aqueous solutions of halogens and halides. We add the halogen to the halide and observe any change of colour. We repeat this with different combinations of halogens and halides.

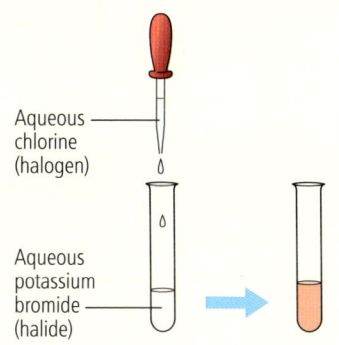

Figure 13.3.3 Aqueous chlorine reacts with aqueous potassium bromide.

The table shows the colour changes when halogens are added to different halides. A cross shows that there is no colour change. In aqueous solution chlorine is very light green, bromine is orange and iodine is brown.

halogen	halide		
	potassium chloride	potassium bromide	potassium iodide
chlorine	X	turns orange	turns brown
bromine	X	X	turns brown
iodine	X	X	X

In these reactions the colour changes show that chlorine displaces bromine (orange) from potassium bromide, and iodine (brown) from potassium iodide. Bromine has only displaced iodine from potassium iodide. Iodine has not reacted at all. So the more reactive halogen displaces the less reactive halogen from a solution of its halide. Chlorine is the most reactive and iodine is the least reactive of these three halogens. So aqueous iodine will not react with aqueous potassium bromide. This is because iodine is less reactive than bromine. However, bromine can displace iodine from potassium iodide:

$$Br_2(aq) + 2KI(aq) \longrightarrow 2KBr(aq) + I_2(aq)$$

We can write similar equations for the other reactions between halogens and halides.

EXAM TIP

Make sure that you can distinguish between the halogens (elements) and halides (compounds). It is a common error to write chlorine ions instead of chloride ions.

KEY POINTS

- The halogens (Group VII) are non-metals. They have diatomic molecules.
- The reactivity of halogens decreases going down the group and the density increases going down the group.
- A more reactive halogen displaces a less reactive halogen from a solution of its halide.

SUMMARY QUESTIONS

1 Copy and complete using the words below:

chlorine decreases diatomic halogens iodine liquid

We call the Group VII elements _____. Group VII elements have _____ molecules. Their reactivity _____ going down the group. At r.t.p. _____ is a yellow-green gas, bromine is a red-brown _____ and _____ is a grey-black solid.

2 Suggest why aqueous potassium chloride does not react with bromine.

3 Copy and complete this symbol equation:
$Cl_2(aq) + $ ___ $KBr(aq) \longrightarrow$ 2 _____ (aq) + _____ (aq)

167

13.4 Noble gases and periodic trends

LEARNING OUTCOMES

- Describe the noble gases as being unreactive monatomic gases and explain this in terms of electronic structure
- Explain how the position of an element in the Periodic Table can be used to predict its properties
- **S** Identify trends in groups given suitable information

The noble gases

The gases in Group VIII or 0 are called the noble gases. All these gases are chemically unreactive (inert) because they have a full outer shell of electrons. They do not need to gain, lose or share electrons to form compounds. This is why they exist as single atoms – they are monatomic. Many of the noble gases are found in the air.

Periodic trends

The second period starts with lithium. As you go across Period 2 each successive element has one more electron in its outer shell until you reach neon with the maximum number of eight electrons. The third period is similar. The fourth period is complicated by the transition elements which form a block of metals in the middle of the Periodic Table.

For Periods 2 and 3 the melting points increase up to Group IV and then decrease again towards Group VIII (see Figure 13.4.4 on the next page). This reflects the different structures of the elements in each group. So for Period 2: on the left we have metallic structures; in the middle we have a giant covalent structure (carbon) in Group IV; on the right we have non-metals with molecular structures.

As you go across a period there are also differences in chemical properties. We have the basic oxides on the left and the acidic oxides on the right.

Predicting properties

We can predict the properties of elements in the Periodic Table by looking at the trends down the groups and across the periods. For example, we know that phosphorus is in Group V and Period 3. Since the element above it (nitrogen) has a simple molecular structure, we can predict that phosphorus also has a simple molecular structure. So it has a relatively low melting point and does not conduct electricity. Because it is on the right of the Periodic Table, we can also predict that its oxide is acidic.

2
He
4
10
Ne
20
18
Ar
40
36
Kr
84
54
Xe
131
86
Rn
222

Figure 13.4.1 The noble gases are found in Group VIII of the Periodic Table.

Figure 13.4.2 Helium is used to fill balloons because it is less dense than air. It is also inert (unreactive).

Unit 13: The Periodic Table

Supplement

Trends in other groups

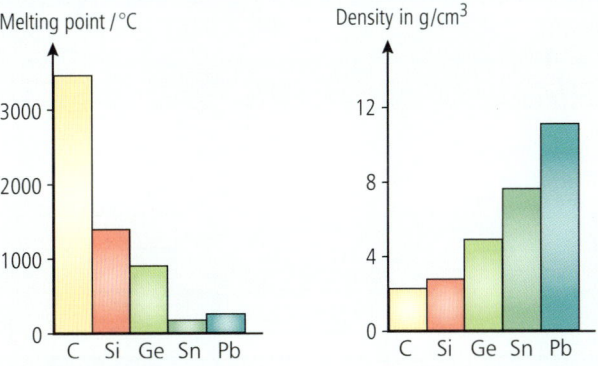

Figure 13.4.3 Trends in the melting point and density of Group IV elements.

We can identify trends in physical and chemical properties in most groups in the Periodic Table. You do not have to remember these trends but you should be able to identify them when given suitable information. In Groups III to VI these trends are often associated with a change from non-metals to metals down the group. Sometimes these trends are general with one element spoiling an otherwise regular pattern.

Sometimes the differences in the chemical properties within a group can be very marked. In Group VI the non-metal, oxygen, at the top of the group has totally different properties from the metal polonium at the bottom of the group.

Group II metals have some trends similar to Group I. For example, they show increased reactivity with water going down the group. But they do not show a very regular trend in density and melting points. There are other trends though. For example, the solubility of the hydroxides increases going down Group II.

SUMMARY QUESTIONS

1 Copy and complete using the words below:

**eight   electrons   energy   Group   inert   outer
noble   remove**

The elements in _____ VIII of the Periodic Table are called the _____ gases. Apart from helium they have a complete outer shell of _____ electrons. They are generally _____ because it takes too much _____ to add, _____ or share _____ from the _____ shell.

2 Describe the trend in the boiling point of the noble gases going down the group.

3 Suggest three properties of sulfur by referring to its position in the Periodic Table.

EXAM TIP

It is better to write that the noble gases are unreactive 'because the electronic structure of a full outer shell is energetically stable' than to write 'the outer shells have eight electrons'.

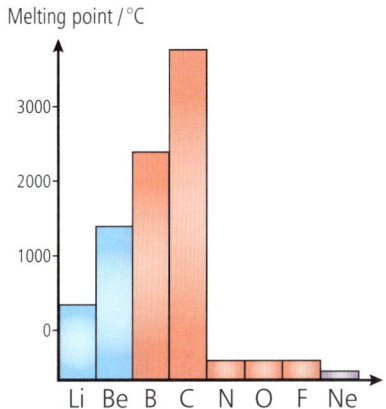

Figure 13.4.4 How melting points vary across Period 2.

EXAM TIP

Be ready to describe the trends from data that you have been given. Look for increases or decreases in the values of the numbers and make sure that you can deal with negative numbers.

KEY POINTS

- Noble gases are inert because their electronic configuration (full outer shell of electrons) is energetically stable.
- The noble gases show trends in physical properties going down the group.
- We can predict the properties of an element from its position in the Periodic Table.

13.5 Transition elements

LEARNING OUTCOMES

- Describe transition elements as metals having high densities, high melting points, forming coloured compounds and acting (as elements and compounds) as catalysts
- **S** Recognise that transition elements have variable oxidation numbers

Transition elements and the Periodic Table

The **transition elements** form a block of elements in the middle of the Periodic Table. They are all metals with typical metallic properties. They all conduct thermal energy and electricity, are malleable and ductile, and are shiny and sonorous.

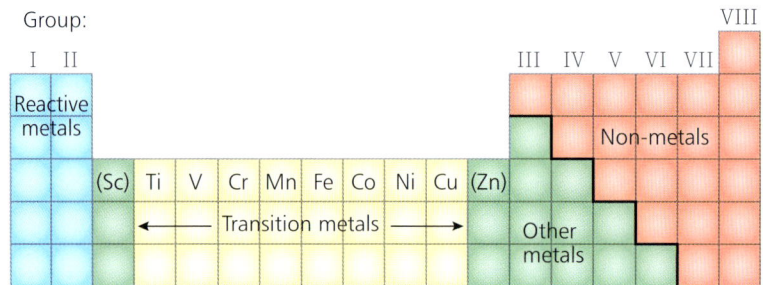

Figure 13.5.1 The position of the transition elements in the Periodic Table.

Many of the properties of transition elements are similar but they are very different from the metals in Groups I, II and III. We can see this by comparing some of their physical and chemical properties. Note that scandium, Sc, and zinc, Zn, are not true transition elements. They do not have some of the properties shown by transition elements, e.g. they do not form compounds where Sc or Zn have different oxidation numbers.

Physical properties of transition elements

We use the following properties to distinguish transition metals from metals in Groups I, II and III.

Transition elements:

- have very high melting points. For example, the melting point of chromium is 1857°C. The melting point of potassium in Group I is only 63°C.
- have very high densities. Compare the densities of chromium and potassium: chromium 7.2 g/cm³; potassium 0.86 g/cm³.
- are stronger and harder than Group I metals.
- form coloured compounds. For example, iron(II) salts are often light green in colour but iron(III) salts are yellow or brown. Salts of Group I and II metals are usually colourless.

Figure 13.5.2 This jet engine has strong turbine blades made from transition metal alloys that can withstand high temperatures

EXAM TIP

It is a common error to suggest that transition elements are highly coloured. It is the *compounds* of transition elements that have a range of colours.

EXAM TIP

When answering questions about transition elements, make sure that you know which properties are physical and which are chemical.

Chemical properties of transition elements

Many of the chemical properties of the transition elements set them apart from other metals:

- Many transition elements and transition element oxides are good catalysts. Iron is a catalyst for the Haber process and vanadium(V) oxide is the catalyst used in the manufacture of sulfuric acid.

Unit 13: The Periodic Table

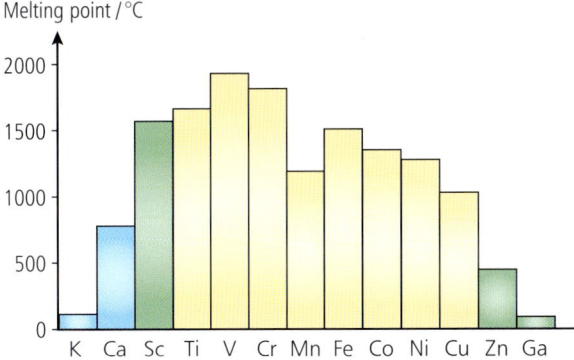

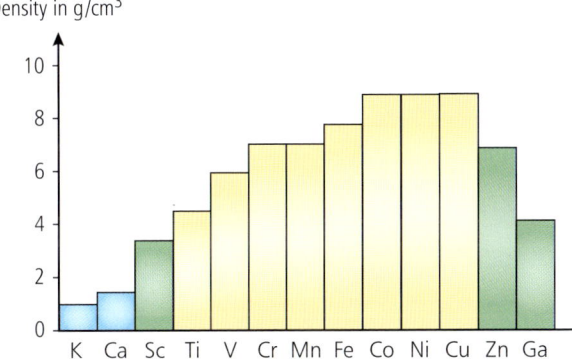

Figure 13.5.3 Comparing the melting points and densities of the metals in Period 3.

- They are less reactive than metals from other groups. They do not react with cold water, although some react with steam.
- They have more than one oxidation number in compounds. For example, iron can form iron(II) ions, Fe^{2+}, or iron(III) ions, Fe^{3+}. Many transition elements have a very wide range of oxidation numbers. For example, manganese can exist in positive oxidation numbers ranging from +1 to +7. This does not mean that you can get ions with a 7+ charge. You cannot get ions with more than a 3+ charge.

EXAM TIP

Remember that oxidation number does not always refer to the charge on the ions. For example, in potassium manganate(VII), $KMnO_4$, the oxidation number of manganese is +7 but the manganese ion with the highest charge is Mn^{2+}.

SUMMARY QUESTIONS

1 Copy and complete using the words below:

block   catalysts   coloured   densities   high   middle

The transition elements form a _____ of metals in the _____ of the Periodic Table. Most transition elements have _____ melting and boiling points and high _____ compared with the metals in Groups I, II and III. Transition elements form _____ compounds and are good _____.

2 Describe two differences in chemical properties that distinguish transition elements from Group I elements.

3 Refer to the properties of iron(II) chloride and iron(III) chloride to explain why iron is a transition element.

KEY POINTS

- Transition elements have high melting and boiling points and high densities.
- Transition elements and their compounds often act as catalysts.
- Compounds of transition elements are coloured.
- Transition elements have different oxidation numbers when in compounds.

SUMMARY QUESTIONS

1 Copy and complete using words from the list.

alkali   coloured   darker   high   less   melting   middle   oxidation   soft

The _____ metals (Group I) are _____ metals that decrease in _____ point going down the group. Halogens are non-metals that get _____ in colour and _____ reactive going down the group. The transition elements are a block of elements in the _____ of the Periodic Table. They are metals with very _____ melting points and densities. They form _____ compounds in which they have a variety of _____ states.

2 Match the elements on the left with the phrases on the right.

chlorine	an unreactive gas
bromine	a metal with a very high melting point
lithium	the least reactive of the alkali metals
neon	a green poisonous gas
potassium	a reddish-brown liquid
iron	a metal that catches fire when it reacts with water

3 State four differences between an alkali metal and a transition element such as nickel.

4 List the trends in physical properties of:
   a the Group I metals   **b** the halogens

5 Describe the differences between the reactions of lithium and potassium with water.

6 Write word equations for the reaction of:
   a sodium with water
   b sodium with chlorine
   c aqueous chlorine with aqueous potassium iodide

7 Explain why chlorine reacts with aqueous potassium bromide but bromine does not react with aqueous potassium chloride.

S 8 Write symbol equations for the reaction of:
   a lithium with water
   b aqueous bromine with aqueous sodium iodide

9 Describe and explain how metallic character changes across a period.

Practice questions

1 Select the correct statement about the reaction of the Group I elements with water.

   A A solution of halide ions is formed.
   B An acidic solution is formed.
   C Oxygen is given off.
   D An alkaline solution is formed.

(Paper 1)

2 Select the correct statement about transition elements in their compounds:

   A they are magnetic
   B they have variable oxidation numbers
   C they are white in colour
   D they form ions of the type M^{5+} and M^{7+}.

(Paper 2)

3 Some properties of the Group I elements are shown in the table.

element	melting point/°C	boiling point/°C	reaction with water
lithium	180	1330	fairly reactive
sodium	98		reactive
potassium	64	760	very reactive
rubidium		690	

   (a) Suggest values for **(i)** the melting point of rubidium **(ii)** the boiling point of sodium. [2]

   (b) How does the reactivity of rubidium compare with the reactivity of potassium? [1]

   (c) Describe three observations that you can make when potassium reacts with water. [3]

   (d) The solution formed when potassium reacts with water is alkaline.
   (i) How can you show that the solution is alkaline? [2]
   (ii) Write a word equation for the reaction of potassium with water. [2]

(Paper 3)

4 Some elements in the Periodic Table are shown.

Li	Be	B	C	N	O	F	Ne
Na	Mg					Cl	Ar

(a) What determines the order of the elements in the Periodic Table? [1]

(b) To which period does chlorine belong? [1]

(c) Choose from the elements shown in the table above to answer these questions:
   (i) Which elements are halogens? [1]
   (ii) Which elements are noble gases? [1]
   (iii) Which elements react rapidly with cold water to form an alkaline solution? [1]
   (iv) Which element has three electrons in its outer shell? [1]
   (v) Name two elements that react with oxygen to form acidic oxides. [2]

(d) Which alkali metal is least reactive? [1]

(Paper 3)

5 Argon is next to chlorine in the Periodic Table.

(a) Use your knowledge of the structures of argon and chlorine to explain why they are both gases at room temperature. [2]

(b) Explain why argon is unreactive. [2]

(c) Chlorine is a diatomic molecule. What do you understand by the term diatomic? [1]

(d) Chlorine reacts with hydrogen to form hydrogen chloride.
   Copy and complete the equation for this reaction:
   _____ + $H_2 \longrightarrow$ _____ HCl [2]

(e) (i) Write the word equation for the reaction of aqueous chlorine with aqueous potassium bromide. [2]
   (ii) Describe the colour change seen in this reaction. [2]
   (iii) Explain why aqueous bromine does not react with aqueous potassium chloride. [1]

(Paper 3)

6 In Period 3, the elements are arranged in order of increasing atomic number.

(a) Describe how the electronic structure of these elements changes across Period 3. [3]

(b) Describe and explain how metallic and non-metallic character is related to the number of outer shell electrons. [4]

(c) The halogens are a group of elements with similar properties. Describe how the density and reactivity of the halogens changes down the group. [2]

(d) Bromine reacts with aqueous potassium iodide but not with aqueous potassium chloride.
   (i) Explain why bromine does not react with aqueous potassium chloride. [1]
   (ii) Write a symbol equation for the reaction of aqueous bromine with aqueous potassium iodide. Include state symbols. [3]
   (iii) Chlorine reacts with cold sodium hydroxide to form sodium chlorate(I), sodium chloride and water. The formula for the chlorate(I) ion is ClO^-.
   $Cl_2(aq) + 2NaOH(aq) \longrightarrow NaCl(aq) + NaClO(aq) + H_2O(l)$
   Write the ionic equation for this reaction. [2]

(Paper 4)

7 Strontium is below calcium in Group II of the Periodic Table.

(a) Use ideas about atomic structure to explain why strontium is more reactive than calcium. [2]

(b) Calcium reacts with water in a similar way to sodium.
   (i) Suggest two observations that you can make when calcium reacts with water. [2]
   (ii) Write a symbol equation for the reaction of calcium with water. [2]

(c) Silicon is in Group IV of the Periodic Table. Suggest two physical properties of silicon. Explain why you selected these properties. [4]

(Paper 4)

14.1 The metal reactivity series

LEARNING OUTCOMES

- State the order of selected metals in the reactivity series
- Describe the chemical reactions of metals with acids and water
- Explain selected reactions of metals with cold water, steam and hydrochloric acid in terms of the position of the metals and hydrogen in the reactivity series
- Deduce an order of reactivity from experimental results

The metal reactivity series

Some metals are very reactive. The Group I meals react very rapidly with cold water. Other metals are less reactive. The transition elements either do not react with water or react only with steam. We can put the metals in order of their reactivity by investigating how well they react with cold water, steam or dilute hydrochloric acid to produce hydrogen gas. The rate at which hydrogen gas is produced gives us an idea of how reactive the metal is. The faster the reaction, the more reactive the metal is.

The order of reactivity of the metals is shown in Figure 14.1.1. You will notice that we have included hydrogen in the metal **reactivity series** because it is the gas which is produced when we react many metals with water or dilute hydrochloric acid. Metals below hydrogen in the reactivity series do not react with cold water or steam. They do not release hydrogen from hydrochloric acid either. So copper, silver and gold are unreactive metals.

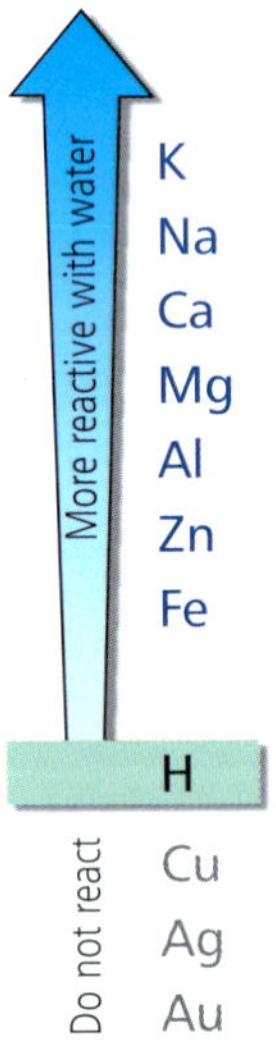

Reaction of metals with water or steam

calcium	reacts rapidly with cold water
copper	no reaction with cold water or steam
iron	does not react with cold water but reacts with steam
magnesium	reacts very slowly with cold water but reacts rapidly with steam
potassium	reacts very rapidly with cold water and catches fire
sodium	reacts very rapidly with cold water
zinc	reacts only when powdered and heated strongly in steam

We can use the information in the table to put these metals in order of reactivity by first comparing how fast the reaction is with cold water, then by comparing how fast the reaction is with steam.

potassium sodium calcium magnesium iron zinc copper

most reactive ⟶ least reactive

We can see that the higher the metal in the reactivity series, the more likely it is to produce hydrogen by reaction with water.

Figure 14.1.1 The metal reactivity series showing the position of hydrogen.

K
Na
Ca
Mg
Al
Zn
Fe
(More reactive with water)
H
Cu
Ag
Au
(Do not react)

Unit 14: Metals and reactivity

PRACTICAL

Reacting iron wool with steam

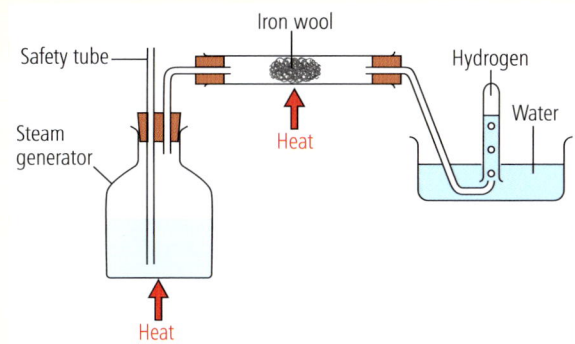

Figure 14.1.2 Reacting iron with steam.

We pass steam over red-hot iron wool. The iron turns black. We collect the gas in a test tube. The gas pops with a lighted splint. The iron has reacted and formed iron oxide and hydrogen.

When a metal reacts with cold water, a metal hydroxide and hydrogen are formed:

$$Ca(s) + 2H_2O(l) \rightarrow Ca(OH)_2(aq) + H_2(g)$$
calcium + water → calcium hydroxide + hydrogen

When a metal reacts with steam, a metal oxide is formed:

$$3Fe(s) + 4H_2O(g) \rightarrow Fe_3O_4(s) + 4H_2(g)$$
iron + steam → iron oxide + hydrogen

Reaction with dilute hydrochloric acid

We can also use dilute hydrochloric acid to compare the reactivity of metals.

The table shows how different metals react with hydrochloric acid.

calcium	very rapid with dilute acid – many bubbles of H_2	most reactive
zinc	slow with dilute acid – bubbles of H_2 produced slowly	↑
iron	very slow with dilute acid – bubbles of H_2 produced very slowly	
copper	no reaction with dilute or concentrated acid	least reactive

EXAM TIP
Metals that react with cold water form metal hydroxides. But when a metal is heated in steam an oxide is formed.

Figure 14.1.3 We can put magnesium (Mg) zinc (Zn) and iron (Fe) in order of reactivity by comparing how rapidly they react with an acid.

KEY POINTS
- Metals can be arranged in a reactivity series by comparing how easily they react with water, steam or hydrochloric acid.
- Only metals above hydrogen in the reactivity series react with water, steam or hydrochloric acid.

SUMMARY QUESTIONS

1 Copy and complete using the words below:

   cold   hydrogen   hydroxide   iron   potassium   steam

   Sodium, _____ and calcium react with _____ water to form a metal _____ and hydrogen. Less reactive meals such as _____ and zinc do not react with cold water but they do react with _____. Metals below _____ in the reactivity series do not react with water or steam.

2 Copper reacts with concentrated nitric acid but gold does not. Suggest the relative position of copper and gold in the reactivity series.

3 Tin is between iron and hydrogen in the reactivity series. Suggest how tin will react with:

   a cold water   b steam

14.2 Metal oxides and their reduction

LEARNING OUTCOMES
- Describe the reaction of metals with oxygen
- Deduce an order of reactivity from experimental results
- **S** Explain the apparent unreactivity of aluminium in terms of its oxide layer

The reaction of metals with oxygen

The table shows how different metals react with oxygen.

copper	reacts very slowly when heated strongly
gold	does not react when heated
magnesium	reacts rapidly and bursts into flames when heated
iron	reacts slowly when heated
sodium	reacts violently and bursts into flames when warmed

We can use the information in the table to put these metals in order of their reactivity:

sodium   magnesium   iron   copper   gold
most reactive ⟶ least reactive

We can see that the higher the metal in the reactivity series, the easier it is to react with oxygen. Some equations for these reactions are shown:

$$2Mg(s) + O_2(g) \rightarrow 2MgO(s)$$
magnesium + oxygen → magnesium oxide

$$3Fe(s) + 2O_2(g) \rightarrow Fe_3O_4(s)$$
iron + oxygen → iron oxide

Figure 14.2.1 Sodium burns rapidly in oxygen to form an oxide of sodium.

Supplement

Why does aluminium seem unreactive?

Aluminium is high in the reactivity series but it does not seem to react with water or acids. It will react with acids only when it is freshly made. This is because, when the surface of freshly made aluminium is left in the air, a thin layer of aluminium oxide quickly forms on its surface:

$$4Al(s) + 3O_2(g) \rightarrow 2Al_2O_3(s)$$

This layer is only about 0.0002 cm thick, but this is enough to make the metal resistant to corrosion. The oxide layer is very unreactive, so the water or acid cannot reach the surface of the metal. The tough oxide layer sticks to the surface of the aluminium very strongly and does not flake off.

EXAM TIP

If a metal quickly forms an oxide layer on its surface, it must be a reactive metal such as sodium or aluminium.

Competing for oxygen

When you heat powdered iron with copper(II) oxide, CuO, the iron displaces the copper:

$$Fe(s) + CuO(s) \xrightarrow{heat} FeO(s) + Cu(s)$$

The iron competes better to 'hold onto' the oxygen. This is because iron is higher in the reactivity series than copper so it is more reactive. The iron is oxidised to iron(II) oxide and the copper(II) oxide is reduced

Unit 14: Metals and reactivity

to copper. The more reactive metal, in this case iron, is the reducing agent. It removes the oxygen from the oxide of the less reactive metal.

Reducing metal oxides with carbon

We can carry out reduction using carbon as a reducing agent. So which metals can carbon reduce? Look at the position of carbon in the reactivity series (Figure 14.2.2). The oxides of the metals below carbon can be reduced to the metal by heating with carbon. Metals more reactive than carbon are generally extracted by electrolysis.

For example, carbon is more reactive than copper. So carbon removes the oxygen from copper(II) oxide when heated.

$$2CuO(s) + C(s) \xrightarrow{heat} 2Cu(s) + CO_2(g)$$

copper(II) oxide + carbon ⟶ copper + carbon dioxide

The table compares the reactivity of metals by looking at how easily the metal oxides are reduced with carbon.

copper(II) oxide	reduced below 800 °C
iron(III) oxide	reduced at 800 °C
titanium(IV) oxide	not reduced at 950 °C
zinc oxide	reduced at 950 °C

We can use the information in the table to put these metals in order of their reactivity:

titanium   zinc   iron   copper
most reactive ⟶ least reactive

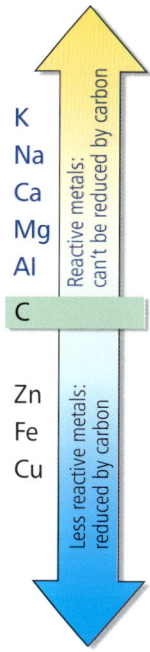

Figure 14.2.2 Oxides of metals below carbon in the reactivity series are reduced to the metal by heating with carbon. Metals above carbon are extracted by electrolysis.

EXAM TIP

Remember that the more reactive a metal the more difficult it is to reduce its oxide.

KEY POINTS

- Most metals react with oxygen to form metal oxides.
- When a more reactive metal is heated with the oxide of a less reactive metal, the more reactive metal acts as a reducing agent.
- Metal oxides below carbon in the reactivity series are reduced by carbon when heated.
- The apparent lack of reactivity of aluminium is due to an unreactive layer of aluminium oxide that forms on its surface.

SUMMARY QUESTIONS

1. Copy and complete using the words below:

    carbon   heated   metals   oxygen   reduced   reducing

    Metal oxides below _____ in the reactivity series are _____ to _____ when they are _____ with carbon. In this reaction carbon is the _____ agent because it removes the _____ from the metal oxide.

2. Write symbol equations for:
    a the reaction of calcium with oxygen
    b the reaction of zinc oxide, ZnO, with carbon to form zinc and carbon dioxide
    c the reaction of magnesium with copper(II) oxide to form magnesium oxide and copper

3. Explain why aluminium does not appear to react with hydrochloric acid.

14.3 From metal compounds to metals

LEARNING OUTCOMES

- Describe bauxite as an ore of aluminium
- Explain why aluminium is extracted by electrolysis
- Describe the ease of obtaining metals from their ores based on the reactivity of metals

Metal ores

Most metals found in the Earth's crust are present as compounds in rocks. Only a small part of the rock may contain useful compounds from which a metal can be extracted. An **ore** is a rock containing enough of a metal compound to extract the metal economically. Most ores are oxides or sulfides. Some important ores are shown in the table:

metal extracted	name of ore	metal compound in the ore
aluminium	**bauxite**	aluminium oxide
iron	**hematite**	iron(III) oxide
zinc	zinc blende	zinc sulfide

Metal extraction and the reactivity series

The way we extract a metal from its ore depends on the position of the metal in the reactivity series. Carbon is used to reduce oxides of metals below it in the reactivity series. In Topic 14.2 we learned that copper(II) oxide can be reduced by heating with carbon.

EXAM TIP

Make sure you know that bauxite is an ore of aluminium and hematite is an ore of iron.

PRACTICAL

Reducing copper(II) oxide with carbon

You put a layer of charcoal powder (carbon) over a layer of copper(II) oxide then heat the tube strongly. When the reaction is over you can see some pinkish-brown copper metal where the two layers of powder meet. This method, however, is not suitable for making copper from copper ores.

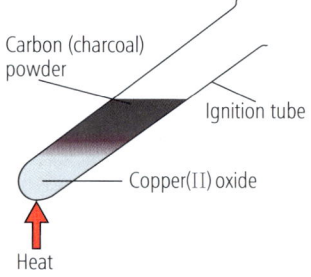

Figure 14.3.1 The reduction of copper(II) oxide with carbon.

Figure 14.3.2 Charcoal kilns like this have been used for centuries to provide carbon for metal extraction.

Oxides of zinc, lead and iron can be reduced by carbon. The carbon is usually used in the form of coke. This is coal from which some impurities have been removed.

$$PbO(s) + C(s) \rightarrow Pb(s) + CO(g)$$

lead(II) oxide + carbon (coke) → lead + carbon monoxide

Carbon monoxide is also a good reducing agent. This is formed when carbon undergoes incomplete combustion:

Unit 14: Metals and reactivity

$$2C(s) + O_2(g) \rightarrow 2CO(g)$$
carbon + oxygen → carbon monoxide

Carbon dioxide is often produced in furnaces used for the extraction of metals:

$$ZnO(s) + CO(g) \rightarrow Zn(s) + CO_2(g)$$
zinc oxide + carbon monoxide → zinc + carbon dioxide

In each of these reactions, we see that the carbon or carbon monoxide takes away the oxygen from the metal oxide. So it reduces the metal oxide. The carbon or carbon monoxide gets oxidised in the reaction.

Reduction with carbon or by electrolysis?

Metals above carbon in the reactivity series are not extracted from their oxides by heating with carbon. This is because the metal bonds to oxygen too strongly and the carbon is not reactive enough to remove it unless extremely high temperatures are used. This is too expensive and needs too much energy. So we use electrolysis to extract metals such as sodium, aluminium, magnesium and calcium.

It is possible to use electrolysis to extract metals less reactive than carbon. This is not usually done because of problems related to the impurities in the ore or difficulties in the extraction of a suitable compound from the ore. For example zinc is often extracted by electrolysis rather than by reduction with carbon since it is cheaper and less polluting.

> **EXAM TIP**
>
> Be prepared to answer questions about oxidation and reduction in metal extraction.

SUMMARY QUESTIONS

1. Copy and complete using the words below:

    bauxite   carbon   extract   iron   ores   zinc

    Rocks from which we can _____ metals are called _____. The main ore of aluminium is _____. Hematite is one of the main ores of _____. We can use _____ to reduce metal oxides such as iron oxide and _____ oxide.

2. Explain why you do not use carbon to extract magnesium from magnesium oxide.

3. Complete these symbol equations. In each case state the element or compound which is oxidised:

    a $SnO_2 +$ ___ $C \rightarrow Sn +$ ___ CO

    b $Fe_2O_3 +$ ___ $CO \rightarrow$ ___ $Fe + 3CO_2$

> **KEY POINTS**
>
> - Bauxite is an ore of aluminium and hematite is an ore of iron.
> - Metal oxides below carbon in the reactivity series are reduced by carbon when heated.
> - Metals more reactive than carbon are extracted by electrolysis.

14.4 More about metal reactivity

Supplement

LEARNING OUTCOMES

- Deduce an order of reactivity from experimental results
- Describe the reactivity of metals in terms of how easy it is to form positive ions in displacement reactions

Displacement reactions

In Topic 13.3 we saw that a more reactive halogen will replace a less reactive halogen in a metal halide. We can think of this as a competition to see which halogen combines the best with the metal ion. We call this type of redox reaction a displacement reaction. Can metals compete in a similar way?

PRACTICAL

The 'thermit' reaction

A mixture of aluminium powder and iron(III) oxide is put into a cone of filter paper. This is placed in a bucket of sand. The magnesium fuse is lit. A vigorous reaction occurs with flames, light and smoke. The result is a lump of iron where the mixture was.

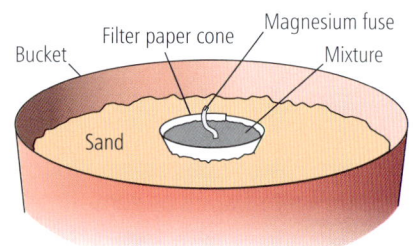

Figure 14.4.1 Setting up the 'thermit' reaction.

In the 'thermit' reaction the aluminium displaces the iron from the iron(III) oxide:

$$Fe_2O_3 + 2Al \rightarrow 2Fe + Al_2O_3$$
iron(III) oxide + aluminium → iron + aluminium oxide

Metal–metal ion displacement reactions

We can carry out similar experiments using metals and solutions of metal ions. When we add excess zinc to a solution of copper(II) sulfate, the zinc gets coated with copper and the solution turns colourless.

Zinc is higher in the reactivity series than copper. So the more reactive zinc displaces copper from the copper(II) sulfate solution. The solution turns colourless because colourless zinc sulfate is formed:

$$Zn(s) + CuSO_4(aq) \rightarrow ZnSO_4(aq) + Cu(s)$$

Another example is:

$$Mg(s) + ZnSO_4(aq) \rightarrow MgSO_4(aq) + Zn(s)$$

Figure 14.4.2 The more reactive metal, zinc, displaces the less reactive metal, copper, from aqueous copper(II) sulfate.

EXAM TIP

You need to know the order of the metals in the reactivity series. You may not be given this in the exam. Make up your own method of memorising it.

Unit 14: Metals and reactivity

The more reactive metal, magnesium, displaces the less reactive metal, zinc, from aqueous zinc sulfate.

From these last two equations we can work out that magnesium is more reactive than zinc and zinc is more reactive than copper.

By carrying out experiments using different combinations of metals and solutions of metal salts, we can arrange all the metals in a metal reactivity series. A more reactive metal will displace a less reactive metal from a solution of its salt.

$$Mg(s) + FeSO_4(aq) \rightarrow MgSO_4(aq) + Fe(s)$$

↑ Mg more reactive   ↑ Fe less reactive

Iron does not react with magnesium sulfate because magnesium is more reactive than iron.

These reactions are redox reactions: we can write half-equations for oxidation and reduction.

Oxidation of magnesium to magnesium ions:

$$Mg(s) \rightarrow Mg^{2+}(aq) + 2e^-$$

Reduction of iron(II) ions to iron:

$$Fe^{2+}(aq) + 2e^- \rightarrow Fe(s)$$

Explaining metal reactivity

In the half-reactions above, we can see that each atom of the more reactive metal loses electrons and each ion of the less reactive metal gains electrons. The more reactive a metal is, the more easily it loses its outer shell electrons. So a more reactive metal forms ions more easily than a less reactive metal.

Figure 14.4.3 The copper coil has reacted with silver nitrate and crystals of silver have formed on its surface.

> **EXAM TIP**
>
> The easier it is to remove the outer shell electrons from a metal atom, the more reactive the metal is.

SUMMARY QUESTIONS

1. Copy and complete using the words below:

    displaces   less   more   outer   solution

    A more reactive metal _____ a _____ reactive metal from a _____ of its salt. This is because the _____ reactive metal loses its _____ shell electrons more easily.

2. Silver is less reactive than copper but copper is less reactive than magnesium. Write full symbol equations and ionic equations for:

    a the reaction of copper(II) sulfate with magnesium

    b the reaction of silver nitrate with copper to form silver and copper(II) nitrate, $Cu(NO_3)_2$

3. Suggest why copper does not react with iron(II) sulfate.

> **KEY POINTS**
>
> - A more reactive metal displaces a less reactive metal from a solution of its salt.
> - A more reactive metal loses its outer shell electrons more easily than a less reactive metal.
> - More reactive metals form ions more easily than less reactive metals.

SUMMARY QUESTIONS

1 Copy and complete using the words below.

**alkali   hydrogen   hydroxides   oxide
potassium   sodium   water**

Metals can be put in order of reactivity using their reactivity with _____ as a guide. Alkali metals such as _____ and _____ react rapidly with water as well as oxygen. The _____ metals react with cold water to form alkali metal _____. Iron reacts with steam to form iron _____ and _____.

2 Use the following reactivity series to answer the questions below.

calcium   magnesium   zinc   iron   copper   gold
most reactive ⟶ least reactive

  a Which metals in the list react with dilute hydrochloric acid?

  b Which metals in the list react with cold water?

  c Which metals in the list react with steam?

  d Where would (i) potassium and (ii) silver come in this list?

3 Match the metals on the left with the phrases on the right.

sodium	a reactive metal that reacts slowly with cold water
copper	a metal that reacts with water to form an alkaline solution
iron	a metal above sodium in the reactivity series
potassium	a pinkish-brown metal that is not very reactive
magnesium	a metal that reacts with steam but not with cold water

S 4 Describe how the reactivity series depends on the ease with which a metal forms a positive ion.

5 Aluminium is a metal high in the reactivity series. Explain why aluminium apparently does not react with dilute hydrochloric acid.

Practice questions

1 Some information about the reaction of three metals with hydrochloric acid is given.

- Metal P: dissolves slowly and a few bubbles are formed.
- Metal Q: dissolves very rapidly and bubbles are formed very rapidly.
- Metal R: dissolves rapidly and bubbles are formed rapidly.

The order of reactivity of these metals, starting with the most reactive, is:

A PQR **C** QRP
B RQP **D** PRQ

(Paper 1)

2 Which one of these equations represents a reaction that is possible.

A $Fe^{2+} + Cu \rightarrow Fe + Cu^{2+}$.
B $Fe + Cu^{2+} \rightarrow Fe^{2+} + Cu$
C $Fe^{3+} + 3Ag \rightarrow Fe + 3Ag^+$
D $Fe + Mg^{2+} \rightarrow Fe^{2+} + Mg$

(Paper 2)

3 Some of the metals in the reactivity series are shown below:

sodium   calcium   magnesium   zinc   iron   copper
most reactive ⟶ least reactive

  (a) Which of these metals will react rapidly with cold water? [1]

  (b) Iron reacts with steam to form iron oxide.

  (i) Name one other element in the list that reacts with steam but does not react with cold water. [1]

  (ii) Copy and complete this equation:
  $\_\_Fe + \_\_H_2O \rightarrow Fe_3O_4 + \_\_H_2$ [3]

  (iii) Explain why this is a redox reaction by referring to the equation. [3]

  (c) Magnesium reacts with hydrochloric acid. The products are $MgCl_2$ and hydrogen.

  (i) Write a word equation for this reaction. [1]

  (ii) Write a symbol equation for this reaction. [2]

(d) Magnesium reacts with black copper(II) oxide when heated.
  (i) Describe the observations during this reaction. [2]
  (ii) Write a word equation for this reaction. [1]
  (iii) Which reactant is reduced in this reaction? Explain your answer. [1]

(Paper 3)

4 (a) Zinc blende is an ore of zinc. What is meant by the term ore? [2]
  (b) Zinc blende contains zinc oxide.
    (i) Copy and complete the equation for the reaction of zinc oxide with carbon monoxide.
    $ZnO + CO \rightarrow \_\_\_\_ + \_\_\_\_$ [2]
    (ii) State the name of the compound which is oxidised in this reaction. Give a reason for your answer. [2]
  (c) Zinc reacts with steam but not with cold water. Potassium reacts with cold water.
    (i) Suggest why zinc does not react with cold water but potassium does. [1]
    (ii) Copy and complete the equation for the reaction of potassium with water:
    $\_\_ K + \_\_ H_2O \rightarrow \_\_ KOH + H_2$ [1]

(Paper 3)

5 Zinc powder reacts with copper(II) oxide on heating:
$Zn + CuO \rightarrow ZnO + Cu$
  (a) Name the reducing agent in this reaction. Explain your answer. [2]
  (b) Describe the direction of electron transfer in this reaction. [2]
  (c) Suggest why the reverse reaction does not occur. [1]
  (d) Both magnesium and zinc are above copper in the reactivity series. Explain, in terms of ease of formation of ions, why magnesium is able to remove oxygen from copper(II) oxide more easily than zinc. [2]

(e) (i) Describe the observations when magnesium reacts with aqueous copper(II) sulfate, $CuSO_4$. Aqueous magnesium sulfate is one of the products. [2]
  (ii) Write an ionic equation for this reaction. Include state symbols. [2]

(Paper 4)

6 Sodium and magnesium are both reactive metals.
  (a) (i) Describe two observations when sodium reacts with cold water. [2]
    (ii) Write a symbol equation to show the reaction of magnesium with steam. Include state symbols. [2]
  (b) Magnesium reacts with iron(II) chloride:
    $Mg(s) + FeCl_2(aq) \rightarrow MgCl_2(aq) + Fe(s)$
    (i) This reaction can be described as either a displacement reaction or a redox reaction.
    State the meaning of the term displacement reaction. [1]
    (ii) Explain in terms of oxidation number changes which atom or ion is oxidised and which is reduced during this reaction. [4]
  (c) Magnesium is more reactive than aluminium.
    (i) Describe how you could use magnesium to extract aluminium from molten aluminium oxide, Al_2O_3. [2]
    (ii) Construct a symbol equation for this reaction. [2]
    (iii) Suggest why this method is not used to extract aluminium industrially. [1]
    (iv) Explain why aluminium containers can be used to store acidic foods even though it is a reactive metal. [3]

(Paper 4)

15.1 Extracting iron

LEARNING OUTCOMES

- Describe the essential reactions in the blast furnace for the production of iron from hematite
- **S** Construct the chemical equations for the extraction of iron from hematite

The blast furnace

Iron is the second most common metal in the Earth's crust. The main ore of iron, hematite, usually contains more than 60% iron. Hematite is largely iron(III) oxide. We extract iron by reduction of the iron(III) oxide with carbon.

The **raw materials** for extracting iron are hematite, limestone and air. **Coke** is also needed to heat the furnace, as well as making the reducing agent. Coke is mainly carbon. It is made by heating coal in the absence of air. The hematite, coke and limestone are added at the top of the **blast furnace**. A strong current of hot air is blown in at the bottom of the furnace. This is why it is called a blast furnace. The temperature of the hot air is between 550 °C and 850 °C. This is high enough to react with the coke.

Figure 15.1.2 A blast furnace used to extract iron from hematite.

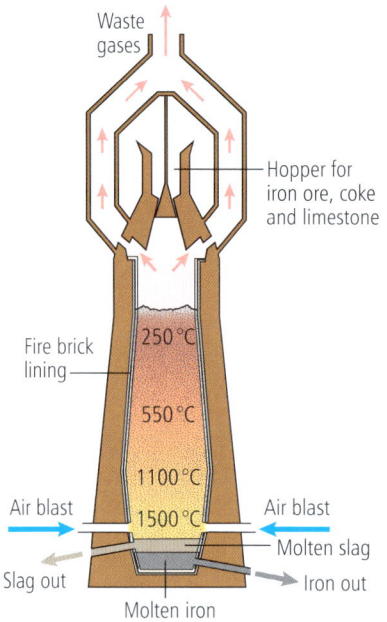

Figure 15.1.1 A blast furnace for extracting iron.

The main reducing agent in the blast furnace is carbon monoxide, but in some parts of the blast furnace carbon also reduces the iron(III) oxide. The temperature in the blast furnace ranges from 1500 °C at the bottom where the air enters to 250 °C at the top.

EXAM TIP

You should be prepared to interpret a diagram of the blast furnace and label the parts where air, iron, slag and limestone enter or exit the furnace.

EXAM TIP

Coke is not a raw material because it has been treated by heating before it is used. Hematite, limestone and air are raw materials because they have not been treated.

EXAM TIP

For Supplement you need to remember the symbol equations shown. For Core you need to know the word equations but you may be asked to complete a given symbol equation.

The chemical reactions in a blast furnace

Reducing the iron(III) oxide

The chemical reactions below result in the production of iron from the iron(III) oxide present in hematite.

- At the bottom of the furnace, the coke burns in the hot air blast to form carbon dioxide. This reaction is exothermic. The thermal energy released heats the furnace.

$$C(s) + O_2(g) \rightarrow CO_2(g)$$

- The carbon dioxide reacts with the coke to form carbon monoxide:

$$CO_2(g) + C(s) \rightarrow 2CO(g)$$

- The carbon monoxide reduces the iron(III) oxide to iron:

$$Fe_2O_3(s) + 3CO(g) \rightarrow 2Fe(l) + 3CO_2(g)$$

iron(III) oxide + carbon monoxide → iron + carbon dioxide

Unit 15: More about metals

Most of the iron is produced in this way. The iron flows to the bottom of the furnace and is removed from time to time as a liquid. It flows into moulds and is left to solidify.

Other reactions may take place in the furnace. In the hotter parts of the furnace, carbon reduces iron(III) oxide directly:

$$Fe_2O_3(s) + 3C(s) \rightarrow 2Fe(l) + 3CO(g)$$

The hot waste gases exiting from the top of the furnace are used to heat the air going into the furnace and so reduce energy costs.

EXAM TIP

Burning the coke has two uses: heating the furnace, and producing carbon dioxide that goes on to form the main reducing agent, carbon monoxide.

Why do we add limestone?

Hematite contains sand (silicon(IV) oxide) as an impurity. Limestone (calcium carbonate) helps remove most of the impurities in the following way.

- The thermal energy from the furnace decomposes the limestone. This is a **thermal decomposition** reaction. Thermal decomposition is the breakdown of a compound into two or more substances by heating.

$$CaCO_3(s) \rightarrow CaO(s) + CO_2(g)$$
calcium carbonate → calcium oxide + carbon dioxide

- The calcium oxide produced reacts with silicon(IV) oxide to form a 'slag' of calcium silicate:

$$CaO(s) + SiO_2(s) \rightarrow CaSiO_3(l)$$
calcium oxide + silicon(IV) oxide → calcium silicate (slag)

- The liquid slag runs down and forms a layer on top of the liquid iron because it has a lower density than molten iron. The slag is run off. The solid slag is used as a building material, especially for building roads.

SUMMARY QUESTIONS

1 Copy and complete using the words below:

air   blast   hematite   limestone   oxide   reduces

We extract iron from iron ore in a _____ furnace. The most common ore of iron is _____. The other raw materials used are _____ and _____. Inside the blast furnace, carbon monoxide _____ the iron(III) _____ to iron.

2 Write word equations for the reactions below which take place in the blast furnace:
   a the reaction of iron(III) oxide with carbon monoxide
   b two equations to show how slag is formed

S 3 Write symbol equations for the reactions in **2a** and **b**.

KEY POINTS

- The raw materials used in the extraction of iron in the blast furnace are iron ore, limestone and air.
- In the blast furnace, carbon monoxide reduces iron(III) oxide to iron.
- The thermal decomposition of limestone produces calcium oxide which reacts with silicon(IV) oxide impurities in the iron ore to form 'slag'.

15.2 The rusting of iron

LEARNING OUTCOMES
- State the conditions required for rusting
- Describe and explain barrier methods for preventing rusting
- S Describe the use of zinc in galvanising (barrier method and sacrificial protection)
- Explain sacrificial protection in terms of metal reactivity and electron transfer

What causes rusting?

Chemicals in the air may attack metals causing the surface to get eaten away. We call this **corrosion**. The corrosion of iron and steel in a particular way is called **rusting**. Only iron and steel rust.

PRACTICAL

What are the best conditions for rusting?

We set up the three tubes as shown, each with a clean iron nail at the bottom. We observe the nails over a period of time. We record how long it takes for each to rust.

The nail in the tube with just air and water rusts quickly. The nails in the other two tubes do not rust.

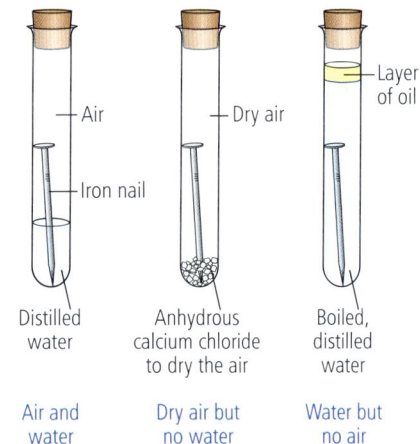

Figure 15.2.1 Both air and water are needed for rusting.

For rusting to occur, both oxygen (from the air) and water are needed. Rusting involves the oxidation of iron:

$$4Fe(s) + 3O_2(g) + 2H_2O(l) \rightarrow 2Fe_2O_3 \cdot H_2O(s)$$
$$\text{iron} + \text{oxygen} + \text{water} \rightarrow \text{hydrated iron(III) oxide}$$

There is no one simple formula for rust. The simplest formula that we can suggest is that of hydrated iron(III) oxide. The amount of water in rust varies with the conditions. You can see that rusting is a redox reaction: the iron increases its oxidation number from 0 to +3, and the oxidation number of oxygen changes from 0 to −2.

A layer of rust is very weak and it soon flakes away from the surface of the iron. The newly exposed surface then starts to rust. Rusting is speeded up by electrolytes such as salt. That is why ships rust very quickly in seawater unless they are treated to prevent rust.

Figure 15.2.2 The hull of a ship has to be painted regularly otherwise the iron will rust when exposed to water and air.

Stopping the rust

Rusting destroys about 20% of the world's iron and steel every year. So it is very important to stop rusting. We can stop rusting by a barrier method. We coat the iron or steel with a layer which keeps out oxygen and water. We can use:

- paint – for bridges and cars
- a plastic coating – wire netting for fencing is often covered with plastic

EXAM TIP

Metals other than iron or steel do not rust. It is a common error to suggest that they do. The other metals corrode.

EXAM TIP

When describing methods that prevent rusting, 'removing water and air' are explanations, not methods. Methods are painting, plating and galvanising.

Unit 15: More about metals

- metal plating – the iron is coated with another metal. This is often carried out by electroplating. Chromium plating is used for bathroom taps. Steel food cans are plated with tin. Iron for roofing is coated with zinc by dipping it into molten zinc – we call this **galvanising**

- greasing and oiling – used for tools and the moving parts of machinery.

Supplement

Galvanising and sacrificial protection

Galvanised steel is steel coated with zinc. The zinc forms a barrier that stops water and oxygen from reaching the iron in steel. But even when the zinc is scratched, the steel underneath will not rust. So there must be another way in which the zinc protects the iron in steel. How does it do this?

Zinc is more reactive than iron. A more reactive metal loses electrons and forms positive ions more easily than a less reactive metal such as iron. When the layer of zinc is scratched, an electrochemical cell is set up in the presence of moisture. The electrons flow from the zinc to the iron.

$$Zn(s) \rightarrow Zn^{2+}(aq) + 2e^-$$

The zinc is oxidised and forms zinc ions. So the zinc corrodes instead of the iron. We say the zinc is a *sacrificial* metal. This method of using a more reactive metal to protect a less reactive metal from corrosion is called **sacrificial protection**. The more reactive metal corrodes instead of the iron or steel.

The iron remains protected because the electrons are accepted on its surface. This keeps the iron in a reduced state. It is not being oxidised to iron(III) ions.

The electrons from the zinc react with hydrogen ions in the water to form hydrogen gas.

$$2H^+(aq) + 2e^- \rightarrow H_2(g)$$

The sacrificial metal does not have to completely cover the surface of the iron for the method to work. Pipelines, oil rigs and ships can be protected from corrosion by attaching bars of zinc or magnesium in direct contact with the iron.

Figure 15.2.3 Zinc or magnesium strips are attached to iron prevent the iron from rusting by sacrificial protection.

KEY POINTS

- Both water and oxygen are needed for iron to rust.
- You can prevent rusting by barrier methods (painting, plating, coating with plastic or greasing).
- In sacrificial protection a more reactive metal is in contact with iron. The more reactive metal corrodes instead of the iron because the more reactive metal loses electrons more easily.

SUMMARY QUESTIONS

1. Copy and complete using the words below:

   flakes   hydrated   oxygen   rust   steel   surface

   Rusting occurs when _____ and water react with iron or _____. Rust is _____ iron(III) oxide. The layer of rust easily _____ off the _____ of the iron. The fresh iron surface then starts to _____ again.

2. Explain why iron plated with tin does not rust.

3. Explain why a piece of magnesium in contact with a steel pipeline prevents the pipeline rusting.

15.3 Alloys

LEARNING OUTCOMES

- Describe the general physical properties of metals
- Recognise that alloys can be harder and stronger than pure metals
- Identify representations of alloys from diagrams of their structure
- **S** Explain why alloys can be harder and stronger than pure metals

Physical properties of metals

In Topic 3.5 we learned that most metals:
- are good electrical conductors
- are good thermal conductors
- are malleable and ductile
- are lustrous (shiny)
- have higher melting and boiling points than most non-metals.

What are alloys?

A metal such as iron is rarely used on its own because it rusts easily. Pure copper is not very strong so cannot be used for parts of machines that are constantly in motion. We can change the properties of a metal to make it stronger, harder or more resistant to corrosion. We do this by mixing it with another metal or with a non-metal.

An **alloy** is a mixture of a metal with one or more other elements. The other elements can be metals such as chromium or non-metals such as carbon. An alloy is not just a mixture of metal crystals. The atoms of the second element form part of the crystal lattice. Figure 15.3.1 shows a representation of an alloy. In this diagram you will notice that:

- two different types of atoms are present – one of them will be a metal
- the atoms are arranged in layers
- there is no particular pattern in the arrangement of the different atoms in each layer
- the two types of atoms have different sizes.

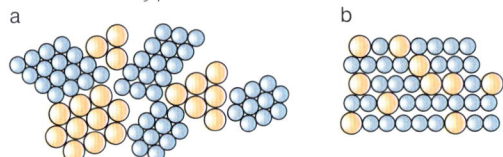

a A mixture of metal crystals   b An alloy

Figure 15.3.1 A mixture of metal crystals is not the same as an alloy.

EXAM TIP

It is a common error to think that all metals are hard and have very high melting points. Remember that Group I metals are soft and have relatively low melting points.

EXAM TIP

Make sure that you can you can identify an alloy from a diagram such as the one in Figure 15.3.1.

PRACTICAL

Tin, lead and solder

Solder is an alloy of lead and tin. It is used to join wires in electrical circuits. You put small pieces of tin, lead and solder on the steel 'tin' lid and heat the centre. You record the time taken for each metal to melt. The solder melts before the tin and lead. This shows that the alloy has lower melting point than the tin or lead alone. Other physical properties are different as well.

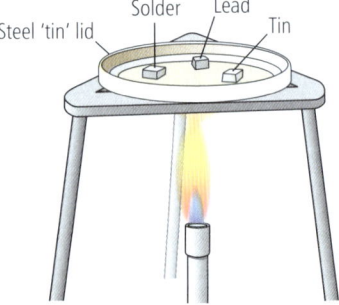

Figure 15.3.2 Which metal melts first?

Unit 15: More about metals

Supplement

Why are alloys harder and stronger?

The atoms in a pure metal are arranged in regular layers. When a force is applied, the layers slide over each other. This explains why metals are malleable and ductile. When a metal is alloyed with a second metal, the different sized metal atoms make the arrangement of the lattice less regular. We say that they disrupt the crystal lattice. This stops the layers of metal atoms sliding easily over each other when a force is applied. So the malleability and ductility decreases. This explains why an alloy is stronger and harder than a pure metal.

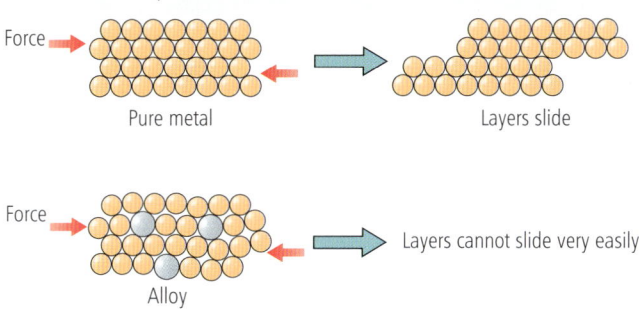

Figure 15.3.3 Alloys are stronger than pure metals because the layers cannot slide past each other as easily.

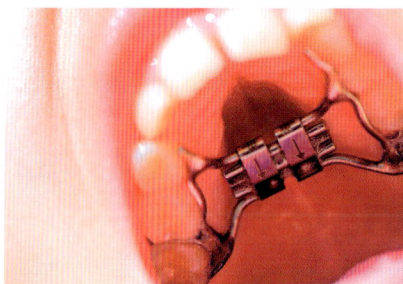

Figure 15.3.4 The alloy 'nitinol' is a mixture of nickel and titanium. It is used in some dental braces. As the alloy warms up, it pulls the teeth into the correct position.

SUMMARY QUESTIONS

1 Copy and complete using the words below:

   alloy   arrangement   mixture   layers   non-metal   stronger

   An alloy is a _____ of metals or a mixture of metals with a _____. By making a metal into an _____ it becomes _____ and harder. In an alloy, the metal atoms are arranged in _____. There is no pattern in the _____ of the different atoms in each layer.

2 Which one of these diagrams, **A**, **B**, **C** or **D**, best shows an alloy of copper, tin and lead?

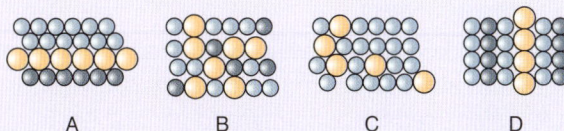

3 Explain why an alloy of aluminium and manganese is stronger than pure aluminium.

KEY POINTS

- Alloys are mixtures of metal atoms with atoms of another element.
- The properties of a metal are changed by making it into an alloy.
- Metals are made into alloys to improve their strength, hardness or resistance to corrosion.
- Alloys are stronger and harder than the metals from which they are made, because the different sized atoms prevent the layers of atoms from sliding over each other easily.

15.4 Uses of metals

LEARNING OUTCOMES

- Describe the uses of aluminium and copper in terms of their physical properties
- State that alloys are often more useful than the pure metals
- Describe the uses of stainless steel

Uses of pure metals

Aluminium

Aluminium is used for making aircraft bodies because of its low density. It is also strong when alloyed. Most aircraft are made from aluminium alloys containing about 90% aluminium and smaller amounts of zinc and copper.

Some food containers and cooking foil are made from aluminium. This is because of its resistance to corrosion. There is an unreactive oxide layer on the surface of aluminium which does not flake off. This oxide layer does not react with the acids that are present in many foods.

Aluminium is used in the manufacture of overhead electrical cables because of its low density and good electrical conductivity.

Copper

Copper is used for electrical wiring because of its high electrical conductivity. It is one of the most malleable and ductile metals so it can easily be shaped and drawn out into wires. It is also used for the base of cooking pans because it is an excellent conductor of thermal energy.

Zinc

Zinc is used to galvanise iron or steel to prevent rusting (see Topic 15.2). To galvanise a steel object, we dip the object into molten zinc. The zinc solidifies and forms a thin coating on the surface of the steel. About a third of the zinc produced in the world is used to galvanise steel. Galvanised steel is used for roofing because it is weather resistant.

Figure 15.4.1 The alloy of aluminium used to make this aircraft has a low density and is also strong.

Steel alloys

Pure iron is too soft and weak to be very useful and the iron from the blast furnace has too much carbon in it to make it useful. It is too brittle to be used to make bridges and frames for buildings.

All steels contain a small, controlled amount of carbon. There are many types of steel. Each of these is used for a particular purpose:

- Mild steel is low carbon steel. It contains about 0.25% carbon. It is soft, malleable and can be drawn into wires easily. We use it to make car bodies and parts of machinery where it will not be worn away. It is also used for buildings and general engineering purposes.

- Low alloy steels contain between 1% and 5% of other metals such as nickel, chromium, manganese and titanium. They are hard and do not stretch much. Nickel steels are used for bridges where strength is needed and for bicycle chains. Tungsten steel is used for high-speed tools because it does not change shape at high temperatures.

EXAM TIP

In most chemistry exams there will be a question about the uses of elements or compounds. Make a list of all the substances in the syllabus whose uses you need to know. Divide the page into two with the names down one side and the uses on the other. Then test yourself.

Unit 15: More about metals

- **Stainless steels** are high alloy steels. They may contain up to 20% chromium. Many stainless steels contain 70% iron, 20% chromium and 10% nickel. They are very strong, hard and resist corrosion. Stainless steel is used to make cutlery (knives, forks and spoons) because it is hard and resistant to rusting. Stainless steel is also used in the construction of pipes and towers in chemical factories. Surgical instruments are made from stainless steel.

Brass and other alloys

Brass is an alloy of 70% copper and 30% zinc. It is stronger than copper and does not corrode. Although it is strong, it can still be easily beaten into shape – it is still malleable. Its gold colour makes it attractive. So it is used to make musical instruments, door handles, ornaments and screws.

Bronze is an alloy of copper and tin which is very hard. So it is used in some parts of machines, where movement might wear away the metal, and in statues and bells.

Figure 15.4.2 Brass is an alloy of copper and zinc. It is used to make musical instruments because of its strength and resistance to corrosion.

Recycling metals

Recycling is the processing of used materials to make new products. There are many advantages of recycling metals.

- it conserves metal ores and other raw materials (natural resources) used
- it saves energy because less fuel is used (extracting and purifying metals needs a lot of energy)
- it reduces pollution arising from extracting and purifying materials
- it reduces waste and problems of disposal of unwanted material (e.g. less landfill)
- it saves land otherwise used for extracting ores.

EXAM TIP

For the examinations you must know about the composition and uses (in terms of their physical properties) of brass and stainless steel.

KEY POINTS

- Aluminium is used to make aircraft because of its low density and for food containers because of its resistance to corrosion.
- The properties of iron are changed by adding controlled amounts of other metals or carbon to make steels.
- Stainless steel is used in making cutlery because it is hard and resists rusting.
- Brass is an alloy of copper and zinc.

SUMMARY QUESTIONS

1. Copy and complete using the words below:

   car   carbon   cutlery   different   mild   stainless

   We use _____ types of steel for different jobs. Low _____ steel, often called _____ steel, is used to make _____ bodies and machinery. We use _____ steel to make _____ and parts of chemical factories.

2. Explain why aluminium is used to make food containers.

3. Give two advantages of using brass instead of pure copper or pure zinc.

SUMMARY QUESTIONS

1 Match each alloy on the left with the elements on the right which they contain.

brass   iron
stainless steel   zinc
mild steel   carbon
    copper
    chromium

2 Draw a diagram of a blast furnace. Label your diagram to show:

P where the solid raw materials are loaded into the furnace
Q where air enters the furnace
R where the iron collects
S where slag collects.

3 Copy and complete using the words below.

air blast calcium coke decomposes impurities monoxide slag

Iron is extracted in a _____ furnace from iron ore using carbon _____ as a reducing agent. The carbon monoxide is formed when _____ burns in a blast of _____. The limestone added to the blast furnace _____ to form _____ oxide. Calcium oxide combines with the _____ in the iron ore to form _____.

4 Match each metal on the left with its use on the right.

aluminium   galvanising iron roofs
mild steel   electrical wiring in the home
stainless steel   aircraft bodies
copper   cutlery
zinc   car bodies

5 Two methods for extracting metals are: (i) heating with carbon and (ii) electrolysis. Which of these methods is best used to extract each of the following metals?

a sodium   d iron
b lead   e aluminium
c calcium

6 Which of these statements about alloys are correct?

A Brass is a compound of zinc and copper.
B An alloy is a mixture of two or more non-metals.
C Stainless steel is harder than iron.
D Stainless steel is a mixture of iron and other elements such as chromium, nickel and carbon.

Practice questions

1 Which one of these statements about the extraction of iron in a blast furnace is correct? [1]

A Limestone is added to combine with excess carbon dioxide.
B A slag of iron(III) oxide forms at the bottom of the furnace.
C Hot air is blown in at the top of the furnace.
D Carbon monoxide reduces iron(III) oxide to iron.

(Paper 1)

2 Which one of these substances is NOT added to the blast furnace for the extraction of iron? [1]

A hematite
B coke
C water
D air

(Paper 1)

3 Which one of these equations does NOT represent a reaction happening in the blast furnace during the extraction of iron? [1]

A $Fe_2O_3(s) + 3CO(g) \rightarrow 2Fe(l) + 3CO_2(g)$
B $FeO(s) + H_2(g) \rightarrow Fe(s) + H_2O(g)$
C $CaCO_3(s) \rightarrow CaO(s) + CO_2(g)$
D $CaO(s) + SiO_2(s) \rightarrow CaSiO_3(l)$

(Paper 2)

4 Iron is extracted in a blast furnace.

(a) Name the three raw materials used to extract iron. [3]

(b) The diagram shows a blast furnace.

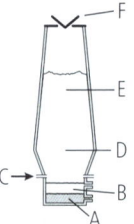

Give the letter in the diagram which shows:

(i) where the solid raw materials are put into the furnace [1]
(ii) where the furnace is hottest [1]
(iii) where the slag is collected. [1]

(c) In the blast furnace, carbon monoxide reacts with iron(III) oxide.
  (i) Write a word equation for this reaction. [2]
  (ii) The carbon monoxide is formed by carbon dioxide reacting with excess carbon. Write a symbol equation for this reaction. [2]

(d) Iron rusts.
  (i) Name the two substances needed for iron to rust. [2]
  (ii) Give two methods that prevent iron from rusting. [2]

(Paper 3)

5 Stainless steel is an alloy.
(a) Give the meaning of the term alloy. [2]
(b) Name two elements present in all types of steel. [2]
(c) Name another metal that is present in stainless steel. [1]
(d) Stainless steel is used for many purposes.
  (i) Give one use of stainless steel. [1]
  (ii) Give two reasons why steel and not pure iron is used to make car bodies. [2]

(Paper 3)

6 The relative reactivity of some metals is shown:

sodium   calcium   zinc   iron   tin
most reactive ⟶ least reactive

(a) State the names of two metals from this list that can be extracted by heating with carbon. [2]
(b) Give two uses of aluminium. Explain each of these uses in terms of the physical properties of aluminium. [4]
(c) Carbon monoxide is used to extract iron from iron(III) oxide.
  (i) State the condition needed for this extraction. [1]
  (ii) Copy and complete the symbol equation for this reaction:
  $Fe_2O_3 + \_\_CO \rightarrow \_\_Fe + 3CO_2$ [2]
(d) Limestone is used in the production of iron.
  (i) Explain what happens to the limestone in the furnace. [2]
  (ii) Explain the purpose of adding limestone to the furnace. [2]

(Paper 3)

7 Iron ore is mainly iron(III) oxide, Fe_2O_3.
(a) State the name of the main ore of iron. [1]
(b) In the blast furnace iron(III) oxide is reduced by carbon monoxide.
  (i) Explain how the carbon monoxide is formed in a two-step process. [3]
  (ii) Write a symbol equation for the reaction of iron(III) oxide with carbon monoxide. Include state symbols. [3]
(c) At high temperatures, iron(III) oxide reacts with carbon to form iron and carbon dioxide. Explain why this is a redox reaction in terms of electron transfer. [2]
(d) Iron rusts when exposed to water and oxygen.
  (i) Give the chemical name for rust. [2]
  (ii) Rusting can be prevented by galvanising iron. Explain two different ways by which galvanising prevents rusting. [4]

(Paper 4)

8 Steel is an alloy of iron.
(a) State the meaning of the term alloy. [2]
(b) Stainless steel contains several elements other than iron. Name two of these elements. [2]
(c) State one use of stainless steel and explain this use in terms of physical properties. [2]
(d) (i) Draw the structure of a typical alloy. [2]
  (ii) Explain why steel is harder than pure iron. [3]

(Paper 4)

16.1 Making ammonia
Supplement

LEARNING OUTCOMES
- State the chemical equation for the Haber process and the sources of the hydrogen and nitrogen needed
- State the essential conditions in the Haber process and explain why these conditions are used
- Predict and explain how the position of equilibrium is affected by changes in reaction conditions

Ammonia has many uses

Millions of tonnes of ammonia, NH_3, are produced every year to make fertilisers, nitric acid, nylon and explosives. Nitrogen combines with hydrogen at high pressure and temperature in the presence of a catalyst to make ammonia. We call this method the **Haber process**.

Sources of hydrogen and nitrogen

The hydrogen for the Haber process is made by heating steam and methane in the presence of a nickel catalyst

$$CH_4(g) + H_2O(g) \rightarrow CO(g) + 3H_2(g)$$
methane + steam → carbon monoxide + hydrogen

Nitrogen for the Haber process is extracted from air, the oxygen having been removed by reaction with hydrogen.

The Haber process

The stages in the Haber process are:

- A mixture of nitrogen (1 volume) and hydrogen (3 volumes) is compressed.

- The compressed gases pass into a large tank called a converter. This contains trays of iron catalyst. The temperature in the converter is 450 °C and the pressure is 20 000 kPa (200 atmospheres). Nitrogen and hydrogen combine:

$$N_2(g) + 3H_2(g) \rightleftharpoons 2NH_3(g)$$

Figure 16.1.1 A chemical plant for making ammonia.

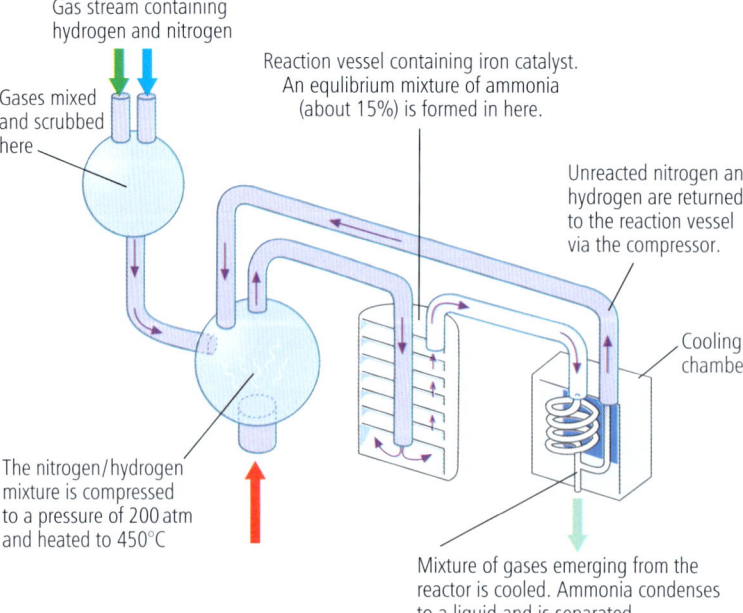

Figure 16.1.2 The production of ammonia by the Haber process.

EXAM TIP

You need to know the conditions used in the Haber process and why these particular conditions are used.

Unit 16: Compounds of nitrogen and sulfur

- The mixture is passed to a cooling chamber. The ammonia condenses here and is removed.
- The unreacted nitrogen and hydrogen are returned to the converter. In this way they are not wasted.

What are the best reaction conditions?

The reaction to make ammonia is a reversible reaction. Figure 16.1.2 shows how the equilibrium yield of ammonia varies with temperature and pressure. The yield is the percentage of ammonia produced in the equilibrium mixture.

The yield of ammonia increases with an increase in pressure. This is because, for a gas reaction, increasing the pressure shifts the equilibrium in the direction of fewer moles of gases (lower gas volume). But high pressure is expensive and there is a danger of the vessels bursting. It costs a lot of money to make strong, safe vessels and a lot of expensive fuel is used to keep the pressure high. So we do not use a pressure above 20 000 kPa (200 atmospheres).

In the Haber process the forward reaction is exothermic. An increase in temperature favours the endothermic reaction where thermal energy is absorbed. So the backward reaction is endothermic and the yield of ammonia decreases with increasing temperature. If we want to increase the yield of ammonia, we should lower the temperature. But if we use too low a temperature, the rate of reaction will be too slow for an economical yield. There is a conflict between the best equilibrium yield and the best rate of reaction. So we use a compromise temperature of 450 °C. This gives quite a good yield with a fast enough rate of reaction.

The catalyst has no effect on the equilibrium. It speeds up the rate of the forward and the reverse reactions equally.

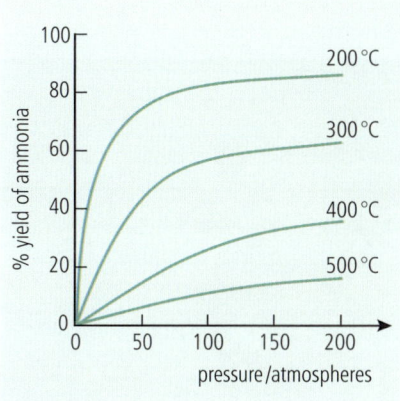

Figure 16.1.3 The yield of ammonia depends on the temperature and pressure.

EXAM TIP

You should be prepared to answer questions about the economics and safety aspects of the Haber process.

KEY POINTS

- Nitrogen for the Haber process is obtained from the air. The hydrogen is obtained from reacting steam with methane.
- The conditions used in the Haber process are 200 000 kPa pressure, a temperature of 450 °C and an iron catalyst.
- A compromise temperature is used in the Haber process which gives a good yield at a fast enough rate of reaction.

SUMMARY QUESTIONS

1. Copy and complete using these words:

   compressed   converter   Haber   iron   kilopascal   nitrogen   rate

   Ammonia is made by the _____ process. Hydrogen and _____ are _____ and reacted together in a _____ at 450 °C and 20 000 _____ pressure. A catalyst of _____ is used to increase the reaction _____.

2. Explain why a compromise temperature of 450 °C is used in the Haber process.

3. State the source of the nitrogen and hydrogen in the Haber process.

16.2 Fertilisers

LEARNING OUTCOMES

- State that ammonium salts and nitrates can be used as fertilisers
- Describe the need for NPK fertilisers

Why do we use fertilisers?

We rely on many plants for our food. Plants need nitrogen to make proteins. Most plants take up nitrogen from the soil through their roots in the form of nitrate ions, NO_3^-, and ammonium ions, NH_4^+.

When farmers harvest their crops, the nitrogen is removed and not returned to the soil. After a few years the amount of nitrogen in the soil is reduced and new crops won't grow very well. Phosphorus and potassium which are also needed for plant growth are removed from the soil as well. So how can we replace these essential elements in the soil?

The answer is to add a **fertiliser**. A fertiliser is a substance added to the soil to replace the elements taken up by plants. Farmers add fertilisers to the soil to improve plant growth so that plants grow faster and crop yields are increased.

Fertilisers containing nitrogen, phosphorus and potassium are added to the soil to replace the essential plant nutrients that have been lost. These fertilisers are called **NPK fertilisers** after the symbols for these three elements. These fertilisers are made of compounds of nitrogen, phosphorus and potassium, not the elements themselves. Typical NPK fertilisers include potassium nitrate, KNO_3, and ammonium phosphate, $(NH_4)_3PO_4$.

Figure 16.2.1 Fertiliser being spread on field by hand.

Making fertilisers

Nearly 10% of the entire chemical industry is involved in making fertilisers.

EXAM TIP

Fertilisers are easily absorbed by plants because ammonium salts and nitrates are soluble in water.

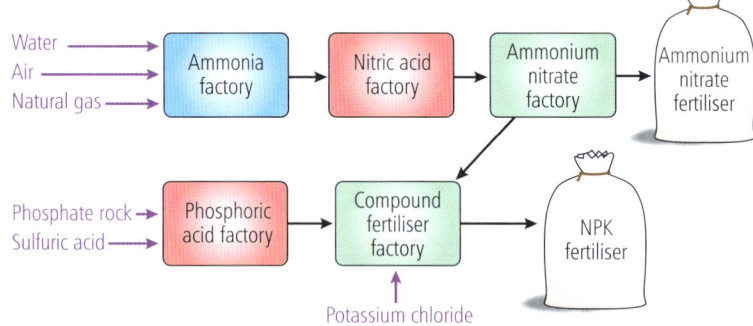

Figure 16.2.2 A flow chart for making an NPK fertiliser.

EXAM TIP

When you write equations for the formation of ammonium salts remember that no water is formed as a product. For example: ammonia + sulfuric acid → ammonium sulfate.

Many fertilisers contain ammonium salts, nitrates and phosphates. These are made by neutralising aqueous ammonia with acids:

$NH_3(aq)$ + $HNO_3(aq)$ → $NH_4NO_3(aq)$
ammonia + nitric acid → ammonium nitrate

$3NH_3(aq)$ + $H_3PO_4(aq)$ → $(NH_4)_3PO_4(aq)$
ammonia + phosphoric acid → ammonium phosphate

Unit 16: Compounds of nitrogen and sulfur

The ammonia for these reactions is made by the Haber process (see Topic 16.1). The sulfuric acid for making phosphoric acid is made by the Contact process (see Topic 16.3). The potassium in fertilisers usually comes from salts, such as potassium chloride.

PRACTICAL

Making ammonium sulfate fertiliser

1. You titrate ammonia with sulfuric acid using methyl orange as an indicator.
2. When the methyl orange turns pink, you record the burette reading.
3. Then you repeat the titration without the indicator, adding the same amount of acid as you did before.
4. You then pour the solution from the flask into an evaporating basin. You evaporate some of the water and allow the fertiliser to crystallise.

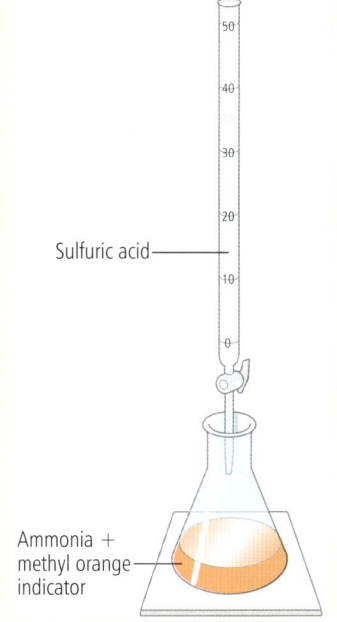

Figure 16.2.3 Making ammonium sulfate.

Releasing ammonia from ammonium salts

Many crop plants do not grow well in acidic conditions. So farmers sometimes add lime (calcium oxide, CaO) to the soil to neutralise the acidity. Lime reacts with water in the soil to form calcium hydroxide. This is alkaline. If too much lime is put on the soil it releases ammonia from ammonium salts in fertilisers. The ammonia may escape as a gas into the air and is lost from the soil:

$$2NH_4Cl(aq) + Ca(OH)_2(aq) \rightarrow 2NH_3(g) + CaCl_2(aq) + 2H_2O(l)$$

ammonium chloride + calcium hydroxide → ammonia + calcium chloride + water

KEY POINTS

- Fertilisers are added to the soil to improve plant growth and to replace essential elements lost when plants are harvested.
- NPK fertilisers are salts which contain nitrogen, phosphorus and potassium.
- Ammonium phosphate and potassium nitrate are important fertilisers.

SUMMARY QUESTIONS

1. Copy and complete using these words:

   elements   fertilisers   harvested   phosphorus   potassium   soil

   Plants need nitrogen, _____ and _____ for healthy growth. Artificial _____ are added to the _____ to replace the essential _____ lost when the crops are _____.

2. Describe how to make crystals of potassium phosphate using phosphoric acid.
3. Write a word equation for making ammonium nitrate from ammonia.

16.3 Sulfuric acid

LEARNING OUTCOMES

- State that aqueous solutions of acids contain H+ ions
- Describe how acids react with metals, metal oxides, metal hydroxides and carbonates
- Predict and explain how the position of equilibrium is affected by changes in reaction conditions

Sulfuric acid is one of the most important industrial chemicals. It has many uses (see Figure 16.3.1).

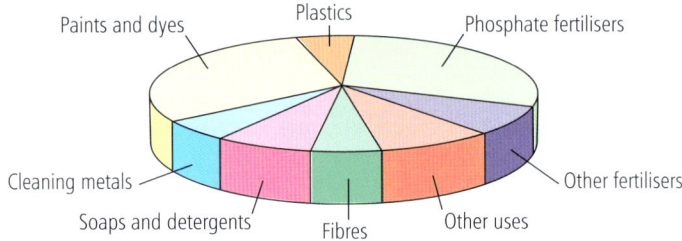

Figure 16.3.1 The uses of sulfuric acid.

Chemical properties of sulfuric acid

Sulfuric acid is a **dibasic acid**. This means that in a neutralisation reaction, two hydrogen ions can be replaced per molecule when a salt is made:

$$H_2SO_4(aq) + 2NaOH(aq) \rightarrow Na_2SO_4(aq) + 2H_2O(l)$$

sulfuric acid + sodium hydroxide → sodium sulfate + water

Dilute sulfuric acid is a strong acid and has typical acidic properties:

- It reacts with metals above hydrogen in the reactivity series to form a salt and hydrogen gas.
- It reacts with bases (metal oxides and hydroxides) to form a salt and water.
- It reacts with carbonates to form a salt, water and carbon dioxide gas.

EXAM TIP

When writing equations for the reaction of sulfuric acid with metals, oxides and hydroxides remember that both hydrogen ions can be replaced.

Supplement

The manufacture of sulfuric acid

Sulfuric acid is manufactured by the **Contact process**. The raw materials needed are sulfur, air and water. The stages in the Contact process are:

- A spray of molten sulfur is burned in a furnace in a current of dry air (which contains oxygen). Sulfur dioxide gas is formed:

$$S(l) + O_2(g) \rightarrow SO_2(g)$$

Sulfur dioxide for the process can also be made by roasting sulfide ores.

- The sulfur dioxide is reacted with excess air in a tower called a converter which contains a catalyst, vanadium(V) oxide. The temperature in the converter is about 450 °C and the pressure is just 200 kPa (2 atmospheres). The product is sulfur trioxide.

Figure 16.3.2 A chemical plant for making sulfuric acid.

Unit 16: Compounds of nitrogen and sulfur

Sulfur trioxide is formed:

$$2SO_2(g) + O_2(g) \rightleftharpoons 2SO_3(g)$$
sulfur dioxide + oxygen ⇌ sulfur trioxide

The reaction is exothermic.

- The sulfur trioxide is absorbed into a 98% solution of sulfuric acid to form a thick liquid called oleum, $H_2S_2O_7$.
- The oleum is mixed with a little water to make concentrated 98% sulfuric acid:

$$H_2S_2O_7(l) + H_2O(l) \rightarrow 2H_2SO_4(l)$$

The reaction conditions for the Contact process

The reaction

$$2SO_2(g) + O_2(g) \rightleftharpoons 2SO_3(g)$$

is an equilibrium reaction. The yield of sulfur trioxide increases with an increase in pressure. This is in the direction of a decreasing number of moles of gas. But a pressure just above atmospheric pressure is used (200 kPa). This is because the yield of sulfur trioxide is very high even at atmospheric pressure.

The reaction is exothermic. So the yield of sulfur trioxide decreases with increasing temperature. This is because an increase in temperature favours the endothermic reaction which is the reverse reaction. But if we use too low a temperature the rate of reaction will be too slow. So we use a compromise temperature of 450 °C. This gives quite a good yield of sulfur trioxide with a fast enough rate of reaction. The catalyst has no effect on the equilibrium – it speeds up the rate of the forward and the reverse reactions equally.

EXAM TIP
You need to know the conditions used in the Contact process and why these particular conditions are used.

SUMMARY QUESTIONS

1 Write word equations for the reaction of sulfuric acid with **a** copper(II) oxide **b** sodium hydroxide.

2 Write symbol equations for the reaction of sulfuric acid with **a** barium hydroxide, $Ba(OH)_2$ **b** calcium carbonate to form calcium sulfate.

3 Copy and complete using these words and formulae:

equilibrium   more SO_2 SO_3 temperature   volume

In the reaction $2SO_2(g) + O_2(g) \rightleftharpoons 2SO_3(g)$ decreasing the pressure shifts the _____ in the direction of _____ moles of _____ (higher gas _____). The reaction is exothermic so an increase in _____ decreases the yield of _____.

KEY POINTS

- Sulfuric acid is a typical strong acid which reacts with metals, metal oxides, metal hydroxides and carbonates.
- In equilibrium reactions, an increase in pressure shifts the equilibrium in the direction of fewer moles of gas.
- A compromise temperature is used in the Contact process which gives a good yield at a fast enough rate of reaction.

16.4 Acid rain

LEARNING OUTCOMES

- Describe the origins of nitrogen dioxide and sulfur dioxide in the atmosphere
- Describe the formation of acid rain and its adverse effects

What is acid rain?

Rain is naturally slightly acidic. This is because carbon dioxide from the air dissolves in rainwater and reacts to form a solution of the weak acid, carbonic acid. However, if the pH of the rain falls below 5, the rain is called acid rain. **Acid rain** is formed when nitrogen dioxide or sulfur dioxide react with water in the atmosphere to form rain with a pH below 5.

How does sulfur dioxide get into the atmosphere?

Fossil fuels contain sulfur and sulfur compounds as well as hydrocarbons. When fossil fuels burn, the sulfur is oxidised to sulfur dioxide gas which escapes into the atmosphere:

$$S(s) + O_2(g) \rightarrow SO_2(g)$$

Volcanoes are also a natural source of sulfur dioxide gas.

How does nitrogen dioxide get into the atmosphere?

Nitrogen forms several oxides. They are all gases. We usually call these by their common names rather than their full chemical names. Nitric oxide is NO (nitrogen(II) oxide) and nitrogen dioxide is NO_2 (nitrogen(IV) oxide). Oxides of nitrogen get into the atmosphere from a number of sources:

- From car engines. The high temperature and pressure inside a car engine cause nitrogen and oxygen to combine. A mixture of nitrogen oxides is formed. This mixture comes out with the exhaust gases from the engine. The mixture is called NO_x to show that several oxides of nitrogen are present.

    formation of nitrogen dioxide: $N_2(g) + 2O_2(g) \rightarrow 2NO_2(g)$

- From high-temperature furnaces. The temperature in these is high enough to allow nitrogen and oxygen to combine.
- The electrical energy in the lightning causes the formation of large amounts of NO_x from nitrogen and oxygen in the air.

How is acid rain formed?

- From sulfur dioxide: sulfur dioxide reacts directly with water in the atmosphere to form sulphurous acid, H_2SO_3(aq). Sulfur dioxide can also be oxidised in the atmosphere to form sulfur trioxide. The sulfur trioxide then reacts with water vapour in the air to form a solution of sulfuric acid:

$$\underset{\text{sulfur trioxide}}{SO_3(g)} + H_2O(l) \rightarrow \underset{\text{sulfuric acid}}{H_2SO_4(aq)}$$

When the acidic water vapour condenses into large enough drops in clouds, acid rain falls.

Figure 16.4.1 These trees have died because of acid rain.

EXAM TIP

It is a common error to suggest that carbon dioxide is responsible for acid rain.

Unit 16: Compounds of nitrogen and sulfur

- From nitrogen oxides: nitric oxide can be oxidised in the atosphere to nitrogen dioxide. Nitrogen dioxide reacts directly with water in the atmosphere to form acids.

$$2NO_2(g) + H_2O(l) \rightarrow HNO_3(aq) + HNO_2(aq)$$
nitrogen dioxide water nitric acid nitrous acid

EXAM TIP
It is a common error to suggest that sulfur rather than sulfur dioxide is responsible for acid rain. Comments such as 'Sulfur dissolves in water to form acid rain' are incorrect.

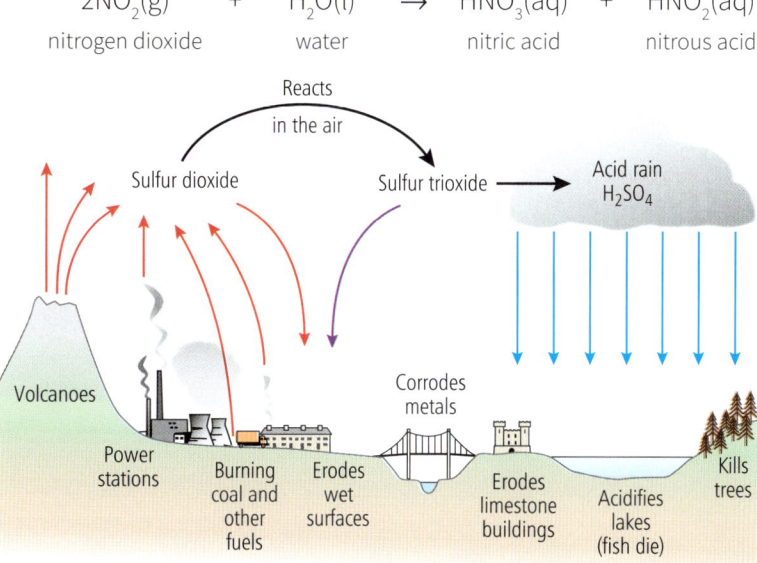

Figure 16.4.2 How acid rain is formed.

The adverse effects of acid rain

Acid rain has many adverse (negative) effects on the environment:

- The leaves of trees can be damaged by the acid. The leaves fall off and the trees die.
- Lakes and rivers become acidic. If they are too acidic, fish and other aquatic life may die.
- Soil may become too acidic to grow crop plants.
- Buildings made from carbonate rocks, such as limestone, will be eroded. The acid reacts with the carbonate to release carbon dioxide and the surface of the building crumbles.
- Metal structures such as bridges and iron railings corrode.

KEY POINTS
- Sulfur dioxide and nitrogen dioxide react with water vapour in the atmosphere to form acid rain.
- The main source of sulfur dioxide is the combustion of fossil fuels.
- The main source of nitrogen dioxide is car engines.
- Acid rain kills trees and acidifies lakes leading to the death of fish. It also acidifies soils leading to reduction in crop growth.

SUMMARY QUESTIONS

1 Copy and complete using these words:

fossil   oxidised   sulfur   sulfuric   sulphurous   water

Sulfur dioxide if formed when _____ fuels containing _____ are burned. Sulfur dioxide reacts with _____ in the atmosphere to form _____ acid. Sulfur dioxide can also be _____ to form sulfur trioxide which reacts with water to form _____ acid.

2 State two adverse effects of acid rain.

3 Write symbol equations for **a** the reaction of sulfur with oxygen **b** the reaction of sulfur dioxide with water.

SUMMARY QUESTIONS

1 Link each compound on the left with the descriptions on the right.

ammonium nitrate	a gas responsible for acid rain
nitrogen dioxide	an alkaline gas used to make fertilisers
methane	a salt used as a fertiliser
ammonia	a gas used to produce hydrogen

2 Copy and complete using the words from the list below:

crops   elements   harvested   nitrate   phosphate   potassium

Farmers use several chemicals to make sure that their _____ grow well. Fertilisers such as ammonium _____ and ammonium _____ are added to the soil to replace the _____ removed when crops are _____. NPK fertilisers contain nitrogen, _____ and phosphorus.

3 State the names of a suitable acid and alkali you can use to make each of the following fertilisers:
   a ammonium sulfate
   b potassium phosphate
   c ammonium nitrate

4 Write word equations for:
   a the reaction of nitric acid with potassium hydroxide
   b the combustion of sulfur
   c the formation of nitrogen dioxide from two elements
   d the reaction of sulfuric acid with calcium carbonate

5 Write symbol equations for the reaction of sulfuric acid with:
   a magnesium oxide
   b zinc
   c sodium hydroxide
   d sodium carbonate

Practice questions

1 Which one of these statements about fertilisers is true?
   A Ammonium nitrate can be used as a fertiliser.
   B Fertilisers contain oxygen, sulfur and iron.
   C Fertilisers are added to the soil to make it more alkaline.
   D Fertilisers are made by combining calcium with oxygen.

(Paper 1)

2 Which one of these statements about the Haber process is completely correct?
   A The raw materials are nitrogen dioxide from the air and methane from the petroleum industry.
   B The temperature used is 450 °C because a higher temperature reduces the yield of ammonia.
   C The temperature is above 800 °C because a higher temperature gives a greater yield of ammonia.
   D The pressure in the reaction vessel is 20 kPa because the reaction is exothermic.

(Paper 2)

3 Fertilisers are spread on fields by farmers.
   (a) Why do farmers use fertilisers? [1]
   (b) State the names of the three elements most commonly found in fertilisers. [3]
   (c) Ammonium sulfate is a fertiliser. Describe how you can make aqueous ammonium sulfate in the laboratory from aqueous ammonia and sulfuric acid. [5]
   (d) Ammonium nitrate is also a fertiliser. Write a word equation to show how ammonium nitrate can be produced from an acid and an alkali. [2]

(Paper 3)

4 Sulfur dioxide and nitrogen dioxide are gases responsible for acid rain.
   (a) Which one of these pH values represents a possible pH of acid rain.
   pH 4.5   pH 6.5   pH 7.0   pH 9.5   pH 12.0 [1]
   (b) Describe a source of each these gases in the atmosphere. [2]

(c) Explain how acid rain is formed from nitrogen dioxide in the atmosphere. In your answer, include a relevant word equation. [2]

(d) Sulfur dioxide can be oxidised in the atmosphere. Copy and complete the symbol equation for this reaction:
_____ SO_2 + _____ → _____ SO_3 [2]

(e) Describe and explain one effect of acid rain on buildings made of carbonate rocks. [2]

(Paper 3)

5 Ammonium phosphate is a fertiliser.
(a) Suggest two substances that you could react together to make ammonium phosphate. [2]
(b) Describe a practical method to make aqueous ammonium phosphate in the laboratory. [5]
(c) Copy and complete these sentences about the crystallisation of ammonium phosphate using words from the list.

**concentrated   dried   evaporated
filtered   phosphate   solution**

A _____ of ammonium _____ is put into an evaporating basin and the water is _____ until the solution is very _____. The solution is then left to form crystals. The crystals are then _____ off and then _____. [4]

(d) Name another two fertilisers that can provide a plant with nitrogen in soluble form. [2]

(Paper 3)

6 The Haber process can be represented by the equation:
$N_2(g) + 3H_2(g) \rightleftharpoons 2NH_3(g)$ $\Delta H = -92$ kJ/mol
(a) State the source of nitrogen for this process. [1]
(b) State one source of hydrogen for this process. [1]
(c) Explain with reference to the equation why the yield of ammonia rises as the pressure increases. [2]
(d) Suggest why a pressure above 20 000 kPa is not used. [2]

(e) Describe how the yield of ammonia varies with temperature. Explain your answer. [3]
(f) Explain why the reaction is carried out at about 450 °C rather than at a higher or lower temperature. [2]

(Paper 4)

7 Sulfuric acid is manufactured by the Contact process. The main reaction is:
$2SO_2(g) + O_2(g) \rightleftharpoons 2SO_3(g)$ $\Delta H = -197$ kJ/mol
(a) Describe and explain how the yield of SO_3 changes when the pressure is decreased. [3]
(b) At just above atmospheric pressure the yield of the reaction is 90%. Suggest why high pressure is not used. [2]
(c) Describe how the yield of SO_3 varies with temperature. Explain your answer. [3]
(d) A catalyst of vanadium(V) oxide is used.
   (i) Explain the purpose of using a catalyst for this reaction and describe the properties of a catalyst. [3]
   (ii) What effect, if any, does vanadium(V) oxide have on the position of equilibrium when the temperature and pressure are constant? [1]
   (iii) Vanadium is a transition element. Describe two physical properties of vanadium which are different from those of a Group I element. [2]

(Paper 4)

8 Sulfuric acid, H_2SO_4, is a dibasic acid.
(a) Suggest what is meant by the term dibasic. [2]
(b) Sulfuric acid is a strong acid. Explain the meaning of the term *strong acid*. [2]
(c) Write a symbol equation for the reaction of dilute sulfuric acid with zinc. Include state symbols. [2]
(d) Write a symbol equation for the reaction of dilute sulfuric acid with calcium carbonate. [2]
(e) A sample of acid rain contains a dilute solution of sulfuric acid. Describe how this acid rain is formed when fossil fuels are burned. [4]

(Paper 4)

17.1 Air pollution

LEARNING OUTCOMES

- State the composition of clean, dry air
- State the source of the air pollutants including carbon dioxide, carbon monoxide and methane
- Describe the adverse effects of air pollutants

The gases in unpolluted air

Oxygen forms 21% of clean, dry air. Oxygen is the reactive gas in the air. It is needed for combustion and rusting, as well as for respiration.

Nitrogen forms 78% of clean, dry air. Nitrogen is an unreactive gas. Processes such as burning and rusting would be much quicker without nitrogen to dilute the oxygen.

The rest of the air – about 1% of it – is largely made up of argon. There are also tiny amounts of other gases. Carbon dioxide is the most important of these. Although only 0.04% of the air is carbon dioxide, this gas has a great effect on our climate because it is largely responsible for global warming.

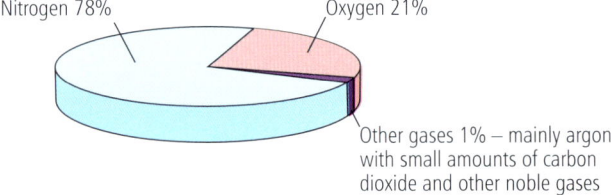

Figure 17.1.1 The gases in clean, dry air.

In addition to these gases, the **atmosphere** contains water vapour. Water vapour usually makes up between 1% and 4% of the atmosphere. Much of the air around us is polluted – it contains unwanted substances. The **pollutants** are dust particles and smoke as well as the gases sulfur dioxide and nitrogen oxides.

Air pollutants and their effects

We burn **fossil fuels** such as coal, oil and gas in power stations and in the home. We use hydrocarbons from the distillation of petroleum for transport. The gases given off when we burn fossil fuels pollute the atmosphere.

Carbon dioxide, CO_2

Fossil fuels contain hydrocarbons. These are compounds containing only carbon and hydrogen. They burn in excess air to produce carbon dioxide and water:

$$\text{hydrocarbon} + \text{oxygen} \rightarrow \text{carbon dioxide} + \text{water}$$

When carbon-containing substances burn completely in excess air we call this complete combustion.

An increased concentration of carbon dioxide in the atmosphere due to human activities causes increased global warming. This leads to climate change.

Figure 17.1.2 Some of the first living things on Earth were bacteria which may have lived in colonies on rocks in the sea. They absorbed sunlight and released oxygen into the early atmosphere of the Earth.

Figure 17.1.3 Burning fossil fuels increases the concentration of carbon dioxide in the atmosphere which increases global warming.

Unit 17: Environmental chemistry

Carbon monoxide, CO

When there is not enough oxygen to burn a carbon-containing fuel completely, carbon monoxide and water are formed. Carbon in the form of soot is also formed. Combustion in a limited amount of air is called incomplete combustion.

Carbon monoxide is a toxic (poisonous) gas. It combines with haemoglobin in the red blood cells that carry oxygen around the body. So if you get carbon monoxide poisoning, it stops respiration. Carbon monoxide is colourless and has no smell. So it is not easy to tell if it is present. It is important that gas fires and boilers are cleaned regularly so that incomplete combustion does not occur.

Particulates

Particulates are tiny particles of solids present in the air. Some come from natural sources but many are formed as a result of burning fossil fuels. An example is soot. Many particulates are so small that they can get deep into the lungs and into the blood. They increase respiratory problems and can cause cancer.

Methane, CH_4

Methane gas is formed by the decomposition of vegetation and from the waste gases formed when animals digest their food. Like carbon dioxide, methane gas increases global warming.

Oxides of nitrogen, NO_x, and sulfur dioxide, SO_2

In Topic 16.4 we learned how sulfur dioxide and oxides of nitrogen are responsible for acid rain and the effects of acid rain on buildings. Nitrogen oxides released from car exhausts also cause **photochemical smog**. Smog is a smoky fog that causes irritation in the throat and nose and can cause increased respiratory problems. The nitrogen oxides react with oxygen in the atmosphere in the presence of sunlight to form ozone, O_3. The ozone reacts with unburned hydrocarbons from vehicle exhausts to form photochemical smog. Nitrogen oxides on their own also harm the lungs and cause breathing difficulties. They also can make asthma worse.

EXAM TIP

A common error when writing about incomplete combustion is to state that the combustion occurs 'in the absence of oxygen' – but no combustion would occur without oxygen. The correct answer is 'there is limited oxygen'.

Figure 17.1.4 Smog is a mixture of SMoke and fOG formed mainly from motor vehicle pollution.

KEY POINTS

- Clean dry air contains 78% nitrogen, 21% oxygen and about 1% argon.
- Carbon dioxide and methane contribute to global warming, leading to climate change.
- Carbon monoxide is formed by incomplete combustion of carbon-containing compounds.
- The adverse effects of oxides of nitrogen include acid rain and photochemical smog.

SUMMARY QUESTIONS

1 Copy and complete using the words below:

   air   argon   climate   dioxide   nitrogen oxygen   warming

   About 78% of air is _____ and 21% is _____. The rest is mainly _____. Carbon dioxide forms 0.04% of unpolluted dry _____. Burning fossil fuels increases the concentration of carbon _____ in the atmosphere and this leads to global _____, resulting in _____ change.

2 State a source of each of these atmospheric pollutants:
   a carbon dioxide   **b** methane   **c** nitrogen oxides

3 Explain how carbon monoxide is formed and describe its adverse effect on humans.

205

17.2 Carbon dioxide in the atmosphere

LEARNING OUTCOMES

- State the sources and word equations for reactions that produce carbon dioxide
- State that photosynthesis uses carbon dioxide
- S State the symbol equations for reactions that produce carbon dioxide
- Describe photosynthesis including a symbol equation

Figure 17.2.1 The increased use of motor vehicles increases the concentration of carbon dioxide in the air.

EXAM TIP

Core candidates will not be asked to write symbol equations for the processes which put carbon dioxide into the air, but may be asked to balance the equations when given the formulae.

How does carbon dioxide get into the atmosphere?

Complete combustion of carbon-containing fuels

Most fuels produce carbon dioxide when they burn. This carbon dioxide escapes into the atmosphere.

$$CH_4(g) + 2O_2(g) \longrightarrow CO_2(g) + 2H_2O(l)$$
methane + oxygen $\longrightarrow$ carbon dioxide + water

Respiration

Respiration is the process by which living things get energy from food. It takes place in the cells of all animals, plants and microbes. We can summarise respiration as:

$$C_6H_{12}O_6 + 6O_2 \longrightarrow 6CO_2 + 6H_2O$$
glucose + oxygen $\longrightarrow$ carbon dioxide + water

Reaction between acids and carbonates

Acids react with carbonates to form a salt, water and carbon dioxide. This is the reaction which occurs naturally when acid rain reacts with carbonate rocks such as limestone (containing mainly calcium carbonate):

$$CaCO_3(s) + 2HNO_3(aq) \longrightarrow Ca(NO_3)_2(aq) + H_2O(l) + CO_2(g)$$
calcium carbonate + nitric acid $\longrightarrow$ calcium nitrate + water + carbon dioxide

Thermal decomposition of carbonates

Carbon dioxide is produced from the thermal decomposition of carbonates.

$$CaCO_3(s) \xrightarrow{heat} CaO(s) + CO_2(g)$$
calcium carbonate $\longrightarrow$ calcium oxide + carbon dioxide

How is carbon dioxide removed from the atmosphere?

Plants remove large amounts of carbon dioxide from the atmosphere by a series of reactions called **photosynthesis**.

The word equation for photosynthesis is:

carbon dioxide + water $\longrightarrow$ glucose + oxygen

Light and chlorophyll are essential for photosynthesis.

Unit 17: Environmental chemistry

Supplement

We can summarise the process of photosynthesis by the equation:

$$6CO_2 + 6H_2O \rightarrow C_6H_{12}O_6 + 6O_2$$
$$\text{carbon dioxide} + \text{water} \rightarrow \text{glucose} + \text{oxygen}$$

The glucose formed is used for plant growth and respiration. The energy for this reaction comes from sunlight. **Chlorophyll**, the green pigment in plants, traps the light and acts as a catalyst.

Changing the balance

The sea, the atmosphere, the rocks and living things all contain carbon. The amount of carbon in each of these has not changed much over millions of years. This is because there is a balance between the release and uptake of carbon. Most of the carbon in the atmosphere is present as carbon dioxide.

Look at the equations for photosynthesis and respiration. You will notice that they are the opposite of one another. Respiration uses oxygen and releases carbon dioxide into the air. Photosynthesis removes carbon dioxide from the air and releases oxygen. These two processes are roughly balanced so that the amount of carbon dioxide in the air remains fairly constant (Figure 17.2.2). In recent years the clearing of large areas of forests (deforestation) has reduced the amount of carbon dioxide removed from the air. This can make the carbon cycle unbalanced so more carbon dioxide is released than absorbed.

The increased burning of fossil fuels over the last 200 years has led to a steady increase in carbon dioxide concentration in the air. This in turn has led to increased global warming.

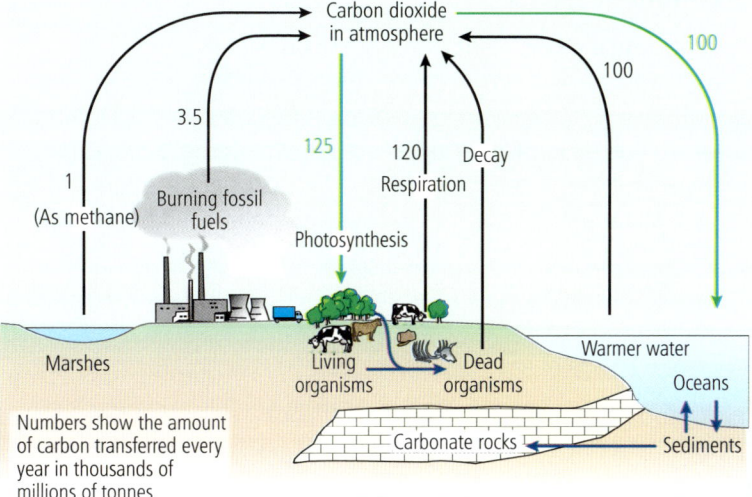

Figure 17.2.2 Sources of carbon dioxide in the atmosphere.

KEY POINTS

- Carbon dioxide is released into the air during the complete combustion of carbon-containing fuels, respiration, thermal decomposition of carbonates and the reaction of acids with carbonates.
- Photosynthesis removes carbon dioxide from the air.
- Burning fossil fuels and deforestation may unbalance the carbon cycle so that the concentration of carbon dioxide in the air increases.

SUMMARY QUESTIONS

1. Describe three processes that put carbon dioxide into the atmosphere.

2. Name the main process that removes carbon dioxide from the atmosphere.

3. Copy and complete using the words below:

   atmosphere carbon glucose oxygen photosynthesis uses water

   The main processes that keep the amount of _____ dioxide in the _____ fairly constant are respiration and _____. Respiration uses _____ from the air to oxidise _____ to carbon dioxide and _____. Photosynthesis _____ carbon dioxide from the air to make glucose and oxygen.

17.3 Global warming and climate change

LEARNING OUTCOMES

- State that carbon dioxide and methane cause global warming which leads to climate change
- **S** Describe how carbon dioxide and methane cause global warming

EXAM TIP

We often use the term 'global warming' inexactly. Remember that climate change is linked to *increased* global warming.

Figure 17.3.1 We can expect more extreme weather and floods with increased global warming.

Global warming and climate change

Carbon dioxide and methane absorb infrared radiation in the atmosphere. This leads to the atmosphere warming up. The more carbon dioxide and methane there is in the atmosphere, the more thermal energy is absorbed. This heating of the atmosphere is called **global warming**. Increased global warming is linked to **climate change**.

Climate change and its results

A warmer atmosphere can affect our climate:

- The temperature of the air increases. This leads to melting of the polar ice-caps and glaciers. This in turn makes sea levels higher and causes flooding of low-lying areas.

- There may be less rainfall in some areas. This results in droughts and the formation of more deserts, so there will be less food production.

- There may be more extreme weather with more storms and stronger winds. This causes more destruction of property and crops.

Supplement

Greenhouse gases

A **greenhouse gas** is a gas that absorbs thermal energy (heat energy) and stops thermal energy escaping into space. The main greenhouse gases are carbon dioxide (CO_2), methane (CH_4) and nitrous oxide (N_2O). These gases are present in the atmosphere in tiny quantities compared with the amounts of nitrogen and oxygen. What are the sources of these greenhouse gases?

- Carbon dioxide is the main greenhouse gas. It is naturally present in the atmosphere. But over the past century the amount of carbon dioxide has been increasing steadily and is now 0.04%. Much of this increase is due to burning fossil fuels in power stations and emissions from vehicle engines.

- Methane is in the atmosphere at a much lower concentration than carbon dioxide. But it can absorb much more thermal energy. Methane is produced by bacteria in the digestive systems of cows, pigs and sheep. The increased number of animals produced for food to feed a growing population means that more and more methane is being released into the atmosphere. The decomposition of vegetation, especially

Unit 17: Environmental chemistry

in swampy areas, also produces methane, as do rice paddy fields.

- Nitrous oxide produced in vehicle and aircraft engines is also a very effective greenhouse gas.

The greenhouse effect and global warming

The **greenhouse effect** has always been with us. Greenhouse gases prevent the Earth from cooling down too rapidly when we are not exposed to the Sun's rays. If we did not have greenhouse gases in the atmosphere, the Earth would be extremely cold.

The greenhouse effect works like this:

- Ultraviolet rays from the Sun get through the Earth's atmosphere easily and are not absorbed by carbon dioxide.
- The ultraviolet rays hit the Earth's surface.
- The Earth's surface absorbs the ultraviolet rays and it heats up.
- Thermal energy is transferred from the Earth's surface as infrared rays. Infrared rays can be absorbed by greenhouse gases.
- Some of the thermal energy is re-radiated (reflected) back to the Earth and some escapes into space.

The more greenhouse gases there are in the atmosphere, the more thermal energy is absorbed by these gases. More thermal energy is re-radiated back to Earth and less is lost into space. So the atmosphere heats up more.

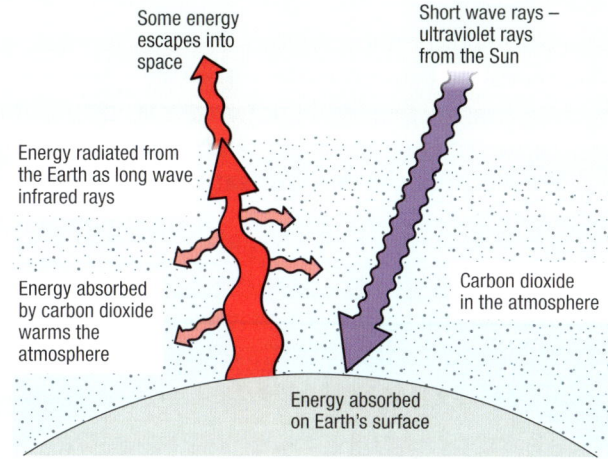

Figure 17.3.2 How global warming happens.

EXAM TIP

You do not have to know the exact details of how global warming works. But you should have an idea of (i) the absorption and reflection of thermal energy in the atmosphere, and (ii) the reduction of thermal energy transferred to space.

SUMMARY QUESTIONS

1 Methane causes global warming. State two sources of methane other than natural gas.

2 Give two effects of global warming.

3 Copy and complete using the words below:

**absorb   atmosphere
global   greenhouse
increase   methane
radiated**

Carbon dioxide and _____ are _____ gases that increase _____ warming. The greenhouse gases _____ thermal energy _____ from the Earth's surface. The more carbon dioxide there is in the _____, the greater the _____ in global warming.

KEY POINTS

- Methane and carbon dioxide are greenhouse gases that cause global warming.
- Increased global warming leads to climate change.
- Greenhouse gases cause an increase in global warming by absorbing thermal energy radiated from the Earth's surface.

17.4 Reducing environmental pollution

LEARNING OUTCOMES

- State and explain how climate change and the emission of greenhouse gases can be decreased
- State and explain how the formation and effects of acid rain can be decreased
- **S** Describe the removal of oxides of nitrogen in car engines by catalytic converters

Lowering the amount of greenhouse gases

How can we decrease the amount of greenhouse gases we make?

- We can decrease the use of fossil fuels by using sources of renewable energy such as solar power, wind power and hydroelectricity.
- We can develop more efficient fuels. Some biofuels burn efficiently. They produce more energy per gram than petrol. However, they still produce carbon dioxide when burned.
- We can plant more trees to absorb carbon dioxide by photosynthesis.
- We can use hydrogen as a fuel. Combustion of hydrogen produces water as the only product. If hydrogen is made from renewable energy sources, we are not putting carbon dioxide into the atmosphere.
- A reduction in livestock farming will decrease the amount of methane released into the atmosphere.

Reducing sulfur dioxide emissions

Coal and petroleum fractions can contain sulfur impurities. When they are burned in power stations or used to heat furnaces, sulfur dioxide is formed. This causes acid rain. The sulfur dioxide can be removed by passing it through a spray of calcium oxide, calcium carbonate or calcium hydroxide in water. This is called **flue gas desulfurisation** (Figure 17.4.1).

$$CaO(s) + SO_2(g) \longrightarrow CaSO_3(s)$$

calcium oxide + sulfur dioxide $\longrightarrow$ calcium sulfite

The calcium sulfite formed is reacted with air and water to form calcium sulfate. This can be used to make plasterboard or to provide sulfur for making sulfuric acid.

Emissions of sulfur dioxide can also be reduced by using low-sulfur fuels. Most of the sulfur in petroleum and natural gas is removed before it is treated further for use as fuel.

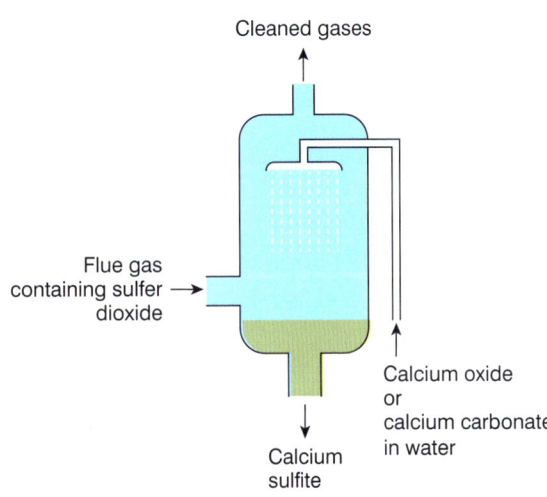

Figure 17.4.1 A flue gas desulfurisation unit for removing sulfur dioxide.

Unit 17: Environmental chemistry

Supplement

Catalytic converters

Oxides of nitrogen are formed in car engines. Nitrogen oxides contribute to both acid rain and global warming. Poisonous carbon monoxide is also formed in the engine by incomplete combustion of the fuel. A **catalytic converter** can be fitted to a car exhaust to remove these gases.

The exhaust gases from the car engine are passed through a 'honeycomb' in a catalytic converter. The surfaces on the honeycomb are covered with a thin layer of catalyst made of platinum, rhodium or palladium or a mixture of these. The gases react on the surface of the catalyst.

Most catalytic converters have two compartments. In the first compartment the metals mainly catalyse the conversion of nitrogen oxides to nitrogen:

$$2NO(g) \longrightarrow N_2(g) + O_2(g)$$
nitric oxide

$$2NO_2(g) \longrightarrow N_2(g) + 2O_2(g)$$
nitrogen dioxide

In the second compartment carbon monoxide is converted to carbon dioxide:

$$2CO(g) + O_2(g) \longrightarrow 2CO_2(g)$$

The reactions in a catalytic converter are redox reactions. The overall reactions are:

$$2NO(g) + 2CO(g) \longrightarrow N_2(g) + 2CO_2(g)$$
nitric oxide   carbon monoxide

$$2NO_2(g) + 4CO(g) \longrightarrow N_2(g) + 4CO_2(g)$$
nitrogen dioxide   carbon monoxide

The nitrogen and carbon dioxide leaving the car exhaust pipe are not poisonous. However, the carbon dioxide contributes to increased global warming.

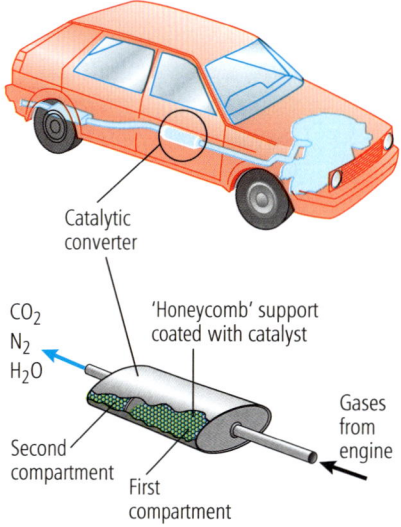

Figure 17.4.2 A catalytic converter.

EXAM TIP

The best equation to remember about reactions in a catalytic converter is the reaction between NO and CO to form N_2 and CO_2.

KEY POINTS

- The emission of greenhouse gases can be decreased by using alternative energy sources, hydrogen as a fuel and decreasing livestock farming.
- Acid rain can be reduced by using low-sulfur fuels, flue gas desulfurisation and the use of catalytic converters.
- Nitrogen oxides and carbon monoxide can be removed from car engine exhausts using catalytic converters.

SUMMARY QUESTIONS

1. State two negative effects of nitrogen oxides on the environment.
2. Describe and explain the term flue gas desulfurisation.
3. Copy and complete using the words below:

   converter exhaust monoxide nitrogen oxidised reduced

   The _____ gases from a car engine include oxides of _____, carbon _____ and unburned hydrocarbons. These gases are passed through a catalytic _____ where nitrogen oxides are _____ to nitrogen and carbon monoxide is _____ to carbon dioxide.

17.5 Clean water

LEARNING OUTCOMES

- State that water from natural sources contains a variety of substances some of which are beneficial
- Describe the purification of the domestic water supply

Water: a valuable resource

Water is all around us in the sea, rivers and lakes as well as in the rocks below the ground and in the air as water vapour. But only 3% of this water is fresh water.

Water is used in the home, in industry and in agriculture:

- In the home we use it for drinking, cooking, washing and cleaning.
- In industry it is used as a solvent for many chemicals and as a coolant to stop industrial processes from getting too hot. It is also used as a cheap raw material for some chemical manufacturing processes. Water is also used to generate electrical power, either in hydroelectric power stations or by turning it into steam to drive turbines.

The benefits of clean water

Unpolluted water from natural sources contains a variety of dissolved substances. The table shows the most common ions present in seawater and fresh water from a lake.

Figure 17.5.1 Clean water is a valuable resource but it needs to be free from bacteria if we are to remain healthy.

EXAM TIP

Be prepared to answer questions about extracting information from tables of data such as this one, even if the information is unfamiliar.

seawater		fresh water	
ion	concentration g/dm^3	ion	concentration mg/dm^3
chloride, Cl$^-$	17	chloride, Cl$^-$	15
sodium, Na$^+$	10	calcium, Ca^{2+}	8
magnesium, Mg^{2+}	9	sodium, Na$^+$	7
sulfate, SO$_4^{2-}$	8	sulfate, SO$_4^{2-}$	2

Table 17.5.1 Comparing seawater and fresh water.

Useful minerals

Minerals are solid chemical compounds which occur in nature, mostly in rocks. They are often metal compounds such as calcium carbonate and magnesium sulfate. Most minerals are slightly soluble in water and so their ions are present in low concentrations in rivers and lakes. They are present in higher concentrations in seawater. Fish and other aquatic organisms need to absorb certain ions from the water to remain healthy. Calcium ions and phosphate ions are needed for bones and cartilage. Magnesium is needed for some enzymes to work.

Dissolved oxygen

Aquatic life depends on oxygen for respiration. The small amount of dissolved oxygen in water allows aquatic life to survive. Rapidly moving water such as fast-flowing rivers and mountain streams contains more oxygen than lakes and ponds. Aquatic plants also

Figure 17.5.2 These fish depend on the dissolved oxygen in water to keep alive.

Unit 17: Environmental chemistry

put oxygen into the water by the process of photosynthesis. The amount of oxygen in water also depends on the temperature. Warm water dissolves less oxygen than cold water.

Water purification

Water from rivers or from the ground contains dirt, pieces of dead animals and plants, dissolved compounds and bacteria, many of which are harmful to health. Cholera and typhoid are just two of the many diseases caused by bacteria in untreated water. These diseases kill millions of people each year. So it is important that the water we use is treated to remove these impurities.

A modern water treatment plant removes insoluble materials by **sedimentation** and filtration, and kills bacteria by **chlorination**.

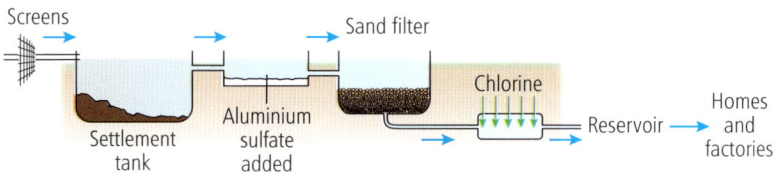

Figure 17.5.3 Treatment of water removes insoluble particles and kills bacteria. Not all the stages are shown.

EXAM TIP

You do not need to know all the details about water treatment. Sedimentation, filtration, use of carbon and chlorination are the learning objectives on the syllabus.

Treating water is a complex process and there are many stages:

- The water passes through metal screens which collect objects such as twigs and leaves.
- In the settlement tank, particles of solids such as soil particles fall to the bottom of the tank because they are denser than water. This is called sedimentation.
- Aluminium sulfate is added to make small particles in the water stick together. The particles then fall to the bottom of the tank.
- The water then undergoes filtration. It passes through a filter made of sand and gravel or crushed coal. This removes any small insoluble particles that were not removed in the other tanks.
- Carbon is used to remove bad tastes and bad odours (bad smells).
- Chlorine is added to kill bacteria (chlorination).
- The pH of the water is adjusted and the water is run off and stored or goes directly to homes and factories.

Although treated water appears colourless, it still contains dissolved salts. So it is not pure.

SUMMARY QUESTIONS

1 Copy and complete using the words below:

   bacteria   carbon   chlorine
   drinking   filter   insoluble

   We make water fit for _____ by passing the water through a sand and gravel _____. This removes any small _____ particles. We then add _____ to kill _____. We use _____ to remove bad tastes and smells.

2 State why **a** oxygen **b** dissolved minerals are essential for aquatic life.

3 Suggest why most aquatic life in lakes lives near the surface of the water.

KEY POINTS

- Unpolluted water contains oxygen and minerals that are beneficial to aquatic life.
- Water treatment removes insoluble materials by sedimentation and filtration.
- Bad tastes and odours are removed from water by treatment with carbon.
- Chlorination of water kills bacteria.

213

17.6 Water pollution

LEARNING OUTCOMES

- State that water from natural sources contains a variety of substances some of which are harmful
- Describe the harmful effects of substances found in the water (metal compounds, plastics, sewage, and nitrates and phosphates)

A range of water pollutants

We saw in Topic 17.5 that some substances dissolved in water are essential for aquatic organisms to survive. There are many other substances which pollute seas, rivers and lakes. These pollutants reduce the quality of the water and harm aquatic life. Examples include:

- metal compounds
- sewage and harmful microbes
- nitrates and phosphates
- plastics.

Toxic metal compounds in water

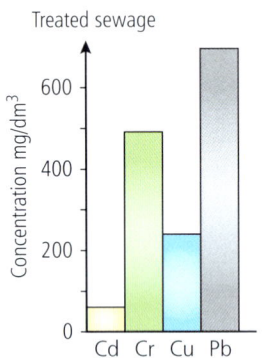
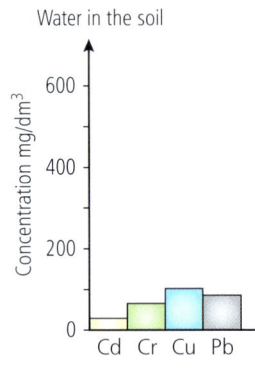

Figure 17.6.1 Treated sewage has a higher concentration of toxic metal ions than water in the soil.

Products such as old televisions, phones and car batteries can be disposed of in household waste sites. These products contain compounds of 'heavy metals' such as mercury, lead, cadmium and arsenic. These metal compounds are toxic. 'Heavy metals' are also present in the ground where old factories once stood. Ground water can dissolve these compounds and remove them from the soil. We call this **leaching**. These toxic compounds can get into rivers and lakes and then into the sea. They can poison aquatic animals. They can also get into drinking water.

The sewage problem

Sewage is a mixture of waste solids and water produced by humans. It comes from basins, showers, washing machines and dishwashers, as well as from toilets. Sewage also contains metal ions, unused medical drugs and phosphates from detergents. Sewage contains harmful **microbes** (bacteria and other microorganisms) which can cause disease. So if sewage gets into rivers, lakes and seas it can be harmful for other organisms, including us!

Sewage is treated to make it less harmful. The water from the treated sewage is sufficiently unpolluted that it is sometimes put into rivers. The solid material (sewage sludge) can be put on the soil as a fertiliser but it contains an increased concentration of 'heavy metal ions'.

Dissolved nitrates and phosphates

Nitrates and phosphates present in fertilisers dissolve in water in the soil. When it rains, excess water runs off the land into rivers and lakes taking some of the nitrates and phosphates with it.

Unit 17: Environmental chemistry

Phosphates from detergents also get into rivers from treated sewage. The presence of excess nitrates and phosphates in rivers and lakes leads to the removal of oxygen from the water (**deoxygenation**) and to the death of most aquatic life. This process is called **eutrophication**.

The sequence of events leading to eutrophication is:
- fertilisers from fields run off into rivers
- the increased concentration of nitrates and phosphates leads to a huge growth in algae (algae, like flowering plants, are a group of photosynthetic organisms)
- the algae on the surface of the water block off light to other water plants, which die
- bacteria in the water feed off the dead plants and use up the oxygen
- aquatic animals cannot respire because there is no oxygen, so they die.

Plastics in the ocean

There are millions of tonnes of waste plastic in our oceans. It comes from ships and coastal towns, and is also transported from inland by rivers. Plastic can harm aquatic life in many ways:
- unwanted plastic fishing nets can trap and even strangle larger sea creatures like seals
- aquatic creatures can eat pieces of plastic. These can harm their digestive systems and can even kill them.
- plastic can get trapped in the gills of fish and the oxygen supply to their bodies may decrease
- chemicals washed out when plastics break down can be poisonous
- tiny particles from microbeads or broken-down plastics can get into the bloodstream and our body cells where they cause harm. It is estimated that about 80% of the tap water in the world has a very small amount of tiny plastic particles in it.

Figure 17.6.2 These fish have died because the polluted water did not have enough oxygen for respiration.

EXAM TIP

You do not have to know the details of the eutrophication process for your exam. But you do need to know that nitrates and phosphates in rivers lead to deoxygenation and the death of aquatic life.

SUMMARY QUESTIONS

1 Copy and complete using the words below:

aquatic   deoxygenation   disease   metal   microbes   phosphates   toxic

Polluted water in rivers and lakes may contain _____ compounds which are _____. Sewage contains harmful _____ which can cause _____. Excess nitrates and _____ in the water lead to _____ of the water and death of _____ life.

2 Give the meaning of these terms:
   a sewage   b deoxygenation   c toxic

3 Describe one harmful effect of plastics in the oceans.

KEY POINTS

- Water from natural sources contains some pollutants such as metal compounds, plastics, sewage, harmful bacteria, nitrates and phosphates.
- Some metal compounds are toxic; sewage contains harmful microbes.
- Excess nitrates and phosphates in the water lead to deoxygenation and the death of aquatic life.

SUMMARY QUESTIONS

1 Link the pollutants on the left to their effects on the right.

particulates	photochemical smog and acid rain
nitrogen oxides	stops respiration
methane	cancer
carbon monoxide	increased global warming

2 Explain how carbon monoxide is formed.

3 Copy and complete using the words below.

21%   78%   climate   dioxide   dry global   increased   noble

Pure _____ air contains _____ nitrogen and _____ oxygen. The rest of the air contains _____ gases such as argon. There is a small amount of carbon _____ the air. Over the past 200 years, the concentration of this gas has _____. This has led to _____ warming and _____ change.

4 Link the pollutants on the left with their possible sources on the right.

CH_4	incomplete combustion of C_2H_6
CO	cows and sheep
SO_2	lightning
NO	complete combustion of C_2H_6
CO_2	burning fossil fuel containing sulfur compounds

5 State three distinct sources of carbon dioxide in the atmosphere.

6 Catalytic converters remove nitrogen oxides from the exhaust gases in car engines.
   a Name two oxides of nitrogen.
   b Explain how nitrogen oxides are formed in car engines.
   c Explain how catalytic converters remove oxides of nitrogen.

7 Copy and complete using words from the list:

carbon   energy   global   greenhouse infrared   change   methane   temperature

Thermal energy is lost from the Earth's surface as _____ rays. The _____ from the infrared rays can be absorbed by _____ gases such as _____ and _____ dioxide. The _____ of the atmosphere increases. This is called _____ warming which leads to climate _____.

Practice questions

1 Which one of these statements about carbon monoxide is correct?
   A It is responsible for acid rain.
   B It is a better greenhouse gas than carbon dioxide.
   C It is produced when cattle digest grass.
   D It is formed by the incomplete combustion of methane.

(Paper 1)

2 Which one of these statements about the carbon cycle is correct?
   A Burning fossil fuels is responsible for most of the carbon dioxide in the atmosphere.
   B Methane gets into the atmosphere by burning fossil fuels.
   C Removing many of the forests on Earth will decrease the amount of carbon dioxide in the atmosphere.
   D The amount of carbon dioxide put into the atmosphere by respiration is roughly the same as carbon dioxide removed by photosynthesis.

(Paper 2)

3 Clean air is about four-fifths nitrogen.
   (a) (i) State the names of three other gases present in unpolluted air. [3]
       (ii) For one of these gases, state the name of the gas and the approximate percentage of that gas in air. [1]
   (b) Carbon monoxide is an atmospheric pollutant.
       (i) Suggest a source of carbon monoxide and explain how it is formed. [2]
       (ii) State an adverse effect of carbon monoxide on health. [1]
   (c) Methane is also an atmospheric pollutant.
       (i)   State two sources of methane. [2]
       (ii) Explain how methane may contribute to climate change. [2]
       (iii) State one effect of climate change. [1]

(Paper 3)

4 Carbon dioxide and oxygen are gases in the atmosphere.

(a) Which one of these values gives the approximate percentage of carbon dioxide in clean dry air?
0.0004%   0.04%   4%   40% [1]

(b) Carbon dioxide is produced during respiration. Write a word equation for respiration. [3]

(c) (i) Give two other sources of carbon dioxide in the atmosphere. [2]

(ii) Give the name of the main process which removes carbon dioxide from the atmosphere. [1]

(d) Oxygen dissolves in water. Explain why dissolved oxygen is important for aquatic life. [1]

(e) Name one other group of substances in water which is essential for aquatic life. [1]

(f) Describe and explain one adverse effect when sewage gets into river water or seawater. [2]

(g) Explain the importance of (i) carbon and (ii) chlorine in water purification. [2]

(Paper 3)

5 The release of carbon dioxide into the atmosphere is balanced by its removal by photosynthesis.

(a) Describe the process of photosynthesis. [5]

(b) Write a symbol equation for the overall reaction in photosynthesis. The formula of one of the products is $C_6H_{12}O_6$. [2]

(c) Describe one process, other than respiration, that adds carbon dioxide to the atmosphere. [1]

(d) Complete the equation for respiration.
$C_6H_{12}O_6$ + _____ $O_2 \rightarrow 6CO_2$ + _____ [3]

(e) Carbon dioxide is a greenhouse gas.
(i) State the meaning of the term greenhouse gas. [2]

(ii) Explain why the concentration of carbon dioxide in the atmosphere has increased over the past 100 years. [1]

(Paper 4)

6 Oxides of nitrogen are atmospheric pollutants.

(a) Write the symbol equation for the formation of nitric oxide, NO, from nitrogen and oxygen. [2]

(b) Explain how the poisonous exhaust gases produced in a car engine can be removed using a catalytic converter. [3]

(c) Unburned hydrocarbons from car engines can also react with nitrogen oxides:
$4C_6H_{14} + 38NO_2 \rightarrow 24CO_2 + 28H_2O + 19N_2$
Identify the reducing agent and the oxidising agent in this reaction. Give reasons for your answers. [3]

(d) Nitrogen dioxide and sulfur dioxide contribute to acid rain. Describe and explain how sulfur dioxide can be removed from waste gases in power stations. Include a relevant equation in your answer. [4]

(e) Describe the formation of photochemical smog. [3]

(Paper 4)

7 Carbon monoxide is formed by the incomplete combustion of methane gas, CH_4.

(a) (i) State the meaning of the term *incomplete combustion*. [1]

(ii) Write a symbol equation for the incomplete combustion of methane. Include state symbols. [3]

(b) Methane is a greenhouse gas. State the meaning of the term greenhouse gas. [2]

(c) The concentration of methane in the atmosphere is gradually increasing. Explain why it is increasing. [1]

(e) Methane is a fossil fuel. Describe and explain two strategies to reduce the use of fossil fuels. [4]

(Paper 4)

18.1 A variety of organic compounds

LEARNING OUTCOMES

- Identify the type of compound present from a chemical name ending, e.g. -ene, -ol
- Define the term 'homologous series'
- **S** Describe the general characteristics of a homologous series

Figure 18.1.1 Some organic chemicals like sugar char (go black) when heated.

EXAM TIP

Remember that the chemical properties of the compounds in a homologous series are similar. It is a common error to suggest that they are 'the same'.

Organic compounds

About 200 years ago the Swedish chemist Jöns Jakob Berzelius divided chemicals into two main groups:

- organic chemicals which either burn or char (go black) when heated
- inorganic chemicals which vaporise without burning when heated.

We now know that there are some exceptions to these rules.

All organic compounds contain carbon. They usually contain hydrogen and may contain other elements as well. Millions of organic compounds are known so we have to make rules for naming them. Fortunately for us, many organic compounds can be put into groups. A group of organic molecules with similar chemical properties is called a **homologous series**.

- A homologous series is a family of similar compounds with similar chemical properties due to the presence of the same **functional group**.
- A functional group is an atom or group of atoms that gives a compound particular chemical properties, e.g. —OH or —COOH.

Two homologous series are **alcohols** and **carboxylic acids**. Here are the names and formulae of some compounds in these two homologous series:

alcohol homologous series

methanol	CH_3OH
ethanol	C_2H_5OH
propanol	C_3H_7OH

carboxylic acid homologous series

methanoic acid	$HCOOH$
ethanoic acid	CH_3COOH
propanoic acid	C_2H_5COOH

You can see that all the alcohols have an —OH functional group and that all the carboxylic acids have a —COOH functional group. Carboxylic acids have different reactions to alcohols, but all carboxylic acids have very similar chemical properties to each other.

A range of functional groups

We can tell which homologous series a compound belongs to by the ending of its name. For example, the members of the alcohol homologous series all end in – ol. The members of the **alkene** homologous series all end in – ene. Table 18.1.1 gives a list of some of these endings. The functional group is usually attached to a carbon atom, but in some simple compounds it is attached to a hydrogen atom, e.g. —COOH in methanoic acid, HCOOH.

Unit 18: Organic chemistry and petrochemicals

homologous series	name ending	functional group	example
alkane	-ane	no functional group	ethane, H_3C-CH_3
alkene	-ene	$\diagdown C=C \diagup$	ethene, $H_2C=CH_2$
alcohol	-ol	$-O-H$	ethanol, CH_3CH_2-O-H
carboxylic acid	-oic acid	$-C\diagup^O_{O-H}$	ethanoic acid, $CH_3-C\diagup^O_{O-H}$, CH_3-COOH

Table 18.1.1 Summary of four common homologous series.

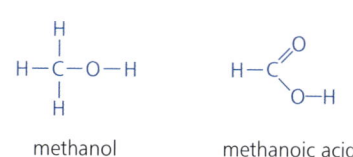

Figure 18.1.2 The smallest alcohol molecule (methanol) and the smallest carboxylic acid molecule (methanoic acid)

More about homologous series

- Each homologous series has a **general formula** which applies to all its members. For example, **alkanes** have the general formula C_nH_{2n+2}, where n is the number of carbon atoms. The alkane with five carbon atoms is called pentane. Its formula is $C_5H_{(2\times5)+2}$, which is C_5H_{12}. All members of the alkene homologous series have the general formula C_nH_{2n}.
- The members of a homologous series have very similar chemical properties because they all have the same functional group.

Supplement

As well as having the same functional group and the same general formula, a particular homologous series has these characteristics.

- As the number of carbon atoms in a homologous series increases by one, the number of hydrogen atoms increases by two. For example: CH_3OH, C_2H_5OH, C_3H_7OH. Each differs from the next by a CH_2 group.
- The physical properties in a homologous series change in a regular way as the number of carbon atoms increases. For example, the boiling points of the alkanes.

KEY POINTS

- A homologous series is a family of similar compounds with similar chemical properties due to the presence of the same functional group.
- A functional group is an atom or group of atoms that gives a compound particular chemical properties.
- **S** A homologous series has general characteristics such as the same general formula, trends in physical properties and a difference of $-CH_2-$ between successive members.

SUMMARY QUESTIONS

1. Copy and complete using these words:

   atom   chemical   compound   ethane   functional   homologous

   Methane and _____ belong to the same _____ series. They have the same _____ group. A functional group is an _____ or group of atoms that gives a _____ its particular _____ properties.

2. Name the homologous series and draw the functional group of:

   a propanol
   b butene
   c pentanoic acid

3. **S** State three characteristics of a homologous series other than 'same functional group'.

18.2 Formulae of organic compounds

LEARNING OUTCOMES

- Name and draw the displayed formulae of methane, ethane, ethene, ethanol and ethanoic acid
- Draw and interpret displayed formulae and molecular formulae

Different types of formula

The displayed formula shows all of the atoms and all of the bonds in a molecule. Look at the displayed formulae of methane, ethane, ethene, ethanol and ethanoic acid:

$$
\underset{\text{methane}}{\begin{array}{c}H\\|\\H-C-H\\|\\H\end{array}} \quad \underset{\text{ethane}}{\begin{array}{c}H\ \ H\\|\ \ \ |\\H-C-C-H\\|\ \ \ |\\H\ \ H\end{array}} \quad \underset{\text{ethene}}{\begin{array}{c}H\ \ \ H\\|\ \ \ \ |\\C=C\\|\ \ \ \ |\\H\ \ \ H\end{array}} \quad \underset{\text{ethanol}}{\begin{array}{c}H\ \ H\\|\ \ |\\H-C-C-O-H\\|\ \ |\\H\ \ H\end{array}} \quad \underset{\text{ethanoic acid}}{\begin{array}{c}H\ \ O\\|\ \ \|\\H-C-C-O-H\\|\\H\end{array}}
$$

You can see that each carbon atom forms four bonds, each oxygen atom forms two bonds and each hydrogen atom forms one bond with other atoms.

The molecular formula is the formula showing the number and type of different atoms in one molecule. The molecular formulae of methane, ethane, ethene, ethanol and ethanoic acid are:

$$\underset{\text{methane}}{CH_4} \quad \underset{\text{ethane}}{C_2H_6} \quad \underset{\text{ethene}}{C_2H_4} \quad \underset{\text{ethanol}}{C_2H_6O} \quad \underset{\text{ethanoic acid}}{C_2H_4O_2}$$

Hydrocarbons

A hydrocarbon is a compound that contains only carbon and hydrogen atoms.

The alkanes and the alkenes are two important homologous series of hydrocarbons. Alkanes have only single covalent bonds. Alkenes have one or more double bonds between their carbon atoms.

EXAM TIP

When drawing <u>displayed</u> formulae for alcohols and carboxylic acids, don't forget the bonds in the functional groups. It is a common error to write —OH instead of —O—H.

EXAM TIP

You may see formulae such as C_2H_5OH and CH_3COOH showing the functional groups. Remember that these are <u>not</u> molecular formulae because the H and the O have been written more than once.

Figure 18.2.1 Natural rubber is a very useful hydrocarbon that comes from the sap of certain trees.

Naming alkanes

What do all compounds with names starting with meth- have in common? The answer is that they have only one carbon atom. A compound beginning with eth- has two carbon atoms in its chain. A compound with three carbon atoms has a name beginning with prop-. The prefix (beginning part of the word) tells us how many carbon atoms there are in the chain.

Unit 18: Organic chemistry and petrochemicals

The names of the first six alkanes are shown in the following table.

prefix	number of carbon atoms	name and molecular formula	full structural formula
meth-	1	methane, CH_4	$H-\underset{\underset{H}{\vert}}{\overset{\overset{H}{\vert}}{C}}-H$
eth-	2	ethane, C_2H_6	$H-\underset{\underset{H}{\vert}}{\overset{\overset{H}{\vert}}{C}}-\underset{\underset{H}{\vert}}{\overset{\overset{H}{\vert}}{C}}-H$
prop-	3	propane, C_3H_8	$H-\underset{\underset{H}{\vert}}{\overset{\overset{H}{\vert}}{C}}-\underset{\underset{H}{\vert}}{\overset{\overset{H}{\vert}}{C}}-\underset{\underset{H}{\vert}}{\overset{\overset{H}{\vert}}{C}}-H$
but-	4	butane, C_4H_{10}	$H-\underset{\underset{H}{\vert}}{\overset{\overset{H}{\vert}}{C}}-\underset{\underset{H}{\vert}}{\overset{\overset{H}{\vert}}{C}}-\underset{\underset{H}{\vert}}{\overset{\overset{H}{\vert}}{C}}-\underset{\underset{H}{\vert}}{\overset{\overset{H}{\vert}}{C}}-H$
pent-	5	pentane, C_5H_{12}	(displayed formula of pentane)
hex-	6	hexane, C_6H_{14}	(displayed formula of hexane)

Table 18.2.1 The first six members of the homologous series of alkanes.

Alkyl groups

When we remove a hydrogen atom from an alkane chain, we have a group called an alkyl group. So the alkyl group from ethane, C_2H_6, is C_2H_5-. The alkyl group from butane, C_4H_{10}, is C_4H_9-. The general formula for an alkyl group is C_nH_{2n+1}. Alkyl groups are named after the hydrocarbons by changing the -ane ending of the hydrocarbon to –yl. So we call C_2H_5- an ethyl group and C_4H_9- a butyl group.

EXAM TIP
When defining a hydrocarbon the word 'only' is very important. The answer 'a compound containing carbon and hydrogen' is not sufficient.

EXAM TIP
You need to know the names and structures of first four alkanes (or the first two if you are doing the Core paper). But it is good know the other names because you are certain to come across them in questions where you have to handle data.

SUMMARY QUESTIONS

1 Copy and complete using these words:

   butane   chain   members   number   pent-   prefixes
   prop-   three

   The different _____ of a homologous series can be identified by the _____ meth-, eth-, _____, but-, _____ and so on. These prefixes show the _____ of carbon atoms in the main _____ of the compound. For example, _____ has four carbon atoms in its carbon chain and propane has _____.

2 Name:
   a the straight-chained alkane with four carbon atoms
   b the alkene with three carbon atoms

3 Draw the displayed formulae for **a** ethanol **b** ethanoic acid.

KEY POINTS
- A hydrocarbon is a compound containing carbon and hydrogen only.
- The prefixes meth-, eth-, prop-, amongst others, tell us the number of carbon atoms in the main chain of an organic compound.
- The displayed formula shows all of the atoms and all of the bonds in a molecule.

221

18.3 Structural formulae and isomerism

Supplement

LEARNING OUTCOMES

- Draw structural formulae for organic compounds
- Define structural isomerism
- Name and draw structural and displayed formulae of alkanes and alkenes containing up to four carbon atoms per molecule

Structural formulae

A **structural formula** is an unambiguous (clearly defined) description of the way the atoms in a molecule are arranged. It shows us each carbon atom in the molecule as well as the hydrogen atoms and functional groups which are bonded to each carbon atom in the molecule. No bonds are shown except for double (and triple) bonds. Look at the displayed formulae and structural formulae below to see the relationship between them.

Displayed formula:

H H H H   H H   H   H H H
 | | | | | | | | | |
H—C—C—C—C—H   H—C—C=C—C—H   H—C—C—C—O—H
 | | | | | | | | | |
H H H H   H H   H   H H H

Structural formula: $CH_3CH_2CH_2CH_3$ $CH_3CH=CHCH_3$ $CH_3CH_2CH_2OH$

butane   but-2-ene   propanol

Structural isomers

The carbon chain in alkanes and other organic compounds can be branched.

EXAM TIP

When writing the structural formula of branched-chain isomers, the side-chain goes in brackets. For 2-methylpropane the structural formula is $CH_3CH(CH_3)CH_3$.

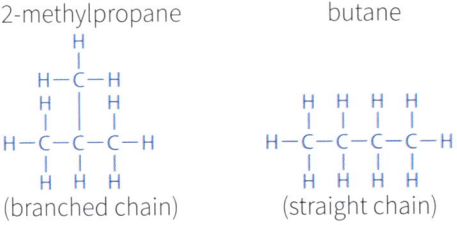

2-methylpropane (branched chain)   butane (straight chain)

2-Methylpropane has four carbon atoms and has the same molecular formula as butane, C_4H_{10}. But it is not butane because the carbon atoms are arranged differently. Compounds with the same molecular formula but with a different structural formula are called **structural isomers**. We say that the **isomer** of butane with the CH_3– group sticking out has a branched chain. Structural isomers may have similar chemical properties but they have different physical properties. The boiling point of straight-chain butane is 0 °C but the branched-chain isomer has a boiling point of −12 °C.

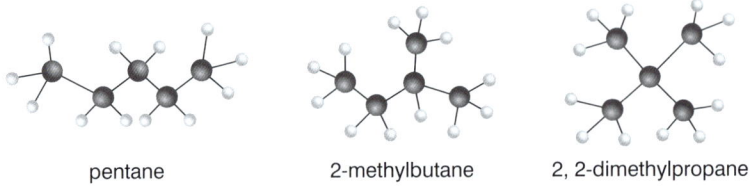

pentane   2-methylbutane   2, 2-dimethylpropane

● Carbon atom   ○ Hydrogen atom

Figure 18.3.1 There are 3 structural isomers of the alkane with 5 carbon atoms.

Unit 18: Organic chemistry and petrochemicals

Rules for naming branched-chain alkanes

We will use the compound opposite as an example:

- Find the longest carbon chain and name the compound after the number of carbon atoms in the longest chain. There are four carbon atoms in the longest chain. So it is named after butane.

- Look for the alkyl side-chain. In this case it is a methyl group. So the compound is methylbutane.

- Number the alkyl group side-chain by counting the numbers of the carbon atoms from one end of the carbon chain. You count from the end of the carbon chain that gives you the lowest number. In this case counting from the left, the alkyl group is on the second carbon atom. So the compound is 2-methylbutane.

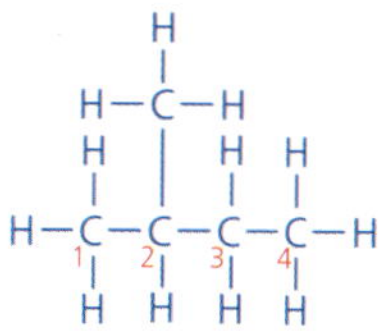

Structural isomerism in alkenes

All members of the alkene homologous series with four or more carbon atoms in a chain can form isomers where the position of the double bond changes. This type of structural isomer is called a position isomer.

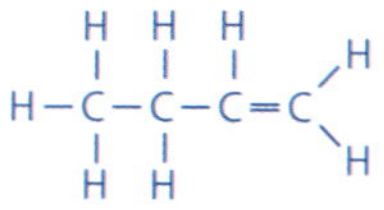

but-1-ene 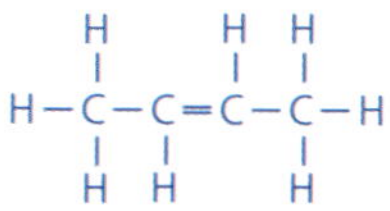   but-2-ene

> **EXAM TIP**
>
> Make sure that you do not draw the same isomer twice. These two compounds are the same.
>
> $CH_3-CH_2-CH-CH_3$
> $\quad\quad\quad\quad\quad |$
> $\quad\quad\quad\quad\quad CH_3$
>
> $CH_3-CH-CH_2-CH_3$
> $\quad\quad |$
> $\quad\quad CH_3$

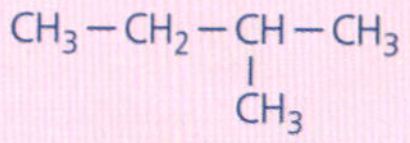

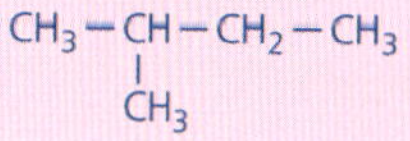

SUMMARY QUESTIONS

1. Write the structural formulae for:
   a. propene
   b. ethanoic acid
   c. the straight-chain alkane with six carbon atoms

2. Name these compounds:
   a. $CH_3CH(CH_3)CH_3$
   b. $CH_3CH=CH_2$

3. Draw the structural formula for the two isomers of butane.

KEY POINTS

- A structural formula is an unambiguous description of the way the atoms in a molecule are arranged.
- Structural isomers are compounds with the same molecular formula but different structural formulae.

18.4 Fuels

LEARNING OUTCOMES

- Name the fossil fuels coal, natural gas and petroleum
- Name methane as the main constituent of natural gas
- Describe the separation of petroleum into fractions containing various hydrocarbons
- Describe the physical properties of the fractions, e.g. relative volatility

The fossil fuels coal, petroleum (crude oil) and natural gas all contain hydrocarbons. We cannot use petroleum (crude oil) as a fuel because it is a sticky black liquid that is difficult to set alight. When it does burn, it produces clouds of poisonous black smoke. Petroleum is a mixture of many types of hydrocarbons with different carbon chain lengths. Some of the chains are branched and there may even be compounds containing rings of carbon atoms.

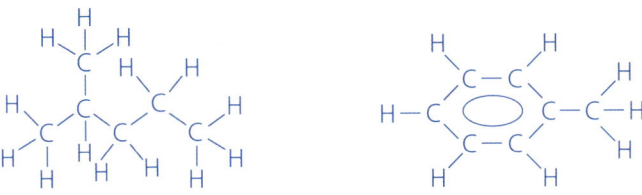

Figure 18.4.1 There are a variety of hydrocarbons in petroleum.

Fractional distillation is used to separate the hydrocarbon molecules in petroleum into groups that have similar boiling points. These groups of molecules are called **fractions**. Each fraction contains hydrocarbons with a certain range of carbon atoms. Many of these fractions are used as fuels:

fraction	number of carbon atoms	type of fuel
refinery gas	1–4	methane, propane, butane for gas cylinders
gasoline (petrol)	5–10	petrol for cars
kerosene (paraffin)	10–16	for aircraft
diesel oil (gas oil)	16–20	diesel for cars and larger vehicles
fuel oil	20–30	for ships and home heating

Figure 18.4.2 Crude oil from porous rocks underground is a mixture containing hydrocarbons and other compounds.

EXAM TIP

Don't get confused between petroleum and petrol. Petroleum is crude oil. Petrol is gasoline, a fraction obtained when we distil petroleum.

The harmful effects of burning fossil fuels are discussed in Topics 17.1, 17.2 and 17.3.

Each fraction has particular properties. The table below shows some of these properties. The properties show a trend from refinery gas (which comes off at the top of the distillation column) to fuel oil (which comes off near the bottom of the column).

fraction	number of C atoms	state	boiling point	viscosity	flammability
refinery gas	1–4	gas	increases ↓	increases ↓	decreases ↓
gasoline	5–10	liquid			
kerosene	10–16	liquid			
diesel oil	16–20	liquid			
fuel oil	20–30	liquid			

The hydrocarbons within each fraction show trends in boiling point, **viscosity** and flammability. Viscosity tells us about the ease of liquid flow. Kerosene flows easily. It has a low viscosity. Fuel oil

Unit 18: Organic chemistry and petrochemicals

flows more slowly. It has a high viscosity. The volatility decreases as the boiling point increases from refinery gas to fuel oil. Volatility is the ease with which a liquid forms a vapour.

Burning hydrocarbon fuels

Hydrocarbons contain carbon and hydrogen. So when they are burned in excess oxygen or excess air, carbon dioxide and water are formed.

PRACTICAL

What is formed when fuels burn?

Figure 18.4.3 CO_2 and H_2O are produced when fuels burn.

We test the products formed when a fuel burns using this apparatus. We burn the fuel under the funnel. The gases produced are sucked through the apparatus by a pump. Water collects in the U-shaped tube. You can test that this is water using white anhydrous copper(II) sulfate which turns blue if water is present. The limewater turns milky showing that carbon dioxide gas is also produced.

What makes a good fuel?

There are several things we take into account when we choose a fuel for a particular job.

- Is it polluting? Coal is much more polluting than other hydrocarbon fuels. But all these fuels produce the greenhouse gas carbon dioxide when burned.
- How much thermal energy does it release per gram burned?
- Is it easy to use?
- Is it readily available, cheap and easy to transport?

EXAM TIP

Candidates taking the Supplement papers will be expected to construct chemical equations for the complete combustion of a given hydrocarbon. Core candidates will be expected to complete a given equation.

KEY POINTS

- Coal, natural gas (methane) and petroleum are fossil fuels.
- The fractions produced by the fractional distillation of petroleum provide us with a variety of fuels.
- There are trends in the physical properties of the fractions from the top to the bottom of the distillation column.
- The products of the complete combustion of a hydrocarbon fuel are carbon dioxide and water.

SUMMARY QUESTIONS

1 Copy and complete using these words:

excess   fractional   petroleum   viscous   volatile   water

Many of the fractions produced by the _____ distillation of _____ are useful fuels. Fractions with shorter chain hydrocarbons are more _____ and less _____ than fractions with longer chains. When you burn a hydrocarbon in _____ air, carbon dioxide and _____ are formed.

2 Write word equations for the complete combustion in oxygen of:

   a methane   b hydrogen

3 Complete this symbol equation: ____C_2H_6 + ____O_2 → ____CO_2 + $6H_2O$

18.5 Petroleum fractionation

LEARNING OUTCOMES

- Describe the fractional distillation of petroleum
- Describe the difference in hydrocarbon chain length in the fractions from the top to the bottom of the distillation column
- Name the uses of petroleum fractions

Fractional distillation of petroleum

In an oil refinery the mixture of hydrocarbons in petroleum is separated into smaller groups. Each of these groups with a limited range of carbon atoms is called a fraction. For example, the gasoline fraction contains hydrocarbons with about five to ten carbon atoms.

The hydrocarbon fractions are separated by fractional distillation. We sometimes call this fractionation. Fractional distillation separates the hydrocarbons using the difference in their boiling points. Hydrocarbons with longer chains have higher boiling points than hydrocarbons with shorter chains.

The petroleum is first heated so that all the hydrocarbons are present as gases. The petroleum is then fed into a tall tower called a fractionating column. The column is kept hot at the bottom (about 350 °C) but it is cooler at the top. So there is a range of temperatures in the column.

Near the bottom of the column those hydrocarbons with higher boiling points and longer hydrocarbon chains condense. Hydrocarbons with lower boiling points are still gases. These move further up the column. As they move up the column, each hydrocarbon condenses at the point where the temperature in the column falls just below the boiling point of the hydrocarbon.

Hydrocarbons with similar boiling points are collected as fractions. Some of the hydrocarbons do not condense. They come off as gases at the top of the column. These are the refinery gases such as methane, ethane, propane and butane. In many oil refineries these are removed from the petroleum before fractionation.

The useful fractions

Fractional distillation separates petroleum into different fractions with a range of boiling points. Each fraction has a particular use.

Figure 18.5.1 Petroleum undergoes fractional distillation in an oil refinery.

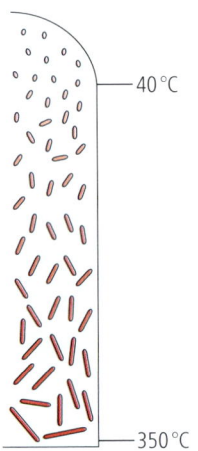

Figure 18.5.2 Shorter-chain hydrocarbons with lower boiling points move further up the fractionating column.

Unit 18: Organic chemistry and petrochemicals

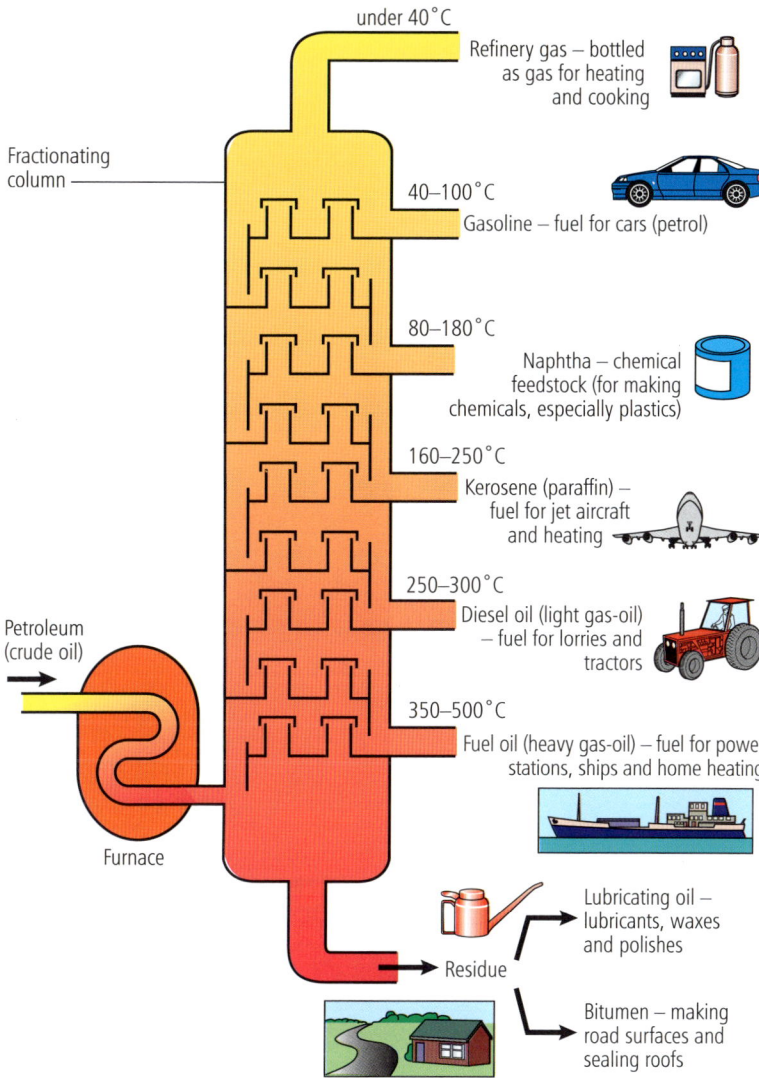

Figure 18.5.3 The fractions from petroleum distillation and their uses.

EXAM TIP

It is a common error to suggest that the fractions come off one by one at the top of the column. A better answer is to state that the fractions come off at different heights in the column.

EXAM TIP

You do not have to remember the boiling range of each fraction. But you do have to know the uses of each fraction and where they condense in the fractionating column.

SUMMARY QUESTIONS

1 Copy and complete using these words:

boiling   condense   fractions   higher   hydrocarbons   tower

Petroleum is separated into different _____ in an oil refinery. Each fraction is a group of _____ with similar _____ points. The hydrocarbon molecules move up the fractionating _____. Hydrocarbons with _____ boiling points _____ lower in the tower.

2 Describe the sequence of the main stages in the fractional distillation of petroleum.

3 Give one use for each of these fractions:

   a fuel oil   b kerosene   c naphtha

KEY POINTS

- Petroleum is separated into different fractions by fractional distillation.
- Each fraction has a specific range of boiling points.
- Fractions with higher boiling points have longer hydrocarbon chains.
- Each fraction obtained from petroleum has a particular use.

227

SUMMARY QUESTIONS

1. Match the petroleum fractions on the left with their use on the right.

naphtha	fuel for diesel engines
bitumen	jet fuel
diesel	surfacing roads
kerosene	making chemicals

2. Copy and complete using these words:

   **alcohols   alkanes   alkenes   ethane
   functional   homologous**

   Methane and _____ belong to the _____ series called the _____. A homologous series is a family of similar compounds with the same _____ group. For example, _____ always have the —OH functional group and _____ have the C=C functional group.

3. Put the following fractions in order of increasing boiling point:

   bitumen; fuel oil; kerosene; naphtha; refinery gas.

4. Match the words on the left with the descriptions on the right.

methane	a group of molecules with a similar range of boiling points
coal	a thick liquid mixture of hydrocarbons
petroleum	a solid fuel that often contains sulfur
fraction	the main constituent of natural gas
hydrogen	a gaseous fuel that forms only water when it burns

5. Give the formula of the functional group present in:

   a alcohols
   b alkenes
   c carboxylic acids

6. a State the meaning of the term structural isomer.

   b (i) Draw isomers of an alkane having four carbon atoms. (ii) Draw position isomers of an alkene with four carbon atoms and one C=C bond.

7. State three general characteristics of a homologous series.

Practice questions

1. Which one of these molecules is an alkene?

   A CH_3COOH
   B $CH_3CH=CH_2$
   C $CH_3CH_2CH_3$
   D $CH_3CH_2CH_2OH$

   (Paper 1)

2. Which one of these statements is a correct description of a homologous series?

   A There is trend in chemical properties as the number of carbon atoms increases.
   B As the number of carbon atoms increases by one, the number of hydrogen atoms also increases by one.
   C Each member of the same homologous series has the same molecular formula.
   D The physical properties change in a regular manner as the number of carbon atoms increases.

   (Paper 2)

3. Petroleum is a mixture of hydrocarbons which are separated into different fractions by fractional distillation.

   (a) State the meaning of the terms:
   (i) petroleum fraction [2]
   (ii) hydrocarbon. [2]

   (b) Explain how fractional distillation separates hydrocarbons into different fractions. [3]

   (c) Kerosene is a fraction obtained from the distillation of petroleum.
   (i) State one use of kerosene. [1]
   (ii) Name two other petroleum fractions. For each of these fractions give one use. [4]

   (d) Copy and complete the following sentences about petroleum fractionation using words from the list. (Not all words are used.)

   **condense   evaporate   fractions   mass
   higher   longer   lower   shorter**

   Hydrocarbon _____ higher in the distillation column have _____ hydrocarbon chains and _____ relative molecular _____ than hydrocarbons lower in the column. The fractions with higher boiling points _____ lower in the column. [5]

(Paper 3)

4 When fuels burn they release energy.
   (a) State the name given to a chemical reaction that releases thermal energy. [1]
   (b) State the name of:
      (i) a gaseous fuel other than hydrogen [1]
      (ii) a liquid fuel [1]
      (iii) a solid fuel. [1]
   (c) Many fuels are alkanes.
      (i) Define the term alkane. [2]
      (ii) Write the molecular formula for ethane. [1]
      (iii) Draw the displayed formula for methane. [1]
   (d) The alkanes are a homologous series of compounds. Define the term homologous series. [2]

(Paper 3)

5 The diagram shows four organic compounds.

[Structural formulas of compounds A, B, C, D]

   (a) State which two of these compounds are hydrocarbons. [1]
   (b) State which compound is an alkene. [1]
   (c) Name the homologous series which compound D belongs to. [1]
   (d) Name compound B. [1]
   (e) Write the molecular formula for compound B. [1]
   (f) Write the displayed formula for another member of the same homologous series as compound A. [1]
   (g) Write the formula for the functional groups present in (i) compound B [1]
      (ii) compound D. [1]

(Paper 3)

6 The diagram shows a column for the separation of petroleum fractions.

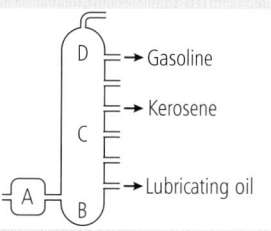

   (a) Give the letter where the temperature is lowest. [1]
   (b) Give the letter where the petroleum turns to vapour. [1]
   (c) Which fraction labelled in the diagram has the lowest boiling point? Explain why. [2]
   (d) State the names of two other fractions that are not shown on the diagram. For each of these fractions state (i) where they condense in the column, and (ii) a use of the fraction. [6]
   (e) Describe how (i) the volatility and (ii) the viscosity of the fractions changes with the position of the fractions in the column. [2]

(Paper 3)

7 The alkanes and the alkenes are both homologous series.
   (a) Give the meaning of the term homologous series. [2]
   (b) Write the general formula for the alkene homologous series. [1]
   (c) Write the molecular formula for the fifth member of the alkane homologous series. [2]
   (d) (i) Draw the displayed formula of 2-methylpropane. [1]
      (ii) Write the structural formula of 2-methylpropane. [1]
   (e) (i) Define the term isomers. [1]
      (ii) Draw the three isomers of butene. [3]

(Paper 4)

19.1 Alkanes

LEARNING OUTCOMES
- State that alkanes are saturated compounds
- Describe alkanes as being generally unreactive except for combustion and substitution by chlorine
- S Describe the substitution reaction of alkanes with chlorine

Bonding in alkanes

Alkanes are saturated hydrocarbons because they have only single covalent C–C and C–H bonds in their structures. We call them saturated because no more atoms can be added to their molecules.

Physical properties of the alkanes

Alkanes are colourless gases, liquids or solids. The first four members of the alkane homologous series are gases at room temperature and pressure. Alkanes with 5 to 17 carbon atoms in their chains are liquids. Alkanes with more than 17 carbon atoms in their chains are solids. The boiling points of the alkanes vary in a regular pattern.

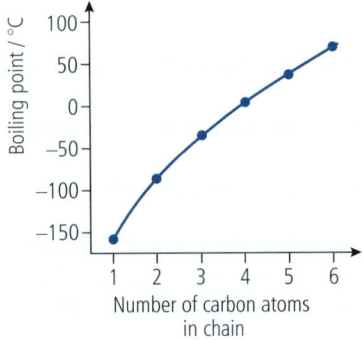

Figure 19.1.2 The boiling points of the alkanes vary in a regular pattern.

You can see that as the carbon chain gets longer, the boiling points of the alkanes increase. You can predict the boiling point of other alkanes by following this trend. For example, using the graph, you might predict the boiling point of heptane (which has 7 carbon atoms in its chain) to be about 96–99 °C. The actual boiling point is 98 °C.

Figure 19.1.1 Candles are generally made from alkanes.

EXAM TIP

You should be prepared to predict values of melting points, densities and other physical properties using data provided.

Chemical properties of the alkanes

Alkanes are generally unreactive compounds except for combustion and substitution by chlorine. In a **substitution reaction** one atom or group of atoms is replaced by another atom or group of atoms. Alkanes do not react with acids or alkalis.

Alkanes burn with a clean blue flame if there is excess oxygen or air present. We describe this reaction as complete combustion. Carbon dioxide and water are formed:

$$C_3H_8(g) + 5O_2(g) \rightarrow 3CO_2(g) + 4H_2O(l)$$
propane + oxygen → carbon dioxide + water

If there is not enough oxygen present, combustion is incomplete. Carbon monoxide gas is formed, and perhaps even black soot (carbon particles).

Unit 19: Some homologous series of organic compounds

$$2C_3H_8(g) + 7O_2(g) \rightarrow 6CO(g) + 8H_2O(l)$$

propane + oxygen → carbon monoxide + water

> **EXAM TIP**
>
> For the Core paper you do not have to know the details of substitution with chlorine.

Supplement

The reaction of alkanes with chlorine

One chemical that alkanes will react with is chlorine – but only under certain conditions. Alkanes do not react with chlorine in the dark. However, if we mix chlorine with an alkane in a sealed tube and keep it in bright sunlight, the green colour of the chlorine disappears. This is a **photochemical reaction**. In a photochemical reaction the ultraviolet light (UV light) in sunlight provides the activation energy for the reaction. The type of reaction that takes place is a substitution reaction. A chlorine atom replaces a hydrogen atom in the alkane.

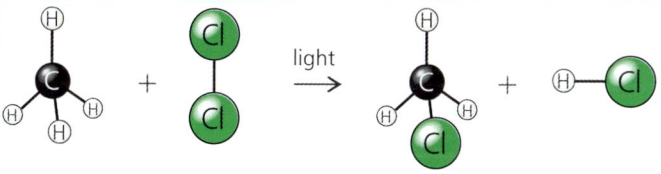

Methane + Chlorine → Chloromethane + Hydrogen chloride

Figure 19.1.3 In UV light a chlorine atom substitutes (replaces) a hydrogen atom in methane.

$$CH_4(g) + Cl_2(g) \xrightarrow{UV\ light} CH_3Cl(g) + HCl(g)$$

methane + chlorine → chloromethane + hydrogen chloride

You will notice that the acidic gas hydrogen chloride is produced. This turns damp blue litmus paper red.

If we use excess chlorine we can substitute more hydrogen atoms:

$$CH_3Cl + Cl_2 \rightarrow CH_2Cl_2 + HCl$$
$$CH_2Cl_2 + Cl_2 \rightarrow CHCl_3 + HCl$$

If enough chlorine is present all four hydrogen atoms can be replaced by chlorine atoms.

We can carry out similar reactions with other alkanes and other halogens, for example:

$$C_3H_8 + Cl_2 \xrightarrow{UV\ light} C_3H_7Cl + HCl$$

> **EXAM TIP**
>
> You will be expected to draw the structural and the displayed formulae of the products of a photochemical substitution reaction, e.g. the structural formula of a product from the substitution of ethane, such as CH_3CH_2Cl.

SUMMARY QUESTIONS

1. Copy and complete using these words:

   carbon   excess   hydrocarbons   single   substitution   unreactive   water

   Alkanes are _____ having only _____ covalent bonds. They are _____ apart from combustion and _____ by chlorine. When alkanes burn in _____ oxygen _____ dioxide and _____ are formed.

2. State the meaning of the term saturated hydrocarbon.

3. Draw the displayed formula of the compound formed when one hydrogen atom in ethane is substituted by chlorine. **S**

KEY POINTS

- A saturated hydrocarbon contains only single covalent bonds.
- Alkanes are generally unreactive except for burning and substitution by chlorine.
- **S** Alkanes react with chlorine in the presence of UV light by a photochemical substitution reaction.

19.2 Cracking alkanes

LEARNING OUTCOMES

- Describe the manufacture of alkenes and hydrogen by cracking alkanes
- Describe the reasons for the cracking of larger alkanes

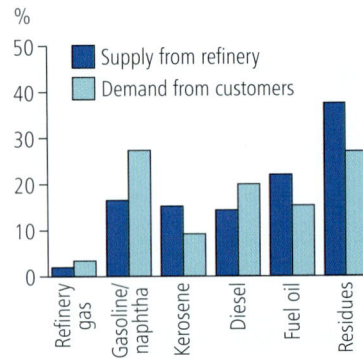

Figure 19.2.1 Supply of and demand for petroleum fractions.

What is cracking?

All the fractions we get from the distillation of petroleum are useful. However, some are more useful than others – there is a greater demand for them. We use more gasoline (petrol) and diesel than can be supplied by the fractional distillation of petroleum.

Oil companies solve this problem by breaking down larger hydrocarbons into smaller, more useful hydrocarbons. This is called **cracking**. Cracking is the thermal decomposition of alkanes. A catalyst is often used. Longer-chain alkanes are cracked to form a mixture of shorter-chain alkanes and alkenes. For example:

$$C_{10}H_{22} \xrightarrow{\text{heat} \atop \text{catalyst}} C_5H_{12} + C_2H_4 + C_3H_6$$
$$\text{decane} \qquad\qquad \text{pentane} \quad \text{ethene} \quad \text{propene}$$

From the cracking, we not only get shorter-chain alkanes which are useful for petrol, we also get alkenes. Alkenes are very useful for making a variety of chemicals including plastics.

Hydrogen can also be produced by cracking. For example:

$$C_2H_6 \longrightarrow C_2H_4 + H_2$$
$$\text{ethane} \qquad \text{ethene} \quad \text{hydrogen}$$

PRACTICAL

Cracking an alkane

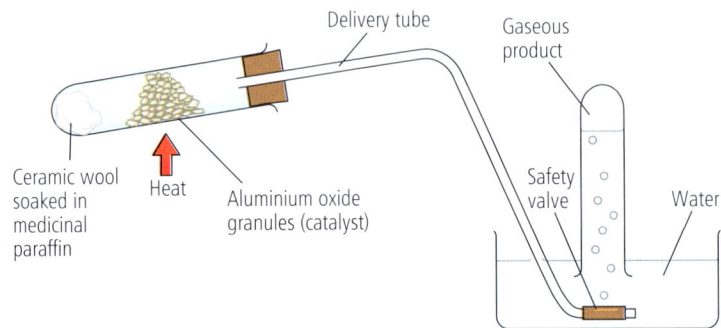

Figure 19.2.2 Aluminium oxide catalyses the cracking of alkanes.

1. You can use medicinal paraffin as your alkane. You heat the aluminium oxide catalyst strongly and then heat the paraffin. The paraffin vapour passes over the aluminium oxide which is kept hot.

2. You collect the gases from the cracking in the test tube. You can tell if the gas contains alkenes by carrying out the bromine water test (see Topic 19.3).

EXAM TIP

When describing cracking you must state that (i) large or long-chain hydrocarbon molecules are broken down into smaller or short-chain ones, and (ii) alkenes are formed.

Unit 19: Some homologous series of organic compounds

Cracking petroleum fractions on a large scale

Cracking is often carried out on a large scale using a catalyst. The huge tank where this takes place is called a catalytic (cat) cracker. The vapour from the gas-oil or kerosene fractions is passed through a catalyst of silicon(IV) oxide and aluminium oxide at 400–500 °C. The catalyst is a fine powder which has to be continuously recycled to the cat cracker through a regenerator tank. This frees the catalyst from any carbon deposited on its surface.

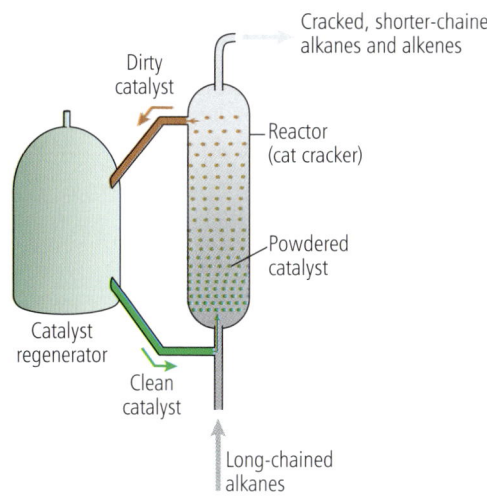

Figure 19.2.3 How a catalytic cracker works.

Figure 19.2.4 A cat cracker like this one is used to split long-chain hydrocarbons into short-chain hydrocarbons and alkenes.

The longer-chain alkanes in the gas-oil or kerosene fractions which are less useful are broken down into shorter-chain hydrocarbons:

- the shorter-chain alkanes are used for petrol and very small alkanes are used for fuel (liquid petroleum gases – LPG)
- the alkenes can be used to make a wide variety of chemicals, including plastics
- hydrogen may also be formed which can be used for making ammonia or as a fuel.

Catalytic cracking is not the only type of cracking. Long-chain alkanes can be cracked at a high temperature without a catalyst. A temperature between 450 °C and 800 °C is used. This type of cracking produces a greater percentage of alkenes.

EXAM TIP

You do not have to remember any of the equations for cracking but will be expected to write or complete equations for cracking when given suitable information.

SUMMARY QUESTIONS

1. Copy and complete using these words:

   alkanes   alkenes   catalyst   high   long   silicon

   We break down _____-chain alkanes into shorter-chain _____ and _____ by cracking them. We carry out cracking at a _____ temperature using a _____ of aluminium oxide and _____ oxide.

2. Complete these equations describing cracking:

   a $C_{12}H_{26} \rightarrow C_8H_{18} +$ _____

   b $C_3H_8 \rightarrow C_3H_6 +$ _____

3. Describe the reasons for cracking long-chain alkanes.

KEY POINTS

- Cracking is used to break long-chain alkanes into short-chain alkanes and alkenes.
- Cracking is carried out because the demand for some short-chain petroleum fractions is greater than the supply.
- Cracking is carried out using a high temperature and a catalyst.

19.3 Alkenes

LEARNING OUTCOMES

- Recognise that unsaturated hydrocarbons have at least one C=C bond or C≡C bond, with alkenes having at least one C=C bond
- Distinguish between saturated and unsaturated hydrocarbons by their reaction with aqueous bromine
- Describe the reaction of alkenes with steam
- **S** Describe the addition reactions of alkenes with bromine and with hydrogen

Unsaturated hydrocarbons

The alkenes are a homologous series of hydrocarbons whose names end in -ene. We call them unsaturated hydrocarbons because they have one or more C=C double covalent bonds or C≡C triple bonds. They do not have the maximum number of hydrogen atoms around each carbon atom – more atoms can be added to their molecules.

The first three compounds in the alkene homologous series are ethene, propene and butene. These three alkenes are all colourless gases. These are their displayed formulae (in blue):

ethene, C_2H_4 propene, C_3H_6 butene, C_4H_8

A test for unsaturated compounds

Aqueous bromine (bromine water) is orange in colour. If aqueous bromine is decolourised when mixed with excess hydrocarbon, the hydrocarbon is unsaturated. All alkenes decolourise bromine water. If the bromine water remains the same orange colour, the hydrocarbon is saturated.

PRACTICAL

Is this compound saturated or unsaturated?

1 You take a test tube of the gas or liquid you want to test.

2 You add a few drops of aqueous bromine and shake the tube. Then you observe the colour of the liquid in the tube.

Figure 19.3.1 The fat molecules in butter are largely saturated but vegetable oils contain compounds which are unsaturated because of their C=C bonds.

EXAM TIP

When explaining the test for unsaturation do not use the word 'clear' to mean colourless.

EXAM TIP

The reaction of bromine liquid with ethene forms $BrCH_2-CH_2Br$ but the reaction of aqueous bromine with ethene forms $HOCH_2-CH_2Br$ as the main product.

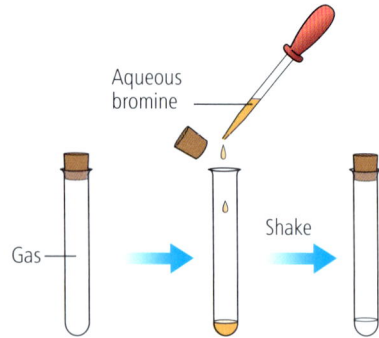

Figure 19.3.2 Aqueous bromine is decolourised by unsaturated compounds.

Unit 19: Some homologous series of organic compounds

Chemical properties of the alkenes

1. The complete combustion of alkenes produces carbon dioxide and water:

   $C_2H_4(g) + 3O_2(g) \longrightarrow 2CO_2(g) + 2H_2O(g)$
   ethene + oxygen $\longrightarrow$ carbon dioxide + water

2. Steam reacts with alkenes to form alcohols.

   A high temperature (about 300 °C) and high pressure (6000 kPa) are needed for this hydration reaction. The steam is passed over a catalyst of concentrated phosphoric acid, H_3PO_4.

   $C_2H_4 + H_2O \xrightarrow{H_3PO_4} C_2H_5OH$
   ethene steam ethanol

Supplement

Addition reactions

The reaction of steam with an alkene is an **addition reaction**. In an addition reaction, two (or more) reactants add together to form only one product. Here are two other examples of addition reactions.

1. The reaction of alkenes with bromine liquid or bromine dissolved in an organic solvent:

   $C_2H_4 + Br_2 \longrightarrow C_2H_4Br_2$
   ethene bromine 1,2-dibromoethane

   $H_2C=CH_2 + Br-Br \longrightarrow H-CBr H-CBr H$ (structural: H₂C=CH₂ + Br–Br → BrCH₂–CH₂Br)

   One of the bonds in the double C=C bond breaks to allow the bromine to be added. Bromine does not react with saturated compounds because they do not have a double bond to break and add on to.

2. Hydrogen reacts with alkenes to form alkanes.

This addition reaction is also called a **hydrogenation** reaction. It can also be classed as a reduction reaction. The reaction is carried out at 60 °C in the presence of a nickel catalyst.

$C_2H_4 + H_2 \xrightarrow{nickel/60\,°C} C_2H_6$
ethene hydrogen ethane

KEY POINTS

- Unsaturated hydrocarbons decolourise aqueous bromine but saturated hydrocarbons do not.
- Unsaturated compounds have one or more C=C double bonds or C≡C triple bonds.
- alkenes react with steam to form alcohols.
- Alkenes undergo addition reactions with bromine, hydrogen and steam. **S**

SUMMARY QUESTIONS

1. Copy and complete using these words:

   bromine colourless hydrocarbon remains saturated shaken

   We can test to see if a _____ is unsaturated or _____ by shaking it with aqueous _____. An unsaturated hydrocarbon turns aqueous bromine _____. The aqueous bromine _____ orange when _____ with a saturated hydrocarbon.

2. Refer to the structure of the compounds $CH_3CH=CH_2$ and $CH_3CH_2CH_3$ to explain which one is saturated and which is unsaturated.

3. Draw the displayed formulae of the compounds formed when: **S**

   a. propene reacts with bromine

   b. but-2-ene, $CH_3CH=CHCH_3$ reacts with hydrogen in the presence of a nickel catalyst

235

19.4 Alcohols

LEARNING OUTCOMES

- Name and draw the structural and displayed formulae of alcohols
- Describe the uses of ethanol and the combustion of ethanol
- **S** Name and draw the structural and displayed formulae of alcohols that are structural (position) isomers
- Describe the formation of ethanoic acid by the oxidation of ethanol

Ethanol

The alcohols are a homologous series having –OH as the functional group. Their names all end with -ol. Their general formula is $C_nH_{2n+1}OH$. The molecular formula of ethanol is C_2H_6O. Its structural formula is CH_3CH_2OH. The displayed formula for ethanol is shown below:

$$\begin{array}{c} \text{H} \quad \text{H} \\ | \quad\; | \\ \text{H}-\text{C}-\text{C}-\text{O}-\text{H} \\ | \quad\; | \\ \text{H} \quad \text{H} \end{array}$$

Ethanol is a colourless liquid that boils at 78 °C. It is miscible with water, meaning it mixes with water.

Ethanol can be manufactured from:

- ethene – by reacting steam with ethene at high pressure and temperature using a phosphoric acid catalyst
- glucose – by fermentation.

Ethanol burns with a clean blue flame in excess air to form carbon dioxide and water:

$$C_2H_5OH + 3O_2 \longrightarrow 2CO_2 + 3H_2O$$

Uses of ethanol

- Solvent: ethanol is used in perfumes and other cosmetics, in printing inks and in glues.
- Fuel: ethanol can be mixed with petrol or used alone as a fuel for cars. It is less polluting than petrol and reduces the reliance on petrol and diesel.
- Making chemicals: it can be used to make **esters** which are used in food flavourings and in many cosmetics.

Figure 19.4.1 Ethanol is a good solvent for perfumes.

EXAM TIP

You will often see the formula of ethanol written as C_2H_5OH. Remember that this is <u>not</u> a molecular formula.

EXAM TIP

When balancing a symbol equation for the combustion of alcohols, remember that the alcohol contains oxygen too.

Supplement

More about the structure of alcohols

The general formula for alcohols is $C_nH_{2n+1}OH$. The displayed formulae (in blue) for the first four alcohols in this homologous series are:

$$\begin{array}{cccc}
\text{H} & \text{H H} & \text{H H H} & \text{H H H H} \\
| & | \;\; | & | \;\; | \;\; | & | \;\; | \;\; | \;\; | \\
\text{H}-\text{C}-\text{O}-\text{H} & \text{H}-\text{C}-\text{C}-\text{O}-\text{H} & \text{H}-\text{C}-\text{C}-\text{C}-\text{O}-\text{H} & \text{H}-\text{C}-\text{C}-\text{C}-\text{C}-\text{O}-\text{H} \\
| & | \;\; | & | \;\; | \;\; | & | \;\; | \;\; | \;\; | \\
\text{H} & \text{H H} & \text{H H H} & \text{H H H H}
\end{array}$$

| CH_3OH | C_2H_5OH | C_3H_7OH | C_4H_9OH |
| methanol | ethanol | propan-1-ol | butan-1-ol |

The number in the name is used so that you can distinguish between different isomers of the alcohols. The numbers are not needed for methanol and ethanol though.

Unit 19: Some homologous series of organic compounds

The isomers with the –OH group at the end of the carbon chain are -1-ols. Some structural (position) isomers of propanol and butanol are shown here:

CH₃—CH₂—CH₂—OH
propan-1-ol

CH₃—CH—CH₃
         |
        OH
propan-2-ol

CH₃—CH₂—CH₂—CH₂—OH
butan-1-ol

CH₃—CH₂—CH—CH₃
              |
             OH
butan-2-ol

           CH₃
            |
CH₃—C—CH₃
            |
           OH
2-methylpropan-2-ol

Oxidising ethanol to ethanoic acid

Ethanol can be oxidised to ethanoic acid by:

1. Oxidation in the air:

Acetobacter is a group of bacteria that are naturally present in the air and on surfaces around us. When we leave a solution of ethanol exposed to the air, enzymes from the bacteria speed up the conversion of ethanol to ethanoic acid. The reaction does not take place in the absence of oxygen:

$$C_2H_5OH + 2[O] \rightarrow CH_3COOH + H_2O$$
ethanol   'oxygen'   ethanoic acid   water

We can write [O] in an equation when the oxidation reaction is complicated but we know that oxygen is involved. This is the reaction that makes vinegar. Vinegar is an aqueous solution of ethanoic acid.

2. Oxidation using acidified potassium manganate(VII):

Potassium manganate(VII) is a good oxidising agent, especially when sulfuric acid is added. We heat the ethanol with potassium manganate(VII) and sulfuric acid. We do this in a flask with a condenser in an upright position. We call this method of heating **refluxing**. This prevents the alcohol, which is very volatile, from escaping. The equation for the reaction can be represented the same as the one above. Other alcohols can be oxidised in the same way.

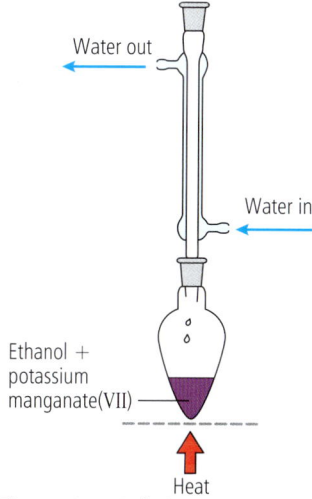

Figure 19.4.2 Refluxing a mixture.

KEY POINTS

- Alcohols are a homologous series with an –OH functional group.
- Ethanol is made by the addition of steam to ethene in the presence of a catalyst or by fermentation.
- Propanol and butanol can exist as isomers with the –OH functional group in different positions.
- Ethanoic acid is made by the oxidation of ethanol.

SUMMARY QUESTIONS

1 Copy and complete using these words:

ethanol   ethene   fermentation   functional   steam

Alcohols have an –OH _____ group. The formula of _____ is C₂H₅OH. Ethanol can be made by reacting _____ with _____ or by _____.

2 Give two major uses of ethanol.

3 Give the structural formula for:
  a propan-1-ol
  b a structural (position) isomer of propan-1-ol

237

19.5 Fermentation

LEARNING OUTCOMES

- Describe the manufacture of ethanol by fermentation
- Describe the manufacture of ethanol by the catalytic addition of steam to ethene
- S Describe the advantages and disadvantages of the two methods of manufacturing ethanol

Enzymes and fermentation

Nearly all chemical reactions in living things are catalysed by enzymes. Enzymes are particular types of **protein** that act as biological catalysts. But unlike inorganic catalysts they are very sensitive to changes in temperature. They work best between 25 °C and 35 °C.

Bacteria and **yeasts** produce enzymes that catalyse fermentation reactions in organic materials. **Fermentation** is the breakdown of organic material with effervescence (bubbles) and the release of thermal energy. For thousands of years, ethanol has been made by fermentation.

Figure 19.5.1 The sugar from this sugar cane can be fermented to make ethanol.

The fermentation of glucose

The most commonly used substances for making ethanol are sugars such as glucose. The bacteria and yeast that cause fermentation are found on the surface of many plants, as well as in the air.

PRACTICAL

Fermenting sugars

1. You add moist yeast to the flask of warm glucose solution.
2. Then you shake the solution so that the yeast is suspended in the solution.
3. You keep the flask warm for about 30 minutes without shaking.
4. Then you 'pour' the gas (not the liquid) from the flask into a boiling tube containing some limewater. The limewater turns milky. This shows that carbon dioxide is a product of fermentation.

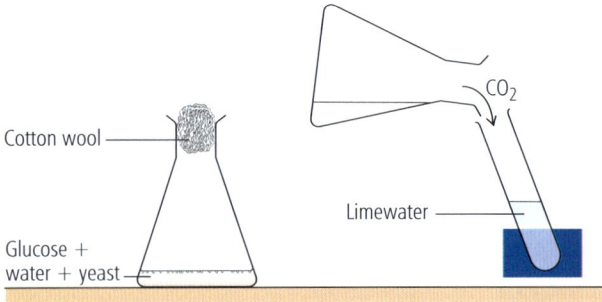

Figure 19.5.2 Fermentation of glucose by yeast.

The fermentation of glucose produces ethanol and carbon dioxide:

$$C_6H_{12}O_6 \xrightarrow{yeast} 2C_2H_5OH + 2CO_2$$

glucose → ethanol + carbon dioxide

Unit 19: Some homologous series of organic compounds

The conditions needed for fermentation are:

- the absence of oxygen or air. The respiration of yeast in the absence of oxygen is called **anaerobic respiration**.
- a temperature of 25–35 °C.
- a pH which is near neutral.

The reaction occurs until about 14% ethanol is present in the mixture. If the ethanol concentration gets much higher the yeast will die. The ethanol is separated from the reaction mixture by fractional distillation.

Supplement

Comparing methods

We can produce ethanol by fermentation or by the reaction of ethene with steam (see Topic 19.3 for ethene reacting with steam). Each of these methods has advantages and disadvantages:

ethanol from fermentation	ethanol from ethene and steam
Simple method	More complex method
Needs low temperature so relatively low energy input	Needs high temperature and pressure so relatively high energy input
Uses a batch process: you have to start again from the beginning once you have removed the solution in the tank	A continuous process: the ethanol is removed continuously and the ethene and steam are fed into the apparatus continuously
Rate of reaction is slow	Rate of reaction is fast
Ethanol needs further purification by distillation	Produces ethanol of high purity
Uses renewable resources	The ethene is made from a non-renewable resource – petroleum

> **EXAM TIP**
> Make sure that you know the difference between conditions and reactants. Conditions are things like temperature and pH which can affect the rate of reaction.

> **EXAM TIP**
> It is a common error to suggest that oxygen is required for the fermentation of glucose to ethanol.

> **EXAM TIP**
> Make sure that you don't confuse the conditions for the two different ways of making ethanol. Remember that for fermentation the yeast is a living organism, so it doesn't survive at high temperatures and pressures.

SUMMARY QUESTIONS

1. Copy and complete using these words:

   carbon   catalyse   enzymes   ethanol   ferments   glucose   yeast

   When you leave a solution of glucose with _____ for a few days it _____. The _____ is broken down to _____ and _____ dioxide. The yeast contains _____ that _____ this reaction.

2. State the conditions needed for fermentation.

3. Give one advantage of making ethanol:
   a. by fermentation
   b. from ethene and steam

> **KEY POINTS**
> - The fermentation of glucose to form ethanol and carbon dioxide is catalysed by enzymes from yeast.
> - The conditions for fermentation are absence of oxygen, temperature between 25 °C and 35°C, and a pH near neutral.
> - There are advantages and disadvantages to the production of ethanol by both fermentation and hydration by steam.

239

19.6 Carboxylic acids and esters

LEARNING OUTCOMES

- Describe the reaction of ethanoic acid with metals, bases and carbonates
- **S** Describe the reaction of a carboxylic acid with an alcohol to form an ester
- Name and the draw the displayed formulae of esters

EXAM TIP

If you are taking the Core paper you only need to know about ethanoic acid.

EXAM TIP

Remember that when naming the carboxylic acids the carbon atom of the –COOH group is included. So CH_3COOH is ethanoic acid because compounds with two carbon atoms have names beginning with eth.

EXAM TIP

For the general properties of acids see Topic 11.2. For those taking the Supplement paper remember that carboxylic acids are weak acids (Topic 11.4).

Figure 19.6.1 The flavour of most fruits is due to chemicals called esters.

Carboxylic acids

The carboxylic acids are a homologous series with –COOH as the functional group. Their names all end in -oic acid. Their general formula is $C_nH_{2n+1}COOH$. The displayed formulae (in blue) of the first four carboxylic acids are shown.

HCOOH — methanoic acid
CH_3COOH — ethanoic acid
C_2H_5COOH — propanoic acid
C_3H_7COOH — butanoic acid

Chemical properties of carboxylic acids

Carboxylic acids are weak acids. They react in a similar way to other acids. The hydrogen of the –COOH group is the only one that can be released as an H^+ ion.

- They react with reactive metals to form a salt and hydrogen:

  $2CH_3COOH + Mg \longrightarrow (CH_3COO)_2Mg + H_2$
  ethanoic acid   magnesium ethanoate

- They react with alkalis to form a salt and water:

  $CH_3COOH + NaOH \longrightarrow CH_3COONa + H_2O$
  ethanoic acid   sodium ethanoate

- They react with metal carbonates to form a salt, water and carbon dioxide:

  $2CH_3COOH + Na_2CO_3 \longrightarrow 2CH_3COONa + H_2O + CO_2$
  ethanoic acid   sodium ethanoate

Note how the salts of ethanoic acid are named by changing the -oic acid to –oate after the name of the metal.

Supplement

The formation of esters

Carboxylic acids react with alcohols when warmed gently in the presence of a catalyst to form compounds called **esters**. The reaction is called esterification.

In this reaction sulfuric acid acts as a catalyst. The $-\overset{\overset{\displaystyle O}{\|}}{C}-O-$ group formed is called an ester linkage. The water given off comes partly from the acid and partly from the alcohol:

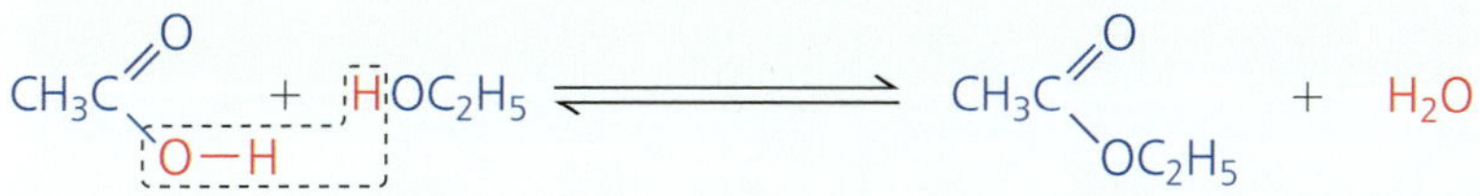

Unit 19: Some homologous series of organic compounds

$$CH_3C(=O)O-H + H-OC_2H_5 \rightleftharpoons CH_3C(=O)OC_2H_5 + H_2O$$

$$\underset{\text{ethanoic acid}}{CH_3COOH} + \underset{\text{ethanol}}{CH_3CH_2OH} \underset{}{\overset{heat/H_2SO_4}{\rightleftharpoons}} \underset{\text{ethyl ethanoate}}{CH_3COOC_2H_5} + H_2O$$

The naming of esters is based on the name of the carboxylic acid and alcohol used to make them.

- The name begins with the alkyl group from the alcohol.
- The name ends with the part coming from carboxylic acid, when –oic acid is changed to –oate.

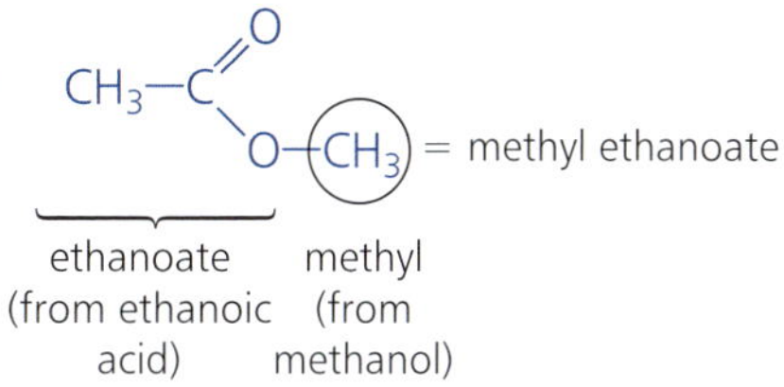

$$CH_3-C(=O)-O-CH_3 = \text{methyl ethanoate}$$

ethanoate (from ethanoic acid)   methyl (from methanol)

So, $C_3H_7COOC_2H_5$ is ethyl butanoate and $HCOOC_3H_7$ is propyl methanoate.

The displayed formulae of two more esters are:

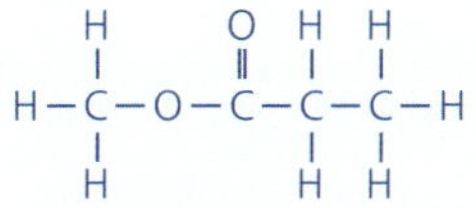

methyl propanoate 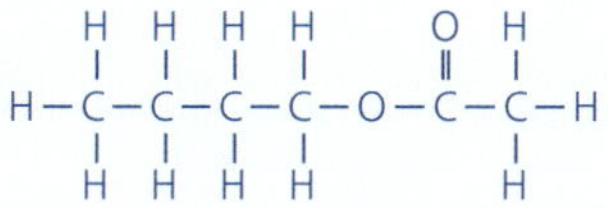   butyl ethanoate

SUMMARY QUESTIONS

1 Copy and complete using these words:

ethanoates functional hydrogen salt water

The _____ group of carboxylic acids is –COOH. Carboxylic acids react with metals to form a salt and _____ and with carbonates to form a _____, carbon dioxide and _____. The salts of ethanoic acid are called _____.

2 Write word equations for the reaction of ethanoic acid with:
  a potassium carbonate
  b zinc

S 3 Write displayed formulae for:
  a sodium methanoate
  b propyl ethanoate

PRACTICAL

Making an ester

1 Put 1 cm³ of ethanoic acid into a test tube.
2 Then add 2 cm³ of ethanol, and then 2 drops of concentrated sulfuric acid.
3 Warm the mixture in a beaker of hot water for 5 minutes.
4 Pour the contents of the test tube into a small beaker of sodium carbonate solution. The smell of the ester will be obvious.

You can repeat the experiment using different carboxylic acids and alcohols.

EXAM TIP

You must be able to name and draw the displayed formulae of esters which can be made from alcohols and carboxylic acids with up to four carbon atoms.

KEY POINTS

- Ethanoic acid reacts with metals, carbonates and hydroxides.
- Carboxylic acids are weak acids.
- Carboxylic acids react with alcohols to form esters and water.

SUMMARY QUESTIONS

1 Match the words on the left with the descriptions on the right.

ethanol	the breaking down of long-chain alkanes to alkenes and shorter-chain alkanes
cracking	organic compounds containing only single C–C bonds
unsaturated	one of the products of the fermentation of glucose
addition	organic compounds containing C=C double bonds
saturated	a reaction in which two or more compounds combine to form only one compound

2 Copy and complete using these words:

added   bonds   bromine   decolourises ethane   ethene   orange

Ethane can be distinguished from ethene by adding aqueous _____ to each compound. Aqueous bromine has an _____ colour. Ethene _____ aqueous bromine but _____ does not. One of the C=C _____ is broken when aqueous bromine is _____ to _____.

3 Copy and complete these symbol equations:

a $C_{10}H_{22} \rightarrow C_2H_4 +$ _____
b $C_2H_4 +$ _____ $\rightarrow C_2H_4Br_2$
c $C_2H_4 + H_2O \rightarrow$ _____
d $C_5H_{12} +$ ___$O_2 \rightarrow$ ___$CO_2 +$ ___H_2O
e $2CH_3COOH + 2Na \rightarrow$ _____ $+$ _____

4 State the type of chemical reaction that occurs in each of these equations:

a $C_3H_6 + H_2O \rightarrow C_3H_7OH$
b $C_2H_5OH + 2[O] \rightarrow CH_3COOH + H_2O$
c $C_7H_{16} + 11O_2 \rightarrow 7CO_2 + 8H_2O$
d $CH_3COOH + C_2H_5OH \rightarrow CH_3COOC_2H_5 + H_2O$

5 What conditions and/or additional reagents are needed for the following reactions:

a   the formation of ethanol from ethene and steam
b   the conversion of ethene to ethane
c   the formation of ethanoic acid from ethanol using inorganic compounds

Practice questions

1 Which one of these statements about cracking alkanes is correct?
   A   Alkanes are cracked using a nickel catalyst and low temperatures.
   B   Long-chain alkanes are broken down to shorter-chain alkanes and alkenes.
   C   Short-chain alkanes are changed into long-chain alkanes and alkenes.
   D   The conditions for cracking are room temperature and an iron catalyst.
(Paper 1)

2 What is the correct name of the ester with the formula $C_3H_7CO_2C_2H_5$?
   A   ethyl butanoate
   B   propyl propanoate
   C   butyl ethanoate
   D   ethyl propanoate
(Paper 2)

3 (a) (i)   Copy and complete the equation for cracking octane:
      $C_8H_{18} \rightarrow C_2H_4 +$ _____   [1]
   (ii)   State the name of the compound C_2H_4 and draw its displayed formula. [2]
(b) State the conditions used for cracking. [2]
(c) Octane is a saturated hydrocarbon. Give the meaning of the terms:
   (i) saturated   [1]   **(ii)** hydrocarbon.   [2]
(Paper 3)

d   the formation of ethyl ethanoate from ethanoic acid

6 Describe how to make ethyl ethanoate starting with ethanol as the only as the organic reactant.

7 Write displayed formulae for:
   a   propene
   b   butanoic acid
   c   a branched-chain alkane with 4 carbon atoms

4 Ethanol is made by either fermentation or reaction of ethene with steam.
   (a) Draw the displayed formulae for
       (i) ethanol [2] (ii) ethene. [1]
   (b) Describe three differences in the conditions used for these two methods of producing ethanol. [6]
   (c) Copy and complete the word equation for fermentation.
       _____ → ethanol + _____ _____ [2]

(Paper 3)

5 Ethene is an unsaturated hydrocarbon. Ethane is a saturated hydrocarbon.
   (a) Describe how you can distinguish between ethene and ethane using a chemical test. [3]
   (b) Draw the displayed formula for ethane. [1]
   (c) Ethene is formed by cracking hydrocarbons.
       (i) Describe what is meant by the term cracking. [4]
       (ii) Give two reasons why cracking is carried out. [2]
   (d) Copy and complete the equation for the complete combustion of ethanol:
       $C_2H_5OH + \_\_\_O_2 \rightarrow \_\_\_CO_2 + \_\_\_H_2O$ [3]
   (e) Give two uses of ethanol. [2]
   (f) Write a word equation for the reaction of ethanoic acid with magnesium. [2]

(Paper 3)

6 Ethanoic acid, CH_3COOH, is a typical weak acid.
   (a) Draw the displayed formula of ethanoic acid. [2]
   (b) Describe two ways in which the properties of 0.1 mol/dm³ ethanoic acid differ from 0.1 mol/dm³ sulfuric acid. [2]
   (c) Write chemical equations for the reaction of propanoic acid, CH_3CH_2COOH, with:
       (i) magnesium [3]
       (ii) potassium carbonate. [3]
   (d) Ethanoic acid can be made in the laboratory by the bacterial oxidation of ethanol.
       (i) Name another oxidising agent that can be used to oxidise ethanol. [1]
       (ii) State the conditions needed for this oxidation. [2]
       (iii) Write a chemical equation for this reaction. Use [O] to represent the oxidising agent. [2]

(Paper 4)

7 Alkanes and alkenes are both hydrocarbons.
   (a) (i) Draw the structural formula for an alkene with four carbon atoms. [1]
       (ii) Draw the displayed formula for an isomer of the alkene you drew in part (i). [1]
   (b) Both alkenes and alkanes react with halogens.
       (i) State the essential condition required for an alkane to react with chlorine. [1]
       (ii) State the type of chemical reaction that occurs. [1]
   (c) (i) Construct an equation to show the reaction of bromine with ethene. [2]
       (ii) State the type of chemical reaction that occurs. [1]
       (iii) State the observations made during this reaction. [2]
   (d) Alkenes react with steam to form alcohols.
       (i) State the conditions needed for this reaction. [3]
       (ii) Write an equation using structural formulae to show the reaction of propene with steam. [2]

(Paper 4)

8 Methane, methanol and methanoic acid each have one carbon atom.
   (a) Write the displayed formula for:
       (i) methanol [1]
       (ii) methanoic acid. [2]
   (b) Methanoic acid reacts with propanol to form an ester.
       (i) State the conditions needed for this reaction. [2]
       (ii) Name and draw the displayed formula of the ester formed. [2]

(Paper 4)

243

20.1 What are polymers?

LEARNING OUTCOMES

- Define the terms monomer, polymer and addition polymerisation with reference to the formation of poly(ethene)
- Describe the environmental issues caused by plastics
- **S** Know that PET can be hydrolysed to its monomers and re-polymerised

Polymers

Plastics are made from **polymers**. Polymers are very large molecules made from many small molecules called **monomers**. The process of joining hundreds or thousands of monomers together to form polymers is called **polymerisation**.

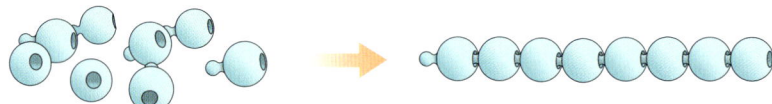

Figure 20.1.1 A model of polymerisation: the 'bead' monomers represent ethene. They join together to form poly(ethene).

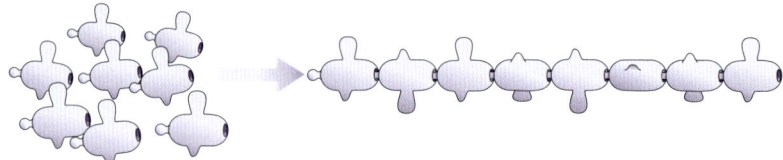

Figure 20.1.2 These 'bead' monomers represent propene. They join together to form poly(propene).

An example of a monomer is ethene, $CH_2=CH_2$. Ethene can take part in addition reactions. Polymers are often formed by addition reactions. We call this type of polymerisation, where monomers join together and no other substance is formed, **addition polymerisation**.

Figure 20.1.3 A wide variety of articles are made from plastics.

Poly(ethene), commonly called polythene, is a plastic that is easy to shape and quite strong. It is made by joining ethene monomers together. So how does this happen?

When alkene molecules react, one of the C=C double bonds breaks and joins with its neighbouring molecule. When poly(ethene) is made, thousands of ethene molecules join together like this to form a long chain.

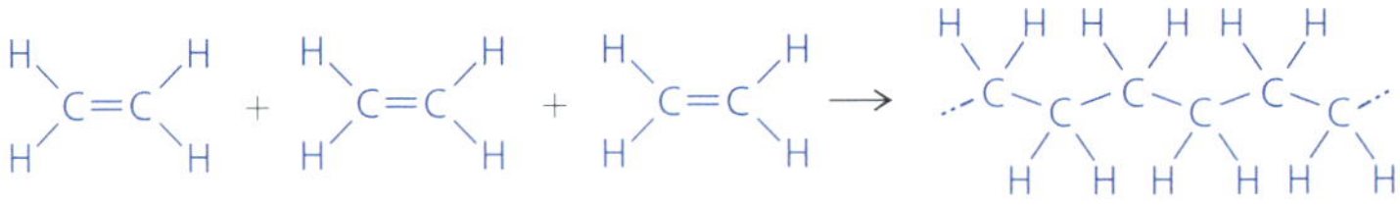

$$\begin{array}{c}H\\ \end{array}C=C\begin{array}{c}H\\ \end{array} + \begin{array}{c}H\\ \end{array}C=C\begin{array}{c}H\\ \end{array} + \begin{array}{c}H\\ \end{array}C=C\begin{array}{c}H\\ \end{array} \rightarrow \cdots -\underset{H}{\overset{H}{C}}-\underset{H}{\overset{H}{C}}-\underset{H}{\overset{H}{C}}-\underset{H}{\overset{H}{C}}-\underset{H}{\overset{H}{C}}-\underset{H}{\overset{H}{C}}- \cdots$$

Figure 20.1.4 Many ethene monomers join together by forming new bonds with the carbon atoms of the molecules next to them.

Plastics and their disposal

Plastics are particular types of polymers that can be moulded. The word plastic simply means something that can have its shape changed. Plastics have a wide range of uses including bottles, buckets, ropes, drainpipes and clothing.

Figure 20.1.5 Tiny particles from these plastics can get into the bodies of sea creatures.

But there is a downside to plastics. Many are **non-biodegradable**. This means that they are not broken down in the soil or water by microorganisms when we throw them away. Non-biodegradable plastics have long chains of C–C bonds which are resistant to **hydrolysis** (the breakdown of compounds by water, acids or alkalis). These plastics build up and cause problems. If plastics get into drains, they can block them and cause flooding. Plastics can kill wildlife by trapping small animals or by blocking the digestive systems of animals and birds that eat the plastic along with their normal food.

More and more plastics are getting into the oceans. Tiny plastic particles have been found in the bodies of sea creatures all round the world. Plastic floating on the surface of the water can block sunlight and oxygen from getting to plants and animals in seas and lakes. Plastic fishing nets can strangle seals and other animals. So how can we deal with unwanted plastics?

- Put them into landfill (waste) sites: These fill up very quickly and use up land that could be used for agriculture or housing.

- Burn them: we can use the thermal energy released to provide electricity or heating. But many plastics form poisonous gases when they burn. PVC produces acidic hydrogen chloride. Plastics containing nitrogen may produce toxic hydrogen cyanide when burned. Many plastics when burned at high temperatures also produce poisonous compounds called dioxins. It is very expensive to put filters on the furnaces used to burn plastics so this is rarely done.

- Recycling: some plastics can be melted and then moulded to make new articles. But not all plastics can be recycled.

- **Biodegradable** plastics: these can be broken down by bacteria in the soil. But it still leaves a problem of tiny harmful hydrocarbon particles in the soil.

Supplement

Recycling and reusing the plastic PET

The polymer PET (polyethylene terephthalate) is used to make plastic bottles. We can recycle PET by melting, remoulding and mixing with other plastics to make carpets. We can also make a higher-grade plastic by hydrolysing the PET polymer to its monomers. The monomers are then re-polymerised to make food containers. In some parts of the world, PET bottles have been used for disinfecting water. UV light can get through the walls of a PET bottle and kill any bacteria present in untreated water.

EXAM TIP

When answering questions about problems with the disposal of plastics remember that they release toxic gases when burned. It is a common error to omit 'when burned'.

KEY POINTS

- Polymers are large molecules built up from small units (monomers).
- In addition polymerisation the double bond between the carbon atoms changes to a single bond when the monomer molecules join together.
- Each method of disposing of a plastic has its disadvantages.
- PET can be hydrolysed to its monomers and then re-polymerised.

SUMMARY QUESTIONS

1 Copy and complete using these words:

   addition ethene molecules monomer polymer

   Small molecules such as _____ can join together by _____ polymerisation. We give the name _____ to the small _____ that join together to form a _____.

2 State two problems that arise from the disposal of plastics.

3 Explain how PET can be recycled to make food containers.

20.2 More about polymer structure
Supplement

LEARNING OUTCOMES

- Identify the repeat units and linkages in addition polymers
- Deduce the polymer structure or repeat unit for a given alkene

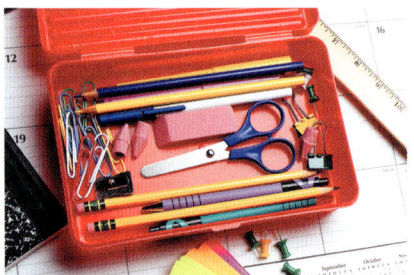

Figure 20.2.1 We make lots of everyday articles from addition polymers.

EXAM TIP

Make sure that you draw the number of repeat units asked for in a question, e.g. for 3 repeat units (from 3 C=C bonds in 3 monomers) draw 6 carbon atoms in the chain.

Naming addition polymers

We can make many different types of addition polymers. Poly(propene), poly(phenylethene) and poly(chloroethene) are just three examples. To name an addition polymer you put the name of the monomer inside brackets and put 'poly' in front. However, some polymers are often called by their common names, so it is not always easy to deduce the structure from the name of the monomer. For example poly(tetrafluoroethene), which is used for non-stick pans, is commonly called 'Teflon'.

From monomer to polymer

Propene is an alkene. The polymer formed from propene is called poly(propene). So how do we write the formula for a polymer when we are given the formula of the monomer?

One way of doing this is to show several **repeat units**:

- write down the formulae for the number of monomer units you want, but change each double bond to a single bond
- draw single bonds between the monomer units
- put 'continuation bonds' at either end of the chain to show that the chain carries on in the same way.

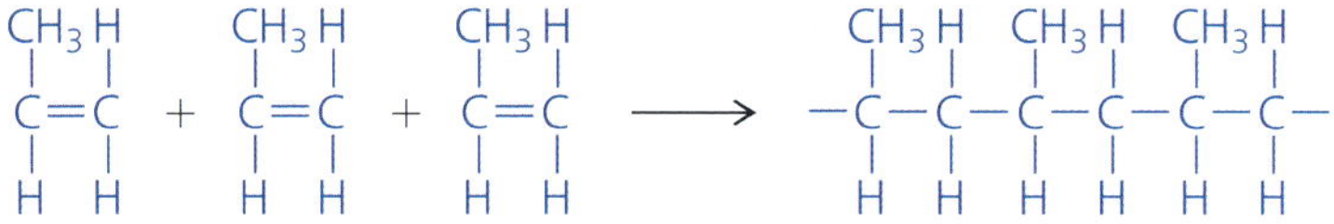

$$\begin{array}{c}CH_3\ H\\ |\ \ \ |\\ C=C\\ |\ \ \ |\\ H\ \ H\end{array} + \begin{array}{c}CH_3\ H\\ |\ \ \ |\\ C=C\\ |\ \ \ |\\ H\ \ H\end{array} + \begin{array}{c}CH_3\ H\\ |\ \ \ |\\ C=C\\ |\ \ \ |\\ H\ \ H\end{array} \longrightarrow \begin{array}{c}CH_3\ H\ \ CH_3\ H\ \ CH_3\ H\\ |\ \ \ \ |\ \ \ \ |\ \ \ \ |\ \ \ \ |\ \ \ \ |\\ -C-C-C-C-C-C-\\ |\ \ \ \ |\ \ \ \ |\ \ \ \ |\ \ \ \ |\ \ \ \ |\\ H\ \ \ H\ \ \ H\ \ \ H\ \ \ H\ \ \ H\end{array}$$

Figure 20.2.2 Three monomer units join to form part of a poly(propene) chain.

We can also write the structure of the polymer in a general form:

- draw the structure of the monomer but change the double bond to a single bond
- put 'continuation bonds' at either end of the 'molecule'
- put square brackets through the continuation bonds
- put an 'n' at the bottom right-hand corner. This means that the unit repeats itself n times.

Unit 20: Polymers

Figures 20.2.3 and 20.2.4 show two more examples.

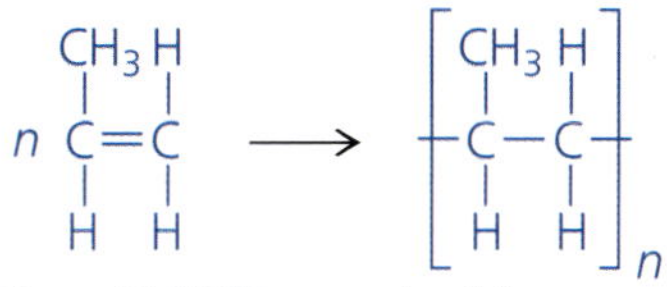

$$n \begin{array}{c} CH_3 \; H \\ | \quad | \\ C=C \\ | \quad | \\ H \quad H \end{array} \longrightarrow \left[\begin{array}{c} CH_3 \; H \\ | \quad | \\ -C-C- \\ | \quad | \\ H \quad H \end{array} \right]_n$$

Figure 20.2.3 The repeat unit for poly(propene) for a large number (n) of propene monomers.

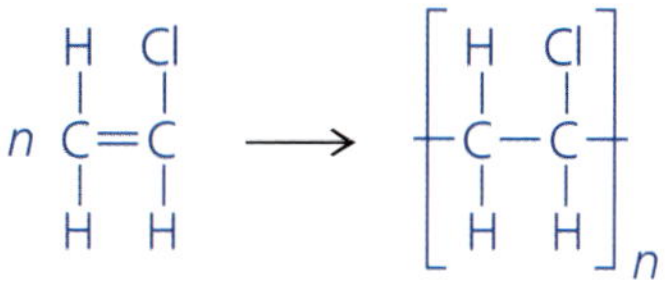

$$n \begin{array}{c} H \quad Cl \\ | \quad | \\ C=C \\ | \quad | \\ H \quad H \end{array} \longrightarrow \left[\begin{array}{c} H \quad Cl \\ | \quad | \\ -C-C- \\ | \quad | \\ H \quad H \end{array} \right]_n$$

Figure 20.2.4 Poly (chloroethene) from chloroethene.

From polymer to monomer

You can work out the structure of the monomer from a diagram of the polymer:

- identify the repeat unit in the polymer and draw this
- ignore the brackets, the continuation bonds and the 'n'
- make the single bond between the carbon atoms in the chain of the polymer into a double bond.

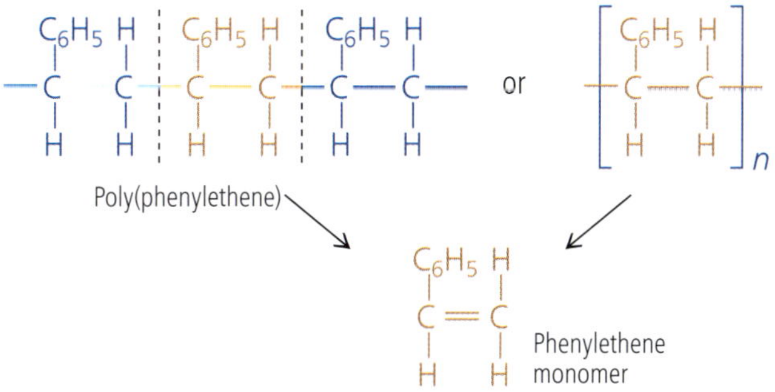

Figure 20.2.5 Deducing the monomer of poly(phenylethene). The brown section between the dashed lines in the polymer is the repeat unit.

EXAM TIP

When writing the formula for an addition polymer don't forget (i) the double bond changes to a single bond and (ii) include the continuation bonds at each end.

SUMMARY QUESTIONS

1 Name the polymers formed from these monomers:
   a butene
   b tetrafluoroethene
   c ethenyl ethanoate

2 Draw part of the chain of poly(but-2-ene), $CH_3CH=CHCH_3$, to show three repeat units.

3 Deduce the structure of the monomer that forms this polymer:

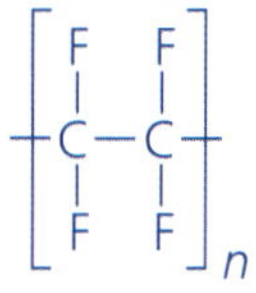

$$\left[\begin{array}{c} F \quad F \\ | \quad | \\ -C-C- \\ | \quad | \\ F \quad F \end{array} \right]_n$$

Figure 20.2.6

KEY POINTS

- We draw the structure of an addition polymer by joining several monomer units and changing the double bonds to single bonds.

- We can draw a shorthand structure for a polymer by placing one repeat unit in square brackets with continuation bonds.

- We can deduce the structure of a monomer from a diagram of a polymer by identifying the repeat units.

20.3 Polyamides and polyesters
Supplement

LEARNING OUTCOMES

- Describe the differences between addition and condensation polymerisation
- Identify the repeat units and linkages in condensation polymers
- Deduce the polymer structure or repeat unit for a given polyamide or polyester

Condensation polymers

We make addition polymers by joining the monomers together to make a long chain. No other substance is formed apart from the polymer. However, many polymers such as nylon and terylene are formed in a different way – by **condensation polymerisation**.

In condensation polymerisation, the monomers react together to form the polymer and another product. The other product is usually a small molecule such as water or hydrogen chloride. We say that the small molecule is eliminated. Condensation polymerisation usually involves two different monomers, each with a different functional group. These groups react with each other, joining monomers together to form the polymer.

Polyamides

Nylon-6,6 is a typical **polyamide**. The two types of monomer that form a polyamide are carboxylic acids and amines. An amine is a compound with an $-NH_2$ functional group. A carboxylic acid reacts with an amine to form an amide. The $-CONH-$ group is called an **amide linkage** (see Figure 20.3.1).

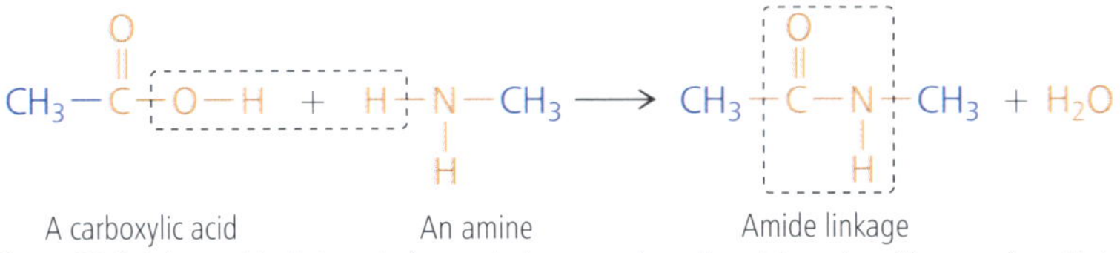

$$CH_3-C(=O)-O-H + H-N(H)-CH_3 \longrightarrow CH_3-C(=O)-N(H)-CH_3 + H_2O$$

A carboxylic acid   An amine   Amide linkage

Figure 20.3.1 An amide linkage is formed when a carboxylic acid reacts with an amine. Water is eliminated.

This type of reaction, where two molecules join together and a small molecule is eliminated, is called a **condensation reaction**.

We can represent the structures of the monomers that react to make a polyamide like this:

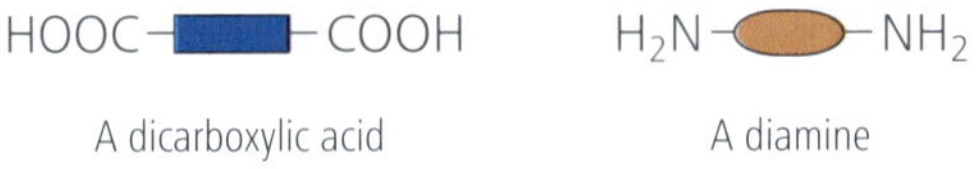

HOOC━━━COOH H_2N━━━NH_2

A dicarboxylic acid   A diamine

Figure 20.3.3 Simplified structures of a dicarboxylic acid and a diamine.

You can see that each monomer has two functional groups. We put the word di- in front of the names to show that there are two of the same functional groups per monomer. Because each end of the monomer has a reactive functional group, a long chain can be formed.

Figure 20.3.2 Nylon is a polyamide which is used to make strings and kites.

EXAM TIP

Make sure that you can identify a repeat unit in a polymer. A repeat unit is the simplest unit which when joined gives the structure of the whole polymer.

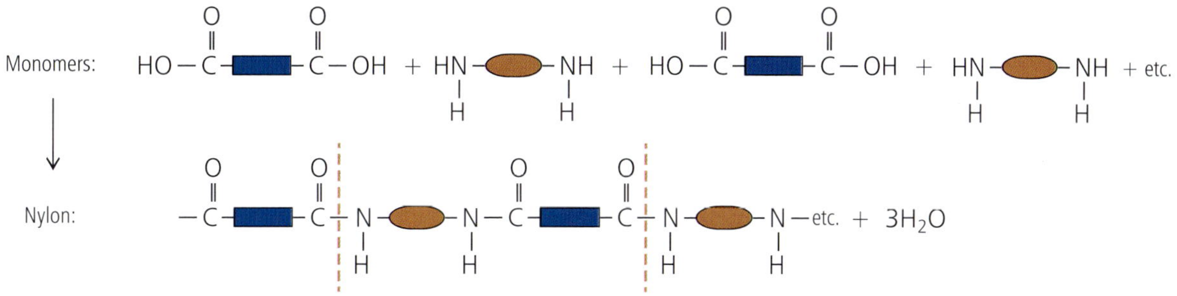

Unit 20: Polymers

Monomers: HO−C(=O)−▬−C(=O)−OH + HN(H)−⬭−NH(H) + HO−C(=O)−▬−C(=O)−OH + HN(H)−⬭−NH(H) + etc.

↓

Nylon: −C(=O)−▬−C(=O)−N(H)−⬭−N(H)−C(=O)−▬−C(=O)−N(H)−⬭−N−etc. + 3H₂O

Figure 20.3.4 Nylon-6,6 is formed from carboxylic acid groups reacting with amine groups. The section between the dashed lines is the repeat unit.

Polyesters

PET is a typical **polyester**. The two types of monomer that form a polyester are dicarboxylic acids and **diols**. An **ester linkage** is formed.

HO−C(=O)−▬−C(=O)−OH   H−O−▬−OH
Dicarboxylic acid   diol

↓

HO−C(=O)−▬−C(=O)−O−▬−OH + H₂O
              Ester linkage

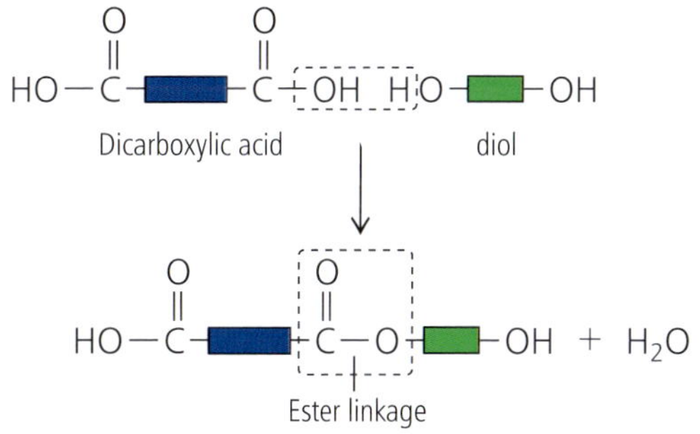

Figure 20.3.5 An ester linkage is formed when a dicarboxylic acid reacts with a diol.

EXAM TIP

When writing the structures of polyamides and polyesters take care to draw the linkages in the correct direction, e.g.

−O−C(=O)−▭−C(=O)−O−

We can represent the equation for polyester formation like this:

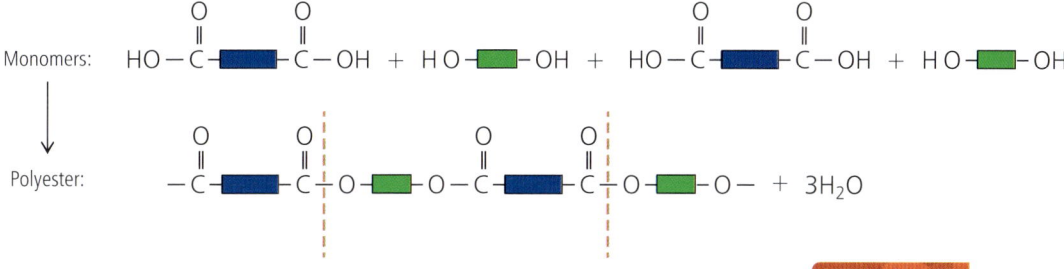

Figure 20.3.6 PET is formed from −COOH groups reacting with −OH groups.

SUMMARY QUESTIONS

1. Copy and complete using these words:

   **condensation diamines eliminated
   monomers polyamide water**

   Nylon is made by _____ polymerisation. The _____ are dicarboxylic acids and _____. These react to form a _____. A small molecule (_____) is _____.

2. Draw the displayed formula for:

   a an amide linkage   **b** an ester linkage

KEY POINTS

- In a condensation reaction the monomers join to form a polymer and a small molecule is eliminated.
- Polyamides such as nylon have amide linkages

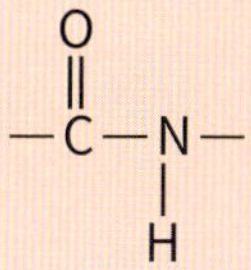

  −C(=O)−N(H)−

- Polyesters such as PET have ester linkages

  −C(=O)−O−

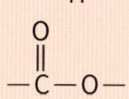

20.4 Amino acids and proteins
Supplement

LEARNING OUTCOMES
- Describe proteins as natural polyamides
- Describe the general structure of amino acids
- Describe the general structure of proteins

Amino acids

Amino acids are the monomers which react together to make proteins. There are 20 naturally occurring amino acids in our bodies. The structures of three amino acids are shown below.

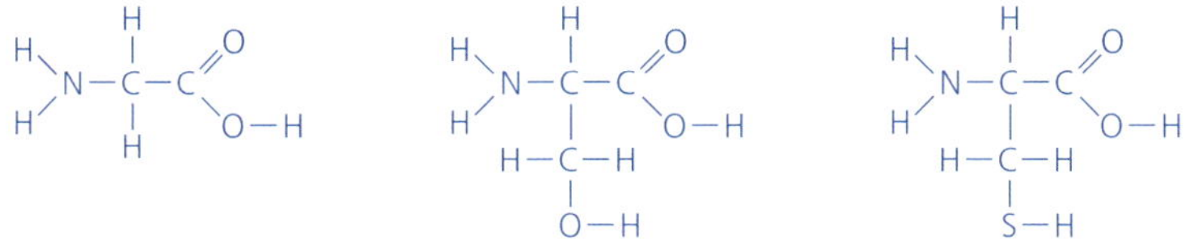

Glycine   Serine   Cysteine

Figure 20.4.1 The structure of three amino acids.

All amino acids have the amine ($-NH_2$) and carboxylic acid ($-COOH$) groups in common. But the side-chains in each of the 20 amino acids are different. The general structure of an amino acid is shown in Figure 20.4.2. The R represents different types of side-chain. These can be alkyl groups or can contain $-OH$, $-COOH$ or $-NH_2$ groups.

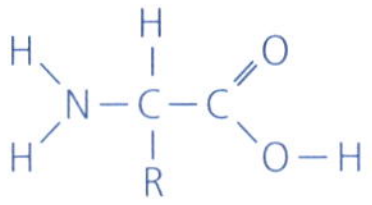

Figure 20.4.2 The general structure of an amino acid.

Figure 20.4.3 Muscle tissue in animals is mainly protein.

Proteins

Proteins, including enzymes, are natural polymers. Proteins have the same amide linkage as nylon. But instead of being made from two different monomers, proteins are made from the 20 different amino acid monomers. The order of the amino acids in proteins does not follow any regular pattern.

When proteins are made, an amide link is formed by the reaction of the amine group of one amino acid with the carboxylic acid group of the next amino acid. This is an example of a condensation reaction. Water is eliminated. The section of the amino acid within the protein chain is called the **amino acid residue**. There is no regular repeat unit in proteins because the amino acid residues are not in any regular order.

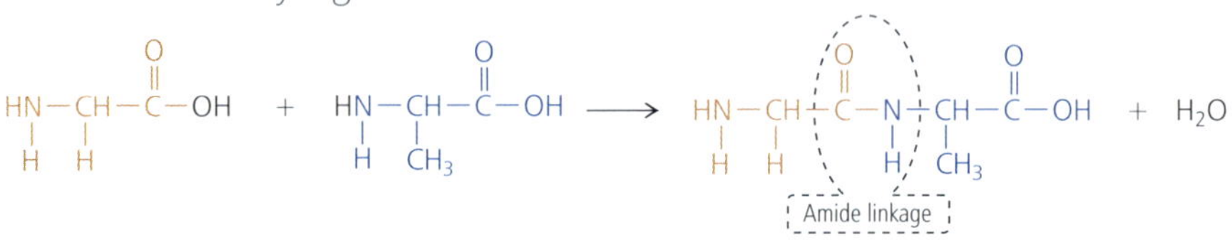

Figure 20.4.4 An amide link is formed in a condensation reaction between two amino acids.

Unit 20: Polymers

We can simplify the structure of an amino acid even further and show what happens when a protein is formed:

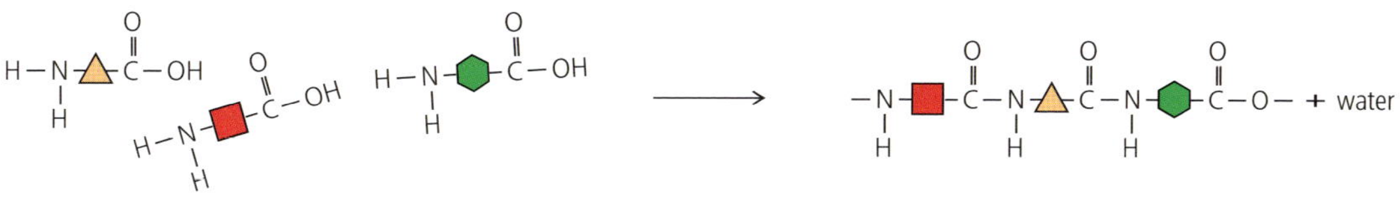

Amino acids → Part of a protein chain + water

Figure 20.4.5 When proteins are made, amino acid monomers form amide linkages with one another by condensation polymerisation.

Comparing proteins and synthetic polyamides

Figure 20.4.6 compares the structures of a protein with a synthetic polyamide such as nylon -6,6.

Note that:

- in a protein, there are many types of amino acid monomers which have been polymerised. In nylon-6,6 there are two types of monomers which have been polymerised.
- in a protein, the order of the amino acid residues is irregular. There is no repeat unit. In nylon-6,6 there is a regular order of repeat units.
- in a protein, the amide linkages are in the same direction: –CO–NH–; –CO–NH–.

  In nylon-6,6 the order alternates: —CO—NH—; —NH—CO—.

EXAM TIP
Make sure that you know the differences in the structures of a protein and a synthetic polyamide.

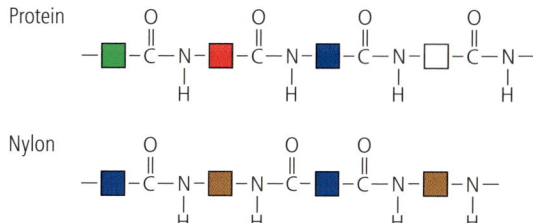

Figure 20.4.6 The structures of a protein and a synthetic polyamide (nylon-6,6) compared.

A synthesis problem

Look at the structure of the amino acid serine in Figure 20.4.1. When we react two serine molecules together we form an ester linkage as well as an amide linkage. The ester linkage is formed by the reaction between the —OH of the side-chain and the —COOH group of another serine molecule.

SUMMARY QUESTIONS

1. State the names of the two functional groups in amino acids that condense to form the amide linkages in proteins.

2. Copy and complete using these words:

   irregular   unit   order   regular   repeat

   In a protein, the _____ of the amino acid residues is _____. There is no _____ unit. In nylon there is a _____ order of the repeat _____.

KEY POINTS

- The general structure of an amino acid is H_2N—$CH(R)$—$COOH$.
- Proteins have the same amide linkage as nylon but the monomer units are different.
- The arrangement of the amide links in proteins is different from those in nylon-6,6.

251

SUMMARY QUESTIONS

1 Copy and complete using these words:

chains   ethene   join   monomers   polymer

At high temperature and pressure _____ molecules combine to form long _____ of poly(ethene). The long chain is called a _____. The small molecules that _____ to form the chain are called _____.

2 Match the words on the left with the descriptions on the right.

monomer	the name of a polymer formed from C_2H_4 monomers
addition	a molecule made by combining monomers
polymer	a simple molecule from which a polymer is made
poly(ethene)	a reaction where two or more molecules combine to form only one product

3 Write down one positive and one negative consequence for each of the following ways of disposing of plastics:

  a burning the plastics
  b recycling the plastics
  c putting the plastics into a landfill site

4 Match each of the polymers on the left with its monomer on the right.

poly(chloroethene)	$C_2H_5CH=CH_2$
poly(butene)	$CH_2=CH_2$
poly(ethenyl ethanoate)	$CH_3CH=CH_2$
poly(propene)	$CH_2=CHCl$
poly(ethene)	$CH_3COOCH=CH_2$

5 State which of these compounds undergoes addition polymerisation.

  A $CH_3CH=CH_2$ B $CH_3CH_2CH_3$
  C CH_3CH_2COOH D $C_6H_5NH_2$
  E $C_6H_5CH=CH_2$

6 Draw simplified diagrams of the monomers used to make **a** PET **b** nylon.

Practice questions

1 Which one of these statements about poly(ethene) is true?
  A Poly(ethene) is formed by the addition polymerisation of ethane.
  B Ethene monomers are used to make poly(ethene).
  C Hydrogen is given off when poly(ethene) is made.
  D Poly(ethene) is made by condensation polymerisation.
(Paper 1)

2 Which statement about the monomers W, X, Y and Z used to make polymers is correct?

$CH_2=CH_2$ (W) $HO-\square-OH$ (X)
$HO_2C-\square-CO_2H$ (Y) $H_2N-\square-NH_2$ (Z)

  A W and X polymerise to form a polyamide.
  B X and Y polymerise to form a polyester.
  C X and Z polymerise to form a polyamide.
  D Y and Z polymerise to form a polyester.
(Paper 2)

3 At high temperature and pressure, ethene molecules combine to form a polymer.
  (a) Draw the displayed formula of an ethene molecule. [1]
  (b) State the general name given to a small molecule that combines to form a polymer. [1]
  (c) Give the name of the polymer formed from ethene molecules. [1]
  (d) State the type of polymerisation that occurs when ethene molecules are polymerised. [1]
  (e) Give the feature of ethene that causes it to polymerise. [1]
  (f) Draw a section of the polymer formed when ethene is polymerised. [2]
(Paper 3)

4 Poly(propene) is a polymer that is made from propene monomers, $CH_3CH=CH_2$.
  (a) State the type of reaction that occurs when propene monomers polymerise. [1]
  (b) Draw the structure of poly(propene) to show three repeat units. [3]

(c) Draw a simplified structure of poly(propene), showing one repeat unit, using the letter '*n*' to show the repetition of this unit. [3]

(Paper 4)

5 A simplified structure of nylon is shown.

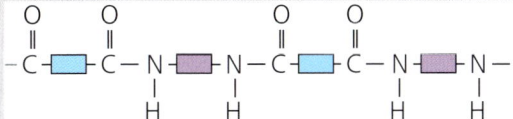

(a) Nylon is made by condensation polymerisation. What do you understand by the term condensation polymerisation? [3]
(b) State the name given to the linkage between the monomer units in nylon. [1]
(c) Write the simplified formulae for the two monomers used to make nylon. [2]
(d) A polymer is made from two monomers having the structures shown:

$H_2N—C_6H_4—NH_2$
$HOOC—C_6H_4—COOH$

(i) Draw the structure of this polymer to show one repeat unit, using the letter '*n*' to show the repetition of this unit. [3]
(ii) State the name of the small molecule eliminated when these two monomers combine. [1]

(Paper 4)

6 Some polymers contain halogen atoms. The structure of the addition polymer poly(dichloroethene) is shown:

$$\left[\begin{array}{c} H \; Cl \\ | \; | \\ -C-C- \\ | \; | \\ H \; Cl \end{array}\right]_n$$

(a) State the meaning of the term addition polymer. [2]
(b) (i) Draw the displayed formula of the monomer used to make this polymer. [1]
(ii) State the name of this monomer. [1]

(c) The polymer 'Teflon' is obtained from the monomer tetrafluoroethene:

$$\begin{array}{c} F \quad\quad F \\ \diagdown \quad \diagup \\ C=C \\ \diagup \quad \diagdown \\ F \quad\quad F \end{array}$$

(i) State the chemical name of this polymer. [1]
(ii) Draw the structure of 'Teflon' to show three repeat units. [2]

(d) State one adverse effect of burning plastics containing halogen atoms. [1]

(Paper 4)

7 Polymer A has the structure:

$$-\overset{O}{\underset{\|}{C}}-C_{10}H_6-\overset{O}{\underset{\|}{C}}-O-CH_2-CH_2-O-\overset{O}{\underset{\|}{C}}-C_{10}H_6-$$
$$\text{(continued)} -\overset{O}{\underset{\|}{C}}-O-CH_2-CH_2-O-$$

(a) State the type of linkage present in this polymer. [1]
(b) Give the name of another polymer with this type of linkage. [1]
(c) (i) Draw simplified structures, using boxes, of the two monomers used to make polymer A. [2]
(ii) State the type of polymerisation that occurs when polymer A is made. Explain your answer. [3]

(Paper 4)

8 The structure of glycine is shown:

$H_2N—CH_2—COOH$

(a) State the name of the group of compounds to which glycine belongs. [1]
(b) Glycine can undergo a condensation reaction in which an amide linkage is formed.
(i) Draw the displayed structure of an amide linkage. [1]
(ii) Name a synthetic polymer that contains an amide linkage. [1]
(c) Glycine can be polymerised to poly(glycine). Draw the partial structure of this polymer to show three repeat units. [3]

(Paper 4)

253

C1 Using and organising techniques, apparatus and material

LEARNING OUTCOMES

- Describe, explain or comment on experimental arrangements and techniques
- Identify sources of error and suggest possible improvements in procedures

EXAM TIP

In (a) make sure that full and correct names are used. 'Cylinder', 'stand' and 'spoon' are not precise enough.

In (b) use a ruler and a sharp pencil and label the diagram clearly. A filter funnel and filter paper both need to be included in the answer.

Naming of apparatus was discussed in Topic 2.1.

Apparatus

Your experience of practical work in chemistry can be gained from working as an individual, from group work or from teacher demonstrations. The recognition of laboratory apparatus and standard equipment, such as Bunsen burners and tripods, is essential.

Example 1

The diagram shows how to make a solution of copper(II) sulfate by reacting excess copper(II) oxide with dilute sulfuric acid.

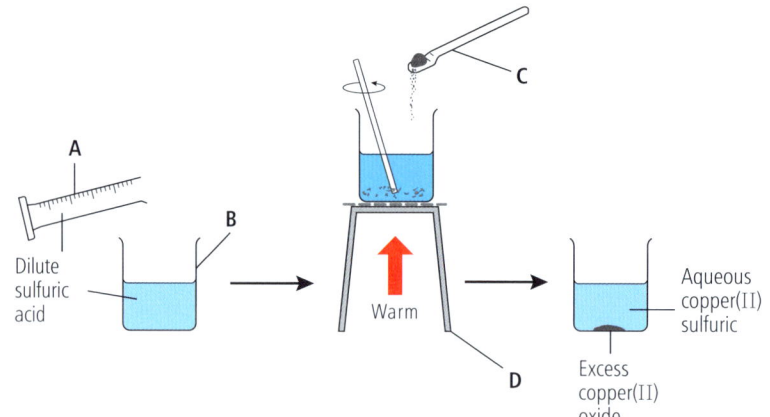

a Name the pieces of apparatus, **A**, **B**, **C** and **D**.

b Draw a labelled diagram of the apparatus needed to separate the excess copper(II) oxide from the aqueous copper(II) sulfate.

Safety

Carry out experiments using safe and efficient procedures. These involve the use of protective clothing, such as goggles, and the awareness of the dangers of flammable, corrosive and harmful chemicals.

Alternative to practical section

Understanding experimental set-ups and identifying problems

Information will be provided to set the context as in the question in Example 2.

Example 2

Chlorine gas is denser than air and slightly soluble in water. A sample of chlorine is prepared by adding hydrochloric acid to manganese(IV) oxide and warming the mixture.

Study the diagram of the apparatus used.

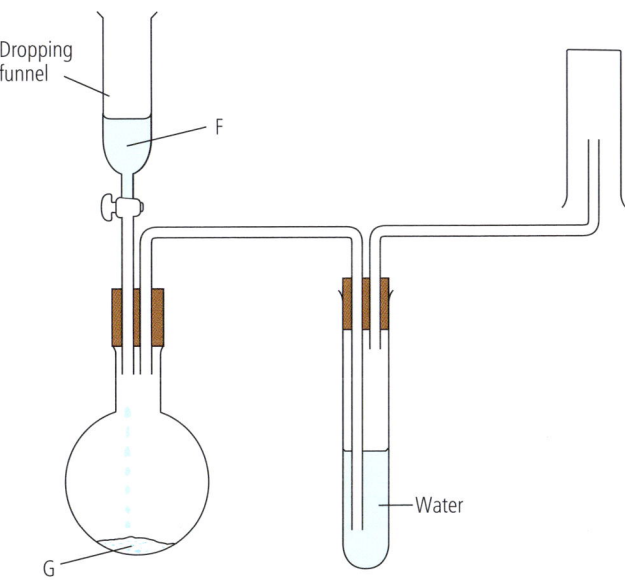

a Identify the chemicals used, **F** and **G**.
b Draw an arrow on the diagram to show where heating is applied.
c Identify two errors in the diagram. Explain your answers.

Reducing errors

Using apparatus that gives greater precision or by repeating the experiments and taking the average of a set of results can help to reduce errors. Improved techniques can also reduce errors. Suggested improvements to reduce the sources or error in experiments should be meaningful and detailed.

- For example, vague statements such as 'Repeat the experiment to obtain more accurate results' and 'Use a more accurate measuring cylinder' are not sufficient.

Good answers include:

- Repeat the experiment and take the average of the results.
- Use a burette/graduated pipette instead of a measuring cylinder.

> **EXAM TIP**
>
> In (a), make sure that you label the chemicals correctly taking into account the physical states.
>
> In the answer to (b) an arrow needs to be positioned correctly with its point touching the flask.
>
> In (c) do not write vague answers, e.g. 'There is no lid on the collecting vessel.'
>
> Methods for collecting gases are discussed in Topic 9.1.

C2 Observing, measuring and recording

LEARNING OUTCOMES
- Record readings from diagrams of apparatus
- Complete tables of data
- Plot graphs

In examination papers, diagrams of apparatus such as thermometers, burettes, measuring cylinders, gas syringes, etc., are used. These show readings at various stages of an experiment and you are expected to fill in a blank table of results.

Example

A student measured the temperature changes when different volumes of acid R react with a fixed volume of alkali S.

The results are shown in the table.

EXAM TIP

Topic 2.1 includes information about accuracy when measuring volumes. Remember that burette diagrams are made to be read to one decimal place.

EXAM TIP

Make sure that you don't misread the temperatures in questions involving thermometer diagrams. The top and bottom values of the temperature scale are not always the same, e.g. compare the top and bottom values at 0°C and 10°C.

Volume of solution R added to 25 cm³ of solution S / cm³	Thermometer diagram	Maximum temperature / °C
0	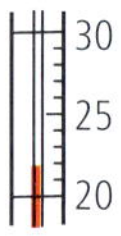 (20–30)	
10	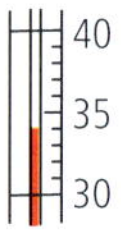 (30–40)	
20	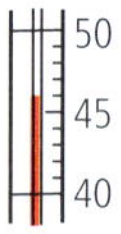 (40–50)	
30	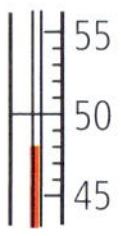 (45–55)	
40	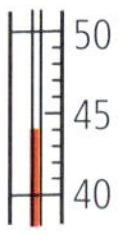 (40–50)	
50	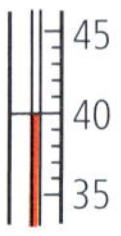 (35–45)	

a Use the thermometer diagrams to record the maximum temperatures reached.

b Deduce the maximum temperature change.

Alternative to practical section

Graphs

Examination questions requiring you to draw a graph will include the grid. Axes may be labelled and scales included. When a blank grid is provided, you need to choose an appropriate scale and axes need to be labelled clearly with units. Each point should be plotted as a small cross using a sharp pencil. Best-fit lines should be drawn leaving out any anomalous points, e.g. the point at 20 seconds in the lowest line on the upper graph. The graphs will be smooth curves or straight lines.

> **EXAM TIP**
>
> When drawing graphs make sure that you use as much of the paper as possible and use a sensible scale, e.g. 10 small squares = 10 or 20 cm^3 of gas.

Example of good practice

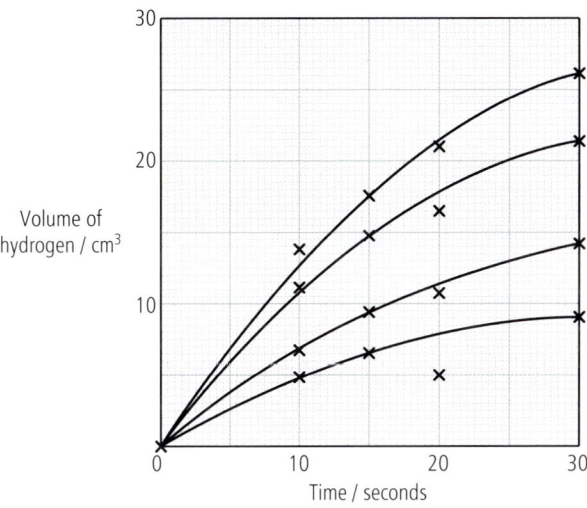

Example of poor practice

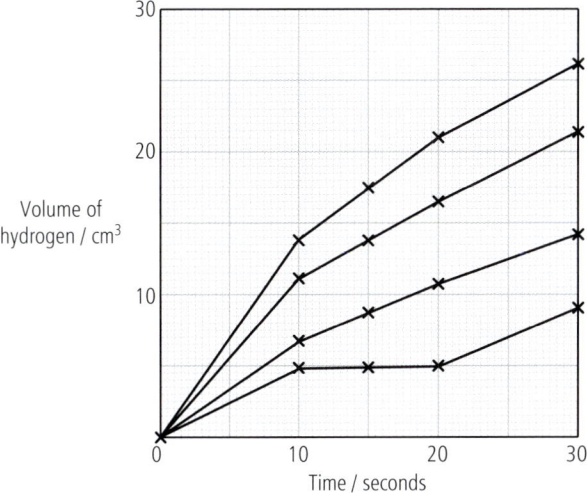

This is an example of poor practice because the points are joined by straight lines. There should be a continuous smooth curve going through or close to each point (and ignoring anomalous points).

C3 Handling experimental data and observations

LEARNING OUTCOMES

- Draw conclusions from information given
- Interpret and evaluate observations and data
- Draw conclusions from tests for gases and ions

EXAM TIP

You can find out more about practical procedures in Unit 2 and Topics 9.1, 9.2 and 9.3.

EXAM TIP

Sample question 1 involves applying experience of common practical procedures to an unfamiliar situation. It also tests knowledge and understanding of chromatography. When drawing chromatography apparatus, make sure that the place where you put the original spot of colour is above the level of the solvent.

You can find out more about chromatography in Topic 2.2.

Making conclusions and evaluating evidence

Making conclusions involves the interpretation of results using knowledge and understanding of chemical facts and concepts, and practical techniques.

Looking for patterns in data

Two titrations were carried out using two different solutions of hydrochloric acid, **A** and **B** and the same volume of aqueous sodium hydroxide. 25.0 cm^3 of acid solution **A** neutralised the same volume of sodium hydroxide as 50.0 cm^3 of acid solution **B**.

Conclusion: Acid **A** is twice as concentrated as acid **B**.

Evaluating an experiment

Evaluation involves asking questions relating to the success or failure of the experiment. For example:

- Was it a fair test?
- Were any of the variables difficult to control?
- Did the repeats of the experiment suggest that the results are reliable?
- Can anything be done to improve the method and/or apparatus used?

Sample question 1

A seaweed was investigated to check the colourings present.

Step 1 The seaweed was ground with sand using a mortar and pestle.

Step 2 Water was added and the mixture was stirred.

Step 3 The mixture was filtered.

Step 4 The filtrate was concentrated.

Step 5 The concentrate was analysed by chromatography.

a State the purpose of Step 1.

b Suggest why the mixture was filtered.

Alternative to practical section

c Describe how Step 4 was carried out.

d Draw a labelled diagram to show the apparatus used in Step 5.

Sample question 2

This example shows a sample answer to a question about qualitative analysis.

Compound **Z** is analysed. The table shows what was done. The observations have been completed for you.

Tests carried out on compound Z	Observation
1 Heat one measure of the mixture gently.	Condensation at the top of the tube
2 The condensation was tested with blue cobalt chloride paper.	Paper turned pink
3 The rest of the mixture was dissolved in plenty of water and the colour observed. The following tests were then carried out on separate portions.	Light violet solution
4 A few drops of aqueous sodium hydroxide were added then sodium hydroxide was added in excess.	Green precipitate forms The green precipitate dissolves
5 A few drops of aqueous ammonia were added then aqueous ammonia was added in excess.	Grey-green precipitate forms The precipitate does not dissolve
6 Dilute nitric acid was added and then aqueous silver nitrate.	White precipitate

a State the conclusion from Tests 1 and 2. Explain your answer. [2]

b State the conclusion from Test 3. Explain your answer. [2]

c State the conclusion from Test 4. Explain your answer. [3]

d State the conclusion from Test 5. Explain your answer. [2]

e State the conclusion from Test 6. Explain your answer. [2]

f Name compound **Z**. [2]

> **EXAM TIP**
>
> You are expected to recall the tests for particular ions, gases and molecules.
>
> You can find information about qualitative tests in Topics 12.5 and 12.6.

> **EXAM TIP**
>
> Make sure that you can distinguish Zn^{2+} ions from Al^{3+} ions and Cr^{3+} ions from Fe^{2+} ions by the use of aqueous sodium hydroxide and aqueous ammonia. The results can seem quite similar.

> **EXAM TIP**
>
> Remember that you can distinguish a precipitate of iron(II) hydroxide from a precipitate of chromium(III) hydroxide by the fact that the green precipitate of iron(II) hydroxide soon turns brown at the surface.

C4 Planning investigations

LEARNING OUTCOMES

- Suggest suitable apparatus and techniques for an investigation

Planning an investigation

In examination, you may be asked a question about planning an investigation.

Details of the investigation should include:

- apparatus/equipment to be used
- conditions to be employed, e.g. high temperature or room temperature
- measurements and observations to be made and recorded
- comparison of experiments to ensure a fair test
- interpretation of results to make conclusions
- evaluation of results.

Sample question 1

You want to compare the energy released when two different liquid fuels **X** and **Y** are burned. Describe how to do this using the apparatus shown below. You can also choose additional apparatus. In your answer describe how you know which fuel releases the most energy.

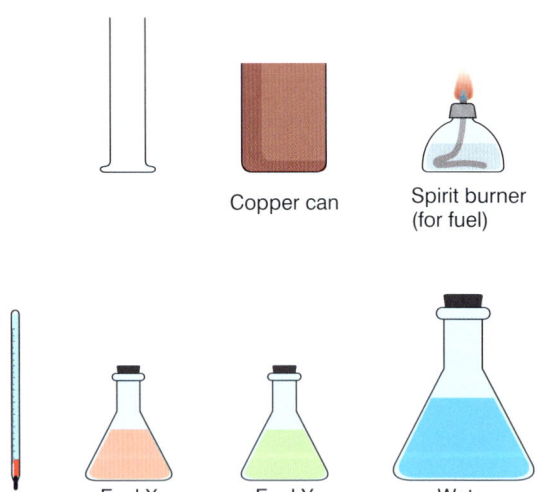

EXAM TIP

When answering questions about the energy released from fuels make sure that:

- every step in the procedure is clearly described
- a clearly labelled diagram is drawn
- when comparing different fuels, the test is fair, e.g. the same distance of the spirit burner from the test tube, the same mass of water used.

You can find out more about comparing the energy released when a fuel burns in Topic 8.4.

Good answers include

- burning each fuel for the same time
- using the same volume of water in the steel can for each fuel
- measuring the initial and final temperatures of the water

Alternative to practical section

- recording the amount of each fuel burned (in grams or moles)
- commenting on thermal energy transfers to the surroundings/attempts to reduce thermal energy transfers to the surroundings
- repeating the experiment
- evaluating the experiment

Sample question 2

This example shows a sample response to a planning question which gains full marks.

The ticks show where the marks have been awarded. Although there are 6 marks for the question, for extended answers such as this there are usually additional points which are worth marks. There are 8 ticks but there is a maximum of only 6 marks which can be obtained.

Chemical drain cleaners contain concentrated sodium hydroxide solution. Plan an investigation to show which drain cleaner contains the highest concentration of sodium hydroxide. [6]

> The acid and alkali can burn your skin so I will wear a labcoat, gloves and goggles. ✓ I will do a titration. First I will dilute 10 grams of each cleaner with 50 cm³ of water. ✓ Then I will put 25 cm³ of one of the drain cleaners in two conical flasks using a volumetric pipette. ✓ I will do a titration using a burette filled with 1 mol/dm³ hydrochloric acid. ✓ I added three drops of methyl orange into the flask with the drain cleaner. ✓ Then I ran acid into the flask until the colour of alkali changes from yellow to red. ✓ Write down the cm³ of acid used. The experiment is repeated with the same volume of cleaner to check the values are the same or nearly. ✓ Then I do exactly the same thing with the other drain cleaner. If less acid was used the alkali is less concentrated. If more acid is used the alkali is more concentrated. ✓ The results from both flasks with the different drain cleaners are compared.

EXAM TIP

When answering this type of question think about these points.

- Are measurements and observations clearly recorded?
- Is there a clear idea of a fair test?
- Has a comparison of the results been made?
- Have appropriate conclusions been made?
- Have safety precautions been considered?
- Has the experiment been repeated to check reliability?

You can find out more about titrations in Topics 6.8 and 11.5.

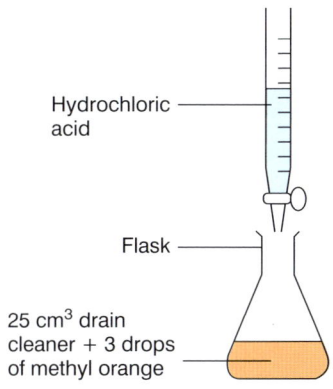

Hydrochloric acid

Flask

25 cm³ drain cleaner + 3 drops of methyl orange

261

Exam skills practice

The following pages contain sample exam questions and a student's answers to these questions. Read the advice and technique boxes for tips on how to approach these questions and the common errors to avoid.

Sample Question 1

Supplementary level questions are in **bold**.

Gallium and aluminium are in Group III of the Periodic Table.

a Deduce the electronic configuration of an aluminium atom. [1]

..... 2 8 3 ✓

b Name the ore of aluminium. [1]

..... Al_2O_3 ✗

> **Exam advice:**
> The answer did not get the mark because the instruction to give a *name* was not applied. The correct answer is 'bauxite'. In addition the student did not realise that an ore is a rock containing a compound which is extracted. Although Al_2O_3 is present in the ore, there are lots of other substances as well.
>
> **Exam technique:**
> Make sure that you read each word in the question carefully. It is a common error to put a formula rather than a name.

c Aluminium is extracted by electrolysis.

State the meaning of the term electrolysis. [2]

..... Breaking down an ionic compound using ✓ an electric current ✓

d Suggest why aluminium is not extracted by heating aluminium oxide with carbon. [1]

..... Aluminium *does not react with carbon* ✗

> **Exam advice:**
> Although the student has some idea that the question is about reactivity, there is no mention of the reactivity series. An answer such as 'aluminium is above carbon in the reactivity series' will get a mark.

e Explain why aluminium is used in the manufacture of overhead electrical cables. [2]

..... It's light ✗ and conducts electricity well ✓

> **Exam advice:**
> The student's answer only got one of the two marks available because the answer 'light' is too vague – it could refer to brightness! The more precise answer 'it has a low density' will get the mark.
>
> **Exam technique:**
> Make sure that you use the correct scientific terms. Scientific terms are always precise. Use the glossary at the end of the book to help you.

f Complete the equation for the reaction of aluminium with chlorine. [2]

... 2 Al ✓   + 3Cl ✗ ⟶ 2AlCl$_3$

> **Exam advice:**
> The answer only got one of the two marks available because the student (i) did not realise that the element chlorine is diatomic, i.e. Cl$_2$, and (ii) did not take notice of the 2 in front of the 2AlCl$_3$. Both marks are given for: 2Al + 3Cl$_2$ ⟶ 2AlCl$_3$
>
> **Exam technique:**
> Make a list of the elements which are diatomic, e.g. Cl$_2$, Br$_2$, H$_2$, O$_2$, N$_2$, and make sure that you use these formulae in balanced equations.

g An ion of gallium is represented by the symbol $^{69}_{31}Ga^{3+}$.

Deduce the number of protons, neutrons and electrons in this ion. [3]

protons 31 ✓   neutrons 38 ✓   electrons 31 ✗

> **Exam advice:**
> The number of neutrons (69 – 31 = 38) was deduced correctly but the student did not take notice of the 3+ charge on the Ga ion. The 3+ charge shows that three electrons have been removed from the atom. So the correct answer is 31 – 3 = 28.
>
> **Exam technique:**
> Make sure that you take the charge of the ion into account. If the ion is negative, the atom has gained electrons, so the number of electrons = proton number + charge number.

Exam skills practice

h A sample of gallium has two isotopes. The symbol and the percentage abundance of these isotopes is shown.

$^{69}_{31}Ga$ abundance = 60.2%; $^{71}_{31}Ga$ abundance = 39.8%

Calculate the relative atomic mass of the gallium in this sample. Give your answer to three significant figures. [3]

69 × 60.2 = 4153.8

71 × 39.8 = 2825.8 ✓

= $\frac{6979.6}{100}$ ✓ = 69.79 ✗

> **Exam advice:**
> This is a good answer which is clearly set out with the relevant working shown. Marks are given for the correct process applied. Although the final answer seems correct, the student (i) Did not round up 69.79 to 69.80 and (ii) did not apply the instruction to give the answer to three significant figures.
>
> **Exam technique:**
> Make sure that you know about significant figures. Remember to apply the instruction in the stem to your calculated answer, for example by highlighting it in some way before starting your answer.

i (i) Deduce the formula of gallium oxide. [1]

..... GaO_3 ✗

> **Exam advice:**
> The student realised that Ga is in Group III of the Periodic Table but did not take into account that the oxide ion has a 2− charge. A comparison with the formula of aluminium oxide, Al_2O_3, gives the correct answer, Ga_2O_3.
>
> **Exam technique:**
> When answering question about unfamiliar metal compounds, find the group that the metal belongs to and then think about a similar compound you know about.

(ii) Gallium oxide is an amphoteric oxide.
Give the meaning of the term amphoteric oxide. [1]

..... It reacts with both an acid and an alkali ✓

j An alloy of aluminium, silicon and iron is stronger than pure aluminium.

   Explain why the alloy is stronger. [3]

   ..... The iron atoms are a different size to the aluminium atoms.
   ✓ They don't move over each other. when a force is applied. But in aluminium they do move over each other.

> **Exam advice:**
> The student's answer only got one of the three marks available because (i) there was no mention of layers of atoms (or ions) for the second mark, and (ii) there was no idea of the layers sliding or not sliding for the third mark. The answer 'atoms move over each other' could just as well apply to a liquid.
>
> **Exam technique:**
> Make sure that you use the idea of 'layers of atoms or ions sliding' when comparing the malleability and hardness of metals.

> **Overall comments**
> This student gained 11 out of a possible 20 marks.
>
> This synoptic question needs knowledge from Units 3, 4, 7 and 15 to gain full marks. Improving your exam technique will help to improve your marks. For example, reading the stem of the question carefully (parts b and h) and using the correct scientific terminology (e and j). Learning to write with more precision will also help to improve your marks.

Sample Question 2

Supplementary level questions are in **bold**.

This question is about iron and compounds of iron

a Iron is a typical transition element.

   Describe three physical properties of iron that are different from metals such as sodium. [3]

   1 ..... high melting point ✓

   2 ..... high boiling point

   3 ..... does not react with water

Exam skills practice

> **Exam advice:**
> The student's answer got only one of the three marks available because (i) high boiling point and high melting point are too similar to be awarded separate marks, and (ii) point 3 did not get a mark because the question asks about the physical properties of iron not its chemical properties.
>
> **Exam technique:**
> When writing about the properties of substances (i) make sure that the properties are sufficiently different, and (ii) that you know the difference between physical properties and chemical properties.

b Iron reacts with sulfuric acid to form the salt iron(II) sulfate. Describe how to prepare pure dry crystals of iron(II) sulfate using excess iron. [4]

..... I will heat the mixture then filter it though a filter paper. ✓ Then I will then heat the iron sulfate to get the crystals. The crystals are filtered off then washed and dried with filter paper. ✓

> **Exam advice:**
> The student realised that the excess iron had to be filtered off and the crystals had to be dried using filter paper. The comment about heating the iron sulfate is too vague. A detailed statement such as 'heat the solution to the point of crystallisation then leave' is needed to get the mark. Specific statements about 'heating the filtrate to evaporate water' or 'washing with' a specific liquid are missing. Note that in these extended questions there are usually more points that can score marks than the number of marks shown in the question.
>
> **Exam technique:**
> Make sure that you write precisely with as much detail as required.

c (i) Iron is extracted from iron ore in a blast furnace. Name two *other* raw materials that are used in the extraction of iron. [2]

..... air ✓   and ..... lime (calcium oxide) ✗

> **Exam advice:**
> Raw materials are substances that have not been processed. Lime is not a raw material because it is made by heating limestone. Limestone is a raw material. Note that coke is not a raw material (as coke is made from coal).
>
> **Exam technique:**
> Make sure that you know the precise meaning of scientific terms.

(ii) One of the reactions in the blast furnace is shown:

$Fe_2O_3 + 3CO \longrightarrow 2Fe + 3CO_2$

Write a word equation for this reaction. [2]

.... iron oxide + carbon oxide ⟶

iron + carbon dioxide ✓

> **Exam advice:**
> A mark has been given for the correct products. Carbon oxide is not precise enough because there is more than one oxide of carbon and we need to distinguish between them; in this case 'carbon monoxide' is needed for the mark.
>
> **Exam technique:**
> Where there are different oxides of the same element, e.g. SO_2 and SO_3, you need to be able to distinguish between them by name. For transition element oxides this can be done using oxidation numbers, e.g. iron(II) oxide and iron(III) oxide.

Exam skills practice

d (i) Iron rusts. Name the two substances needed for rusting. [2]

..... air ✓ and ..... water ✓

(ii) Explain why painting the surface of clean iron prevents rusting. [2]

..... The paint stops the air and water from reacting with the iron ✓

> **Exam advice:**
> The student's answer only got one of the two marks available because there is not enough detail. Their answer does not fully explain what the paint does to prevent rusting. Some idea that the paint forms a barrier is needed for the second mark.
>
> **Exam technique:**
> Note that when a question is worth two marks you need to put down two distinct points in your answer.

e Iron(III) oxide, Fe_2O_3, reacts with magnesium

$$Fe_2O_3 + 3Mg \longrightarrow 2Fe + 3MgO$$

Explain how this equation shows that iron(III) oxide gets reduced. [1]

..... There is less oxygen in the products than in the iron(III) oxide ✗ and there is iron oxide on the left and only iron on the right.

> **Exam advice:**
> To gain the mark, specific reference needs to be made to the iron oxide losing oxygen. The statement that there is 'less oxygen in the products' is incorrect – the total amount of oxygen cannot change. The last part of the student's sentence just repeats the stem of the question.
>
> **Exam technique:**
> Make sure that you write precisely, for example referring to iron oxide losing oxygen, not iron losing oxygen.

f The equation shows a redox reaction.

$$Fe + Cu^{2+} \rightarrow Fe^{2+} + Cu$$

(i) Describe and explain this reaction in terms of electron transfer and oxidation number changes. [3]

..... The iron is oxidised and the copper ✗ is reduced

because iron gains electrons and Cu^{2+} lose electrons.

The oxidation number of iron increases (oxidation) and copper goes from +2 to 0 (reduction) ✓

Exam advice:

The student's answer only got one of the three marks available because there is confusion about which species of iron is being referred to (atoms or ions). The copper is not being reduced. It is the copper <u>ions</u> that are reduced. The loss and gain of electrons is incorrect – the iron is losing electrons.

Exam technique:

When answering questions about redox, you may find it easier to use increase or decrease in oxidation number than to write about gain or loss of electrons. If you refer to electron loss or gain, you must be clear about the species you are referring to, e.g. iron in iron oxide/iron ions/iron atoms.

(ii) Explain which is the reducing agent in this reaction [1]

..... The reducing agent is iron ✓ It is reduced

because iron gains electrons. ✗

Exam advice:

The student has confused reduction with the idea of a reducing agent. A reducing agent reduces another substance and is itself oxidised (syllabus definition). So the iron is the reducing agent because it is getting oxidised. The mark would also have been given for 'because the iron gives (two) electrons to Cu^{2+} ions'.

Exam technique:

When identifying reducing agents it is easier to explain this in terms of reducing another substance than to write about electrons.

Exam skills practice

(iii) Describe a test for reducing agents. [3]

test . . . . . Add acidified potassium mangante(VII) ✓

colour change from . . . . . purple ✓ to transparent ✗

> **Exam advice:**
> The mark for the test has been given despite the incorrect spelling (manganate is correct) because there is no confusion. The colour of the product of the reaction is incorrect – transparent means that you can see through it. The correct term is colourless.
>
> **Exam technique:**
> Make sure that you use the correct spellings and the word 'colourless' to mean no colour.

> **Overall comments**
>
> This student gained 12 out of a possible 23 marks.
>
> This synoptic question needs knowledge from Units 4, 5, 10, 12 and 15 to gain full marks. The marks could be improved by writing answers which are more precise, especially when describing processes (part b) and naming compounds (parts c and f(iii)). Make sure that you practice answering questions requiring longer answers (such as salt preparation in part b) and give sufficient details about the procedure. Marks can also be gained if you have a clear idea of the difference between chemical properties and physical properties (part a). Practice in answering question about redox equations in terms of electron loss and gain in specific species (parts e and f) is essential for those aiming for higher grades.

Sample Question 3

Supplementary level questions are in **bold**.

Zinc powder reacts rapidly with dilute hydrochloric acid.

$$Zn(s) + 2HCl(aq) \longrightarrow ZnCl_2(aq) + H_2(g)$$

a (i) State the name of the salt formed in this reaction. [1]

..... $ZnCl_2$ ✗

> **Exam advice:**
> The student misread the question and just identified the salt rather than naming it. It is essential that you read the stem of the question carefully so that you do not make simple errors like this.
>
> **Exam technique:**
> Make sure that you distinguish between the phrases 'state the name' and 'state the formula'.

(ii) Give the meaning of these symbols: [2]

(aq) aquatic ✗   (g) gas ✓

> **Exam advice:**
> The student's answer to the meaning of (aq) has a significantly different meaning to the correct word 'aqueous'. So the mark was not given.
>
> **Exam technique:**
> Make sure that you remember and use scientific terms in the syllabus accurately. Make a list of them and test yourself or get another student to test you.

b The graph shows how the volume of hydrogen changes with time. The zinc is in excess.

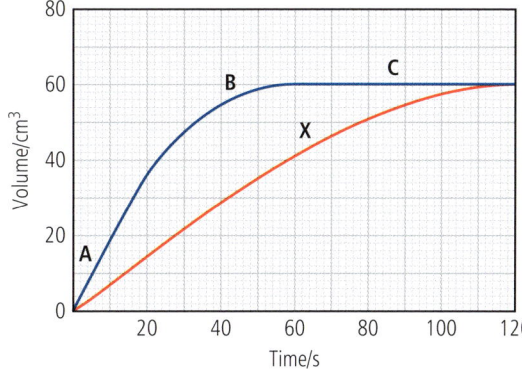

271

Exam skills practice

(i) Explain the shape of the graph at points A, B and C in terms of rate of reaction. [3]

..... At A the reaction is fastest because the gradient is very steep. ✓ At B the rate of reaction is still quite fast. ✗ The reaction has stopped at C where the line is horizontal. ✓

> **Exam advice:**
> The student gained two of the three marks available. The description of the rate at point B is too vague. There is no reference to a 'shallower gradient' or the line being 'less steep'.
>
> **Exam technique:**
> Make sure that you know the meaning of the command words used in the exams, such as 'explain' or 'describe'. 'Explain' means that you have to give a reason for an observation or refer to a particular theory.

(ii) Deduce the volume of gas released 20 seconds from the start of the experiment. [1]

..... 28 cm^3 ✗

> **Exam advice:**
> The student has misread the scale of the graph by counting upwards from 20 cm^3 and thinking that 1 small square = 1 cm^3 (rather than 2 cm^3).
>
> **Exam technique:**
> Before drawing and interpreting graphical data, make sure that you are using the correct scale by working out the value of each small square and each large square on the grid.

(iii) On the grid above draw a line to show how the volume of gas changes when hydrochloric acid of half the concentration is used. Label this line X.

All other conditions stay the same [2]

> **Exam advice:**
> The student's answer (shown on the graph in red) only got one of the two marks available (as the gradient is shallower). The student did not use the information at the beginning of part b which stated that the zinc is in excess. At half the concentration (with all other conditions the same) there must be half the number of moles of hydrochloric acid to start with. So the final volume of gas released will be about half the original value, and the curve should level off at 30 cm^3.
>
> **Exam technique:**
> When answering questions involving calculations or graphical work, make sure that you go back to the stem of the question to get the relevant information.

c State the effect of the following on the rate of reaction. All other conditions stay the same. [2]

Adding a catalyst . . . . . Increases the rate. ✓

Using larger pieces of zinc . . . . . Reaction takes longer time

> **Exam advice:**
> Although the statement about the larger pieces of zinc is true, the second mark was not obtained because the student only referred to time and not to rate. Make sure that you focus on particular words and their meaning when reading the stem of the question.
>
> **Exam technique:**
> When answering questions about rate of reaction you need to remember that rate is change in concentration (or amount of substance) divided by time.

Exam skills practice

d Explain using kinetic particle theory why increasing the temperature increases the rate of reaction. [3]

..... When the temperature is increased, the molecules have more energy and collide with each other more frequently. ✓ There are also more molecules with the higher energy.

> **Exam advice:**
>
> The student's answer only got one of the three marks available because (i) the type of energy there is more of has not been specified. To gain more marks, statements such as 'more kinetic energy' or 'particles move faster' are needed. (ii) The last sentence is not sufficient. There should be reference to 'more molecules with energy greater than the activation energy'.
>
> **Exam technique:**
>
> When answering extended questions make sure that you include specific terms such as kinetic (energy), activation energy and frequency (of collisions).

e In another experiment 0.81 g of zinc oxide reacts with 50 cm^3 of 0.5 mol/dm^3 hydrochloric acid.

$$ZnO + 2HCl \longrightarrow ZnCl_2 + H_2O$$

Show by calculation that the hydrochloric acid is in excess. Use the information in the Periodic Table to help you. [4]

molar mass of ZnO = 65 + 16 = 81 ✓

moles of ZnO = $\dfrac{0.81}{81}$ = 0.01 mol ✓

moles of acid = 0.5 × $\dfrac{50}{1000}$ = 0.025 mol ✓

The acid is in excess because 0.025 is bigger than 0.01

> **Exam advice:**
>
> The student's answer is well set out showing the main steps in the calculation. The fourth mark was not obtained because the 1:2 mole ratio in the equation has not been considered. In order to complete the calculation you need to take this into account, to prove that the hydrochloric acid is in excess. A mark would have been awarded for the idea that 0.025/2 is greater than 0.01 or that 0.01 × 2 is less than 0.025.
>
> **Exam technique:**
>
> In chemical calculations remember that you need to consider the stoichiometry of the equation.

> **Overall comments**
>
> This student gained eight out of a possible 18 marks.
>
> This synoptic question needs knowledge from Units 5, 6 and 9 to gain full marks. More marks could be obtained by writing answers more accurately using the appropriate scientific terms, such as kinetic energy and activation energy (part d). Reading and interpreting the stem of a question are very important (part a(i)). You should also make sure that you return to the start of a question if necessary to get suitable information (part b(iii)). Remember that in chemical calculations you must take account of the stoichiometry of the equation (part e). Your marks can also be improved by studying what each command word means (part b(i)).

Sample Question 4

Supplementary level questions are in **bold**.

We use the symbol ∧ to show that an essential word or piece of information has been omitted.

Calcium carbonate undergoes *thermal decomposition*.

$$CaCO_3 \longrightarrow CaO + CO_2$$

a State the meaning of the term thermal decomposition. [2]

. . . . . It decomposes when heated. ✓

> **Exam advice:**
>
> The student's answer only got one of the two marks available because there are two words in the term 'thermal decomposition' and both need explaining. A second mark can be obtained by defining decomposition as the 'breakdown of a substance'.
>
> **Exam technique:**
>
> When asked to define particular terms, remember that the number of separate points needed is usually equal to the number of marks shown at the end of the question.

Exam skills practice

b Give two **other** sources of carbon dioxide in the atmosphere. [2]

..... respiration ✓ and ..... fossil fuels

> **Exam advice:**
> The second mark was not given because there is no indication that the fuels are being burned. Either 'combustion of fossil fuels' or 'burning fossil fuels' would get the mark.
>
> **Exam technique:**
> When writing about the sources of atmospheric pollutants, remember that many of them (e.g. CO_2, SO_2, NO_2) are produced by combustion reactions of hydrocarbons, sulfur or nitrogen with oxygen.

c State one adverse effect of increasing the concentration of carbon dioxide in the atmosphere. [1]

..... increased global warming ✓

d Describe a test to show the presence of calcium ions in a compound. [2]

Test ..... add acid to calcium carbonate and test the gas produced ✗

Observations ..... brick red flame seen ✓

> **Exam advice:**
> The student understood that a flame test was involved but the exact method of doing this (placing a sample in a non-luminous Bunsen flame) was missing. The test was confused with the test for carbonate ions, possibly because this compound appeared in the stem of the question.
>
> **Exam technique:**
> Make sure that you focus on the question in the section that you are answering. You need to remember the tests for different ions – both flame tests and tests involving precipitation.

e When calcium carbonate is heated in a closed container, the following equilibrium occurs.

$$CaCO_3(s) \rightleftharpoons CaO(s) + CO_2(g) \quad \Delta H = +178 \text{ kJ/mol}$$

State and explain the effect of each of the following on the position of equilibrium: [6]

- the pressure is increased .... the equilibrium moves to the left (to the reactant side) ✓ in the direction of fewer molecules

> **Exam advice:**
> The student's answer got one of the three marks available because (i) 'in the direction of fewer molecules' is not specific enough. To get the mark, gas molecules need to be specified. (ii) There is no explanation of why this is happening, e.g. increasing the pressure pushes the molecules closer so the reaction tries to overcome this by reducing the number of (gas) molecules.
>
> **Exam technique:**
> Make sure that you write with precision using essential words such as 'gas' to highlight the type of particles that are relevant.

- the temperature is decreased .... the equilibrium moves to favour the reactants ✓ because the reaction is endothermic ✓ and shifting the equilibrium to the left heats up the reaction mixture, opposing the decrease in temperature ✓

f When heated in an open container, calcium carbonate decomposes completely.

$$CaCO_3(s) \longrightarrow CaO(s) + CO_2(g) \quad \Delta H = +178 \text{ kJ/mol}$$

Draw and label the reaction pathway diagram for this reaction to show:

- **the reactants and products**
- **the enthalpy change**
- **the activation energy** [4]

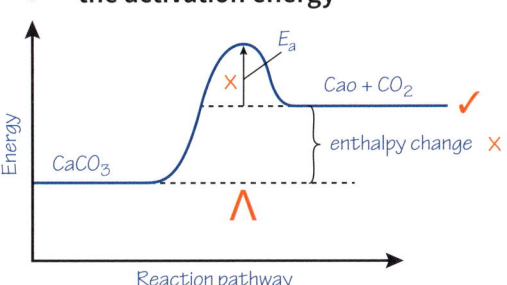

Exam skills practice

> **Exam advice:**
> The student gained a mark for the correct shape of the curve, with the products to the right of the reactants and at a higher energy level. Although the symbol for activation energy is correct, E_A for the backward reaction has been drawn – the arrow should go from the point marked ∧ to the top of the curve. The enthalpy change is unclear – it should be an upward pointing arrow and labelled ΔH or ΔH positive.
>
> **Exam technique:**
> When drawing reaction pathway diagrams make sure that you draw the arrows in the correct direction, and to and from the correct place.

g Explain why calcium carbonate (limestone) is used in the extraction of iron. Include a chemical equation other than the one in part (f). [3]

.... It's used to get rid of impurities in the iron ore like silicon dioxide ✓ by first breaking down to calcium oxide ✓ $CaO + SiO \rightarrow CaSiO$ ✗

> **Exam advice:**
> The student has gained marks for writing about silicon dioxide impurities and the decomposition to calcium oxide. The formula given for calcium silicate is incorrect (it should be $CaSiO_3$). You are also expected to know the formula of silicon dioxide, SiO_2.
>
> **Exam technique:**
> There are formulae and equations in the syllabus that you are expected to recall. Make sure that you know these.

> **Overall comments**
> This student gained 11 out of a possible 20 marks.
>
> This synoptic question needs knowledge from Units 8, 10, 12, 15 and 17 to gain full marks. More marks can be gained by writing your answers more accurately, using appropriate terms such as combustion (part b) and gas molecules (part e). You should also make sure that you learn the tests for specific ions mentioned in the syllabus (part d) and make a list of the specific formulae and equations in the syllabus that you have to recall. Make sure that you know the correct direction and position of the arrows in reaction pathway diagrams (part f).

Sample Question 5

Supplementary level questions are in **bold**.

We use the symbol Λ to show that an essential word or piece of information has been omitted.

This question is about chlorine and compounds of chlorine.

a Chlorine is a yellow-green gas at room temperature.

Describe the arrangement and separation of the molecules in a gas. [2]

arrangement .... *arranged far apart* ✗

separation .... *a long way from each other* ✓

> **Exam advice:**
> The student's answer only got one of the two marks available because they did not understand the word 'arrangement'.
>
> **Exam technique:**
> Make sure that you understand the meaning of specific words in the syllabus which are not necessarily scientific terms. For example, 'arrangement of particles' refers to whether the particles are ordered regularly or randomly in space.

b (i) Complete the chemical equation for the reaction of chlorine with lithium bromide. [2]

$$Cl_2 + 2LiBr \checkmark \longrightarrow 2Br ✗ + 2LiCl$$

> **Exam advice:**
> The answer only got one of the two marks available because the student did not realise that the element bromine is diatomic.
>
> **Exam technique:**
> Make a list of the elements which are diatomic (e.g. Cl_2, Br_2, H_2, O_2, N_2) and make sure that you write these formulae in relevant equations.

Exam skills practice

(ii) ideas of chemical reactivity to explain why bromine does not react with lithium chloride. [1]

.... bromine is lower in the group than chlorine so is less reactive than lithium chloride ✗

> **Exam advice:**
> The student should have compared the reactivity of the halogens (chlorine is more reactive than bromine) instead of the reactivity of the chloride. The statement that 'bromine is lower in the group' tells us nothing about the reactivity.
>
> **Exam technique:**
> When answering questions about the reactivity of halogens with halides, remember that the comparison is between the halogens and not the halides or the metal ion in the halide.

c Sodium chloride is an ionic compound.

Describe two properties of ionic compounds. [2]

1 .... high melting points ✓

2 .... conduct electricity ∧

> **Exam advice:**
> The student realised that ionic compounds have high melting points but essential detail was missing from the conductivity answer. To gain the mark, the student should have included the words 'when molten' or 'in aqueous solution'. Ionic compounds do <u>not</u> conduct when solid.
>
> **Exam technique:**
> When writing about electrical conductivity always mention the physical state of the element or compound, e.g. solid, liquid, aqueous.

d Concentrated aqueous sodium chloride is electrolysed.

  (i) State the names of the products at each electrode. [2]

    Positive electrode .... *chlorine* ✓

    Negative electrode .... *sodium* ✗

> **Exam advice:**
> The student recognised that chlorine is released at the positive electrode, but the product at the negative electrode is hydrogen not sodium. The student did not take note of the important word 'aqueous' in the stem of the question and gave the product for the electrolysis of <u>molten</u> sodium chloride.
>
> **Exam technique:**
> Make sure that you read the question carefully. The products formed during electrolysis depend on the physical state – molten or aqueous.

  (ii) Give the name of the negative electrode. [1]

    .... *cathode* ✓

e When very dilute aqueous sodium hydroxide is electrolysed, oxygen is produced at the anode.

Construct ionic half-equations for the reactions occurring at the anode and cathode. [2]

Anode $2O^{2-}$ ✗ $+ 4e$ ✗ $\longrightarrow O_2$

Cathode $2H^+ + 2e^- \longrightarrow H_2$ ✓

> **Exam advice:**
> The student recognised that hydrogen is released at the cathode but the product at the anode should have come from the breakdown of OH^- ions. The student knew that oxygen is produced but gave the equation for the electrolysis of a molten oxide. In addition the electrons are on the incorrect side of the equation. The correct half-equation is $4OH^- \rightarrow 2H_2O + O_2 + 4e^-$.
>
> **Exam technique:**
> Make sure that you read the question carefully. The products formed during the electrolysis of very dilute aqueous solutions come from H^+ and OH^- ions.

Exam skills practice

f Chlorine reacts with phosphorus to produce phosphorus trichloride, PCl₃.

Construct the chemical equation for this reaction. [2]

$2P + 3Cl_2 \rightarrow 2PCl_3$ ✓

g Draw the dot-and-cross diagram for PCl₃. [2]

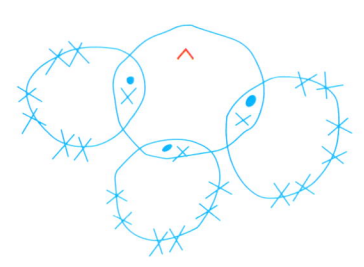 ✓

> **Exam advice:**
> The student's answer only got one of the two marks available because the lone pair of non-bonding electrons on the phosphorus atom should have been drawn at point ∧ on the diagram. The marks were for three bonding pairs of electrons (1) and drawing the rest of the structure correctly (1). The atoms were not labelled but this was not penalised. But it is always good practice to include the correct symbols for the atoms. In the supplement paper this sort of equation may only be allocated 1 mark.
>
> **Exam technique:**
> In dot-and-cross diagrams remember that there should be eight electrons around the outer shell of each atom (except for hydrogen). It is good practice to label each atom with its symbol.

h Compound Y has the structure shown.

Draw the partial structure of the polymer formed when Y is polymerised. Show 3 repeat units. [3]

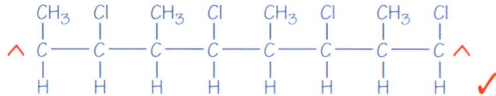 ✓

> **Exam advice:**
> The student has drawn the repeat unit correctly (1 mark) but has drawn four repeat units instead of the three asked for and has not added the continuation bonds at the ends of the chain (∧).
>
> **Exam technique:**
> When drawing polymer structures, remember to add the continuation bonds and draw the required number of repeat units.

> **Overall comments**
>
> This student gained 10 out of a possible 19 marks.
>
> This synoptic question needs knowledge from Units 1, 4, 5, 7, 13 and 20 to gain full marks. More marks could be gained by reading the stem of the question carefully and noting the physical state (solid, molten or aqueous). This is especially important in electrolysis questions (parts d(i) and e). When drawing dot-and-cross diagrams of unfamiliar structures (part g), think of a similar compound that you have come across before. The structure of PCl_3 is rather like NH_3 (N and P are both in Group V). Make sure that you understand the meaning of words in the syllabus such as 'arrangement', 'relative' and 'adverse effect' which are not scientific words.

Sample Question 6

Supplementary level questions are in **bold**.

We use the symbol ∧ to show that an essential word or piece of information has been omitted.

This question is about compounds of nitrogen.

a Nitrogen dioxide, NO_2, is a red-brown gas.

A small jar of NO_2 is placed in a large glass container. The lid of the jar is then removed.

At first a brown colour is only seen around the top of the jar.

After half an hour the brown colour has spread throughout the large glass container.

Explain these observations using the kinetic particle theory. [3]

.... The nitrogen dioxide ∧ moved by diffusion ✓ by going from where their concentration is high to where their concentration is low ✓ down a concentration gradient.

> **Exam advice:**
>
> The student's answer only got two of the three marks available because they did not mention the essential word 'particles' (or 'molecules'). This mark would have been awarded if the student had written 'NO_2 particles move randomly' or 'NO_2 molecules move around'.
>
> **Exam technique:**
>
> When answering questions about kinetic particle theory (diffusion, effect of temperature and pressure, rates of reaction) make sure that you use the word particles (or molecules, atoms or ions, as appropriate).

Exam skills practice

b Oxides of nitrogen are atmospheric pollutants.

   (i) Describe how oxides of nitrogen contribute to acid rain. [2]

      .... They react with water in the atmosphere ✓ to form an acidic solution ✓

   (ii) Give one other adverse effect of oxides of nitrogen. [1]

      .... They are harmful ⋀

> **Exam advice:**
> The student's answer was too vague. A specific effect is needed to get the mark, e.g. 'irritates the lungs' or 'contributes to photochemical smog'.
>
> **Exam technique:**
> When writing about the effects of pollutants give specific effects, e.g. 'irritates the throat' or 'erodes limestone buildings'. Vague statements such as 'harmful' or 'pollutes the air' will not be given marks.

c (i) Explain how oxides of nitrogen are formed in car exhausts. [2]

      .... nitrogen and oxygen from the air combine in the car engine ✓

> **Exam advice:**
> The student's answer only got one of the two marks available because there is no explanation of the conditions. To get the second mark, the student has to refer to the high temperature and pressure in the engine which cause the gases to combine.
>
> **Exam technique:**
> When a question is worth two marks, give two points in your answer.

   (ii) Describe how oxides of nitrogen are removed by catalytic converters. Include a relevant chemical equation in your answer. [3]

      .... The catalyst changes nitrogen oxides to nitrogen ✓ and carbon dioxide is formed. The reactions are redox reactions, carbon is oxidised and nitrogen is reduced.

      $NO_2 + C \rightarrow CO_2 + N_2$

Exam advice:

You are expected to know that carbon monoxide (rather than carbon) is converted to carbon dioxide (1 mark) by reaction with oxides of nitrogen. The third mark is for a relevant equation which must be balanced. The student's equation represents a possible reaction but was not balanced. Note that $2NO + 2CO \rightarrow 2CO_2 + N_2$ would score 3 marks.

Exam technique:

Note that the command word 'describe' indicates that you have to state what happens not why it happens. Writing about redox or how the catalyst works is not necessary.

d Aqueous ammonia is an alkali.

   (i) Give the formula of the ion that is present in all alkalis. [1]

      .... OH ✗

Exam advice:

The student has realised that hydroxide ions are present in alkalis but has not shown the negative charge.

Exam technique:

Make sure that you learn the formulae of the molecules and ions mentioned in the syllabus. You are expected to remember them!

   (ii) Describe the colour change when excess acid is added to aqueous ammonia containing the indicator thymolphthalein. [2]

      from .... blue ✓ to .... red

Exam advice:

The colour when acid is in excess is incorrect. It should be colourless. The student has confused the colour change with that of litmus. Make sure that you know the colours of the indicators mentioned in the syllabus in acidic and alkaline conditions.

Exam technique:

When answering this type of question, make sure that it's clear in your mind whether the acid or the alkali will be in excess after adding one to the other.

   (iii) State the name for the type of reaction that occurs when an acid is added to an alkali. [1]

      .... neutralisation ✓

285

Exam skills practice

(iv) Complete the word equation for the reaction of sodium hydroxide with a chloride salt to produce ammonia. [2]

sodium + *ammonia* ⟶ sodium + ammonia + *water*
hydroxide *chloride* chloride ✓

> **Exam advice:**
> The student made the common error of giving the incorrect name ('ammonia chloride') for the salts of ammonia. Ammonia changes to ammon<u>ium</u> in its salts, e.g. ammonium chloride, ammonium sulfate.
>
> **Exam technique:**
> When naming compounds pay special attention to the word endings, e.g. chlor<u>ide</u>, sulf<u>ate</u>, ammon<u>ium</u>.

e (i) Nitric acid is a strong acid. State the meaning of the term strong in strong acid. [1]

.... An acid which ionises to produce a high concentration of hydrogen ions in solution

> **Exam advice:**
> The student's answer was not precise enough to get the mark. Mention of **complete** dissociation or **complete** ionisation is needed.
>
> **Exam technique:**
> Make sure that you learn the definitions that are in the syllabus. There are quite a few!

(ii) Describe how you can tell the difference between a strong and a weak acid of the same concentration without referring to pH. Explain your answer. [2]

.... See how fast they react with magnesium. The strong acid reacts more rapidly ✓ ∧

> **Exam advice:**
> The student's answer only got one of the two marks available because no explanation was given. For the second mark, some idea of the difference in concentration of H^+ ions is needed. Note that there are two command words ('describe' and 'explain') and each of these needs to be considered.
>
> **Exam technique:**
> Where there are two command words in the stem of a question, make sure that you write something relevant for each.

(iii) Nitrous acid, HNO_2, is a weak acid.

Write a symbol equation to show the ionisation of nitrous acid in aqueous solution. [2]

$$HNO_2 \longrightarrow H^+ + NO_2 \text{ ✗}$$

Exam advice:
The student made two errors. (i) There was no charge given on the NO_2 ion. Remember that when an acid ionises, negative ions are formed to balance the H^+ ions. (ii) An equilibrium mixture is formed when a weak acid ionises. The symbol $\rightleftharpoons$ should be used, not the symbol $\rightarrow$.

Exam technique:
When writing equations involving ions, make sure that the positive and negative charges balance.

Overall comments
This student gained 10 out of a possible 22 marks.

This synoptic question needs knowledge from Units 1, 5, 11, 16 and 17 to gain full marks. Reading questions carefully and noting specific words such as 'particles' (part a) are very important in gaining more marks. Make sure that you know the meaning of particular command words such as describe and explain (parts c(i) and e(ii)). If two different command words are used in a question, make sure that you write something relevant for each. You can also improve your marks by learning or working out the formulae of the ions in the syllabus (parts d(i) and e(iii)) and by avoiding vague answers (part b(ii)).

Exam skills practice

Sample Question 7

Supplementary level questions are in **bold**.

This question is about carbon and organic carbon compounds.

a **(i)** The atomic number of carbon is 6.

State the meaning of term atomic number. [1]

..... the number of protons in an atom. ✓

(ii) Most carbon atoms have a nucleon number (mass number) of 12.

Suggest what extra information this nucleon number gives about an atom of carbon-12. [2]

..... the number of neutrons ✓

> **Exam advice:**
> The student's answer only got one of the two marks available because the question specifically asks about an atom of carbon-12. The second mark is for stating that there are 'six neutrons' in its atoms.
>
> **Exam technique:**
> Make sure you read the question carefully and when a question is worth two marks give two points in your answer.

b Carbon-12 and carbon-13 are isotopes of carbon.

(i) Define the term isotopes. [2]

..... Atoms of the same element ✓ that have different numbers of neutrons ✓

> **Exam advice:**
> There are other ways to express this definition that will also gain full marks. For example, 'Atoms with the same atomic number but different nucleon (mass) numbers'.

(ii) State one similarity and one difference between the atomic structures of a carbon-12 atom and a carbon-13 atom. [2]

Similarity .... *They are both the element carbon*

Difference .... *One atom is heavier than the other because carbon-13 has one more neutron than carbon-12.* ✓

> **Exam technique:**
> The student has lost the mark for a similarity because, although their answer is a correct statement, it does not refer to the atomic structures of the two isotopes as required in the question stem. Their answer for a difference gets a mark, as it mentions both isotopes and their atomic structures in its comparison.

(iii) Explain why carbon-12 and carbon-13 have identical chemical properties. [1]

.... *Because both atoms have the same number of electrons, four, in their outer shells* ✓

c Methane, CH_4, is the first member of the alkane homologous series.

(i) Methane is an example of a hydrocarbon.

Define the term *hydrocarbon*. [2]

.... *A mixture of carbon and hydrogen atoms.* ✗

> **Exam advice:**
> The student's answer could not be awarded any marks because they have incorrectly stated that a hydrocarbon is a 'mixture', but the hydrogen and carbon atoms are bonded together in molecules of compounds. Both marks would be given for answers that also mentioned that hydrogen and carbon are the <u>only</u> elements/atoms in a hydrocarbon.
>
> **Exam technique:**
> Make sure that you learn the key definitions thoroughly. There are some facts you really do need to memorise!

(ii) Draw a dot-and-cross diagram of a methane molecule. Only show the outer shell electrons. [1]

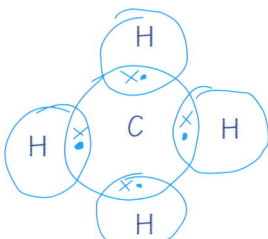 ✓

> **Exam technique:**
> When drawing dot-and-cross diagrams, make sure that both the electrons in a covalent bond are in the area of overlapping outer shells, as in this answer.

Exam skills practice

(iii) Name the second member of the alkane homologous series and give its molecular formula. [2]

..... ethene, ✗ C$_2$H$_6$ ✓

> **Exam technique:**
> 'Ethane' is the correct name – this could have been a careless spelling mistake. 'Ethene' is the first member of the homologous series of alkenes. Read through your answer after you have written it down.

(iv) Draw the displayed formula of the alkane, propane. [1]

 ✓

> **Exam advice:**
> This is a good answer as the 'displayed formula' must show all the atoms (using their symbols) and all the bonds in a molecule. Make sure that you know the number of carbon atoms for the first four members of the alkane homologous series.

d Methane burns in excess oxygen.

CH$_4$(g) + 2O$_2$(g) → CO$_2$(g) + 2H$_2$O(l)

Calculate the minimum volume of oxygen, at room temperature and pressure, required to completely burn 72.0 g of methane.

(1 mole of any gas occupies 24.0 dm^3 at r.t.p.) [3]

..... We have 72.0 g of methane

1 mole of CH$_4$ = 6 ✗ + 4 = 10

Moles of CH$_4$ = $\dfrac{72.0}{10}$ = 7.2 moles

From equation 1 mole of CH$_4$(g) reacts with 2 moles of oxygen.

So 7.2 moles reacts with 7.2 x 2 moles = 14.4 moles of oxygen ✓

If 1 mole of gas occupies 24.0 dm^3 then 14.4 moles will occupy 24.0 x 14.4

= 345.6 dm^3 ✓

.... 345.6 .... dm^3

> **Exam technique**
> The student made a careless mistake when looking up the relative molecular mass of carbon in the Periodic Table, using the atomic number of carbon (i.e. '6') by mistake. However the mistake was only penalised once, and the following use of the wrong relative molecular mass of methane was ignored. The clear working out shown allowed the student to gain two out of the three available marks for the calculation. The correct answer for all three marks is 216 dm^3 of oxygen gas.

e Methane gas from the decomposition of plants escapes into the atmosphere.

   (i) Name one other source of methane in the atmosphere. [1]

   .... From cows digesting grass ✓

   (ii) Give one adverse effect of methane on the environment. [1]

   .... It pollutes the air

> **Exam advice:**
> This answer does not gain a mark as just referring to 'pollute' or 'pollution' is too vague. It does not give a specific effect of methane in the atmosphere, i.e. 'it is a greenhouse gas that contributes to global warming/climate change.'

> **Overall comments**
> This student gained 12 out of a possible 19 marks.
>
> This synoptic question needs knowledge from Units 3, 4, 5, 17, 18 and 19 to gain full marks. Improving your exam technique will help you to gain more marks, for example by reading questions more carefully, checking the spelling of chemical terms and using the Periodic Table to find the correct data. Some marks can be gained quite easily if you spend time learning the basic definitions in the syllabus – in parts b(i) and c(i) straightforward definitions of isotopes and hydrocarbons were asked for. The calculation in part d illustrates the importance of setting out your working clearly to gain marks even when errors are made.

Exam skills practice

Sample Question 8

Supplementary level questions are in **bold**.

We use the symbol ∧ to show that an essential word or piece of information has been omitted.

This question is about carboxylic acids and alcohols.

a Complete the word equation for the reaction of ethanoic acid with sodium carbonate. [3]

ethanoic acid + sodium carbonate ⟶ sodium ethate ✗ + carbon dioxide ✓ + water ✓

> **Exam advice:**
> The student did not give the correct name for the salt 'sodium ethanoate'. The spelling given is too different from the proper name.
>
> **Exam technique:**
> Make sure that you know the correct spelling of the salts mentioned in the syllabus. Remember that the same naming procedure applies to both organic salts and inorganic salts, e.g. nitric ⟶ nitrate, ethanoic ⟶ ethanoate.

b Describe how to find the pH of a solution of ethanoic acid without using a pH meter. [2]

.... Add universal indicator to the acid. ✓ If it goes red or yellow it is acidic and you can work out the pH

> **Exam advice:**
> The student's answer only got one of the two marks available because there is no indication of how the pH value is obtained. A statement such as 'match the colour observed to the colour on the pH colour chart' is needed to get the second mark.
>
> **Exam technique:**
> When answering questions about finding pH using universal indicator, do not write down specific colours unless asked to do so.

c The molecular formula of ethanol is C_2H_6O.

  (i) Define the term *molecular formula*. [1]

  .... The formula showing the number of atoms in a molecule

> **Exam advice:**
> The student's definition is inaccurate because it does not mention the different types of atom present. A better answer is 'the number and type of atoms in one molecule'.
>
> **Exam technique:**
> After writing definitions, read what you have written to see if it makes sense. Make sure that you learn the definitions given in the syllabus. These are the ones which will be used in the exam.

  (ii) Draw the displayed formula of ethanol. [1]

$$\begin{array}{c} \text{H} \quad \text{H} \\ | \quad | \\ \text{H}-\text{C}-\text{C}-\text{OH} \\ | \quad | \\ \text{H} \quad \text{H} \end{array}$$

> **Exam advice:**
> The student's answer did not get the mark because the O–H bond in the functional group was not shown. It is a very common error not to include the bonds in the functional group.
>
> **Exam technique:**
> When you draw a displayed formula make sure that (i) all the atoms and all the bonds are shown including the bonds in the functional group, and (ii) all the hydrogen atoms are shown.

d Ethanol can be manufactured by fermentation.

  (i) Name the reactant and catalyst used. [2]

  reactant glucose ✓   catalyst yeast ✓

  (ii) State two conditions required for fermentation. [2]

  1 anaerobic/presence of oxygen ✗   2 pH neutral ✓

> **Exam advice:**
> The student answered this question well, as they were able to distinguish between reactants and conditions. The mark was not awarded for anaerobic because the statement about the 'presence of oxygen' contradicts this.
>
> **Exam technique:**
> Always read your answers through when finished to make sure that there are no contradictory statements.

Exam skills practice

e Ethanoic acid, CH$_3$COOH, reacts with butanol, C$_4$H$_9$OH, to form an ester.

(i) Name and draw the structural formula of this ester. [3]

Name *ethyl butanoate* ✗

Formula CH$_3$COOC$_4$H$_9$ ✓

> **Exam advice:**
> The student named the ester incorrectly. It should be butyl ethanoate. It is a common error to use the carboxylic acid name and the alcohol name in the wrong order. Remember that the first word in the ester's name comes from the name of the alcohol.
>
> **Exam technique:**
> Make sure that you learn the names of the compounds mentioned in the syllabus.

(ii) Draw the displayed formula of an ester which is an isomer of compound A. [1]

compound A:
```
        O    H H H
        ||   | | |
   H—C—O—C—C—C—H
        | | |
        H H H
```

Student's answer:
```
   H O     H H
   | ||    | |
   H—C—C—O—C—C—H     ✗
   |       | |
   H       H H
```

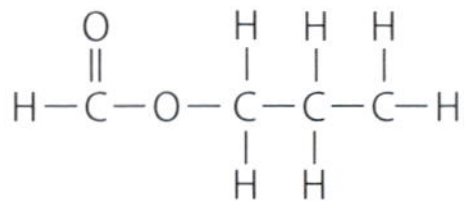

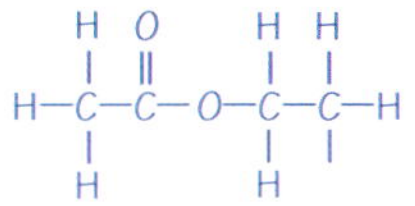

> **Exam advice:**
> The student's answer at first appears correct but a hydrogen atom has been omitted. This is a common error, especially when drawing displayed formulae of hydrocarbons and polymers.
>
> **Exam technique:**
> After drawing a displayed formula, check it to make sure that all the hydrogen atoms have been included and all the bonds between atoms.

(iii) Construct the chemical equation for the reaction of ethanoic acid with magnesium. [2]

$2CH_3COOH + Mg \longrightarrow (CH_3COOH)_2Mg$ ✗ $+ H_2$ ✓

Exam advice:
The student's answer only gets one of the two marks available (for the correct formulae of ethanoic acid and hydrogen). The formula for magnesium ethanoate is incorrect. Although the student realised that two ethanoate ions are required for one Mg^{2+} ion, the H atom has not been removed from the COOH group. The correct formula is $(CH_3COO)_2Mg$.

Exam technique:
After you have written an equation in organic chemistry, especially for the reactions of carboxylic acids, check that the atoms are balanced.

f Alcohols contain carbon, hydrogen and oxygen.

An alcohol has the composition 38.7% of carbon and 9.7% hydrogen by mass.

Deduce the empirical formula of this alcohol. [3]

$$\begin{array}{cc} \text{carbon} & \text{hydrogen} \\ \dfrac{38.7}{12} & \dfrac{9.7}{1} \quad \checkmark \text{ ecf} \\ = 3.225 & 9.7 \end{array}$$

divide by smallest
$$\dfrac{3.225}{3.225} \quad \dfrac{9.7}{3.225}$$
$$1 \quad\quad 3.00 \quad = CH_3 \; \checkmark \text{ ecf}$$

Exam advice:
The student made a careless mistake at the beginning in not calculating the % of oxygen present. However the mistake was only penalised once. The correct processes were carried out to give a sensible answer. The clear working allowed the student to gain two out of the three available marks for the error carried forward (ecf). The correct answer for all three marks is CH_3O.

Exam technique:
When calculating empirical formulae make sure that you read the question carefully and include all types of atom present in the compound.

Exam skills practice

> **Overall comments**
>
> This student gained nine out of a possible 20 marks.
>
> This synoptic question needs knowledge from Units 5, 6, 11 and 19 to gain full marks. Marks can be improved by reading questions more carefully (part f), checking the spelling of chemical terms (part a) and avoiding contradictory statements (part d(ii)). Taking time to learn definitions and how to name substances will also help. The calculation in part f shows the importance of setting out your working clearly, to gain marks even when errors are made.

Glossary

A

Accuracy How close a measurement is to the true value

Acid (Core definition) A substance that forms hydrogen ions when dissolved in water. Acidic solutions have a pH less than 7.

Acid (Supplement) A proton donor

Acid rain Rain with an acidity below about pH 5 due to the reaction of sulfur dioxide or nitrogen dioxide with rainwater

Activation energy, E_a The minimum energy that colliding particles must have in order to react

Addition polymerisation The formation of polymers from monomers (having double bonds) where no substance other than the polymer is formed

Addition reaction A reaction where only one product is formed

Air The mixture of gases present in the atmosphere

Alcohol An organic compound with an –OH functional group

Alkali A soluble base which contains OH⁻ ions in aqueous solution. An alkaline solution has a pH above 7.

Alkali metal The Group I elements in the Periodic Table. They have one electron in their outer shell.

Alkane A hydrocarbon having only single bonds

Alkene A hydrocarbon containing one or more C=C bonds

Alloy A mixture of a metal with other elements

Amphoteric oxide An oxide which reacts with both acids and alkalis separately to form a salt and water

Amide linkage The –CONH– group formed when polyamides are formed from their monomers

Amino acid Molecules having the structure H₂N-CHR-COOH

Amino acid residue The individual repeat units of the amino acids in a protein

Anaerobic respiration Respiration in the absence of oxygen

Anhydrous substance A substance containing no water

Anion A negative ion (which moves to the anode during electrolysis)

Anode The positive electrode in electrolysis

Aqueous solution A solution made by dissolving a substance in water

Atmosphere (air) The layer of gases that surround the Earth

Atom The smallest part of an element that can take part in a chemical change

Atomic number The number of protons in the nucleus of an atom

Avogadro constant The number of defined particles (ions, atoms, molecules or electrons) in one mole of those particles

B

Base (Core definition) A metal oxide or hydroxide that reacts with an acid to form a salt and water

Base (Supplement) A proton acceptor

Basic oxide An oxide that reacts with an acid to form a salt

Bauxite The main ore of aluminium

Binary compound A compound containing just two types of atom or ion, e.g. NaCl

Biodegradable Naturally decaying in the environment with the help of bacteria and fungi

Blast furnace A furnace into which air is blown and in which a metal oxide is reduced using carbon

Boiling The change in physical state from liquid to gas

Bond energy The energy needed to break a mole of given bonds

Bonding The way the atoms or ions are held together in a substance

Brass An alloy of copper and zinc

Burette A piece of glassware for delivering a variable volume of liquid accurately (usually up to 50 cm³)

Burning An exothermic reaction where substances combine and a flame can be seen

C

Carboxylic acid A weak organic acid that contains the –COOH functional group

Catalyst A substance that increases the rate of a chemical reaction and is unchanged at the end of the reaction

Catalytic converter A piece of equipment put on a car exhaust to remove nitrogen oxides and carbon monoxide

Cathode The negative electrode in electrolysis

Cation A positive ion (which moves to the cathode during electrolysis)

Celsius A scale used to measure temperature (degrees Celsius, °C)

Chemical bond The force of attraction between atoms or ions which keeps them together

Chemical change When substances react together and new substances are formed

Chemical equation A balanced equation showing the symbols for all reactants and products

Chemical property A property involving the formation of a new substance

Chlorination The addition of chlorine, especially in water purification

Chlorophyll A green plant pigment that catalyses photosynthesis

Chromatogram A piece of paper showing the separation of substances after chromatography has been carried out

Chromatography The separation of a mixture of soluble substances using filter paper and a solvent

Climate change The change in weather patterns that is linked to an increase in global warming

Closed system A reaction where there is no loss or gain of substances to or from the environment

Coke A form of impure carbon made by heating coal in the absence of air

Collision frequency The rate at which particles collide

Collision theory Using the idea of colliding particles to explain how reaction rates change with surface area, temperature and concentration

Combustion Burning (usually in a reaction with oxygen gas)

Complete combustion Combustion in excess air/oxygen

Compound A substance containing two or more types of atoms chemically combined

Concentration The amount of one type of substance dissolved in a specified volume of another. It is measured in mol/dm³ or g/dm³.

Condensation The change in physical state from gas to liquid

Condensation polymerisation Polymerisation reaction where a small molecule is eliminated

Condensation reaction A reaction where two or more substances combine and a small molecule, such as water or hydrogen chloride, is eliminated (given off)

Condenser A piece of apparatus for cooling a vapour and converting it to a liquid

Conductor A substance that allows thermal energy or electricity to flow through it

Contact process The industrial process for making sulfuric acid

Control variable A variable that is kept constant during an investigation to ensure fair testing

Cooling curve A graph showing how the temperature changes with time when a substance is cooled at a constant rate

Corrosion The 'eating away' of the surface of a metal by a chemical, e.g. by acids

Covalent bond A pair of electrons shared between two atoms, usually leading to a noble gas electronic configuration

Covalent compound A compound having covalent bonds. They can be simple molecules or giant structures.

Cracking The thermal decomposition of an organic compound into smaller molecules by heating (or heating and a catalyst)

Crystallisation The formation of crystals when a saturated solution is left to cool

Crystallisation point The point at which crystals will form very quickly when a drop of saturated solution is placed on a cold surface

D

Decimetre cubed Unit of volume (dm³) often used in chemistry. It has the same volume as 1000 cm³.

Decomposition The breakdown of a substance into two or more products

Delocalised electrons Electrons that are not associated with any particular atom

Density The mass of a substance divided by its volume

Deoxygenation The removal of oxygen

Dependent variable The variable we measure to show the effect of changing the independent variable

Diatomic Molecules containing two atoms

Dibasic acid An acid which contains two atoms of replaceable hydrogen per molecule of acid

Glossary

Diffusion The gradual spreading out and mixing of different particles because of their random motion

Diol A compound containing two –OH functional groups

Discharge series A type of reactivity series showing the relative ease with which ions are changed to atoms at the anode or cathode

Displacement A reaction in which one atom or group of atoms replaces another in a compound

Displaces Replaces

Displayed formula The formula of a substance which shows all of the atoms and all of the bonds

Dissociated (when referring to acids and alkalis) The breaking down of molecules into ions

Distillation A method of separating a liquid from a mixture by boiling the mixture then condensing the vapours

Dot-and-cross diagram A diagram showing the arrangement of the electrons is a molecule or in an ionic structure (usually only the electrons in the outer shells are shown)

Double bond Two atoms share two pairs of electrons between them

Ductile Can be pulled into wires

E

Electrical conductivity The ease with which an electric current can flow through a substance. Conduction is due to moving electrons in metals and moving ions in molten ionic compounds or ionic solutions.

Electrode A rod of metal or carbon (graphite) that leads an electric current into or out of an electrolyte

Electrolysis The decomposition of an ionic compound when molten or in aqueous solution by the passage of an electric current

Electrolyte A molten or aqueous substance that undergoes electrolysis

Electron A negatively charged particle in the atom found in electron shells outside the nucleus

Electron shell The energy levels at different distances from the nucleus where the electrons are found

Electronic configuration The number and arrangement of electrons in the electron shells of an atom (sometimes called the electronic structure)

Electroplating A process that uses electricity to coat one metal with another

Element A substance containing only one type of atom

Elimination reaction A reaction where two molecules react and a small molecule, such as water, is released

Empirical formula The simplest whole number ratio of the different atoms or ions in a compound

End point The point in a titration when the indicator changes colour showing that the reaction is complete

Endothermic A transfer of thermal energy from the surroundings to a reaction mixture leading to a decrease in the temperature of the surroundings

Enthalpy change, ΔH The transfer of thermal energy during a reaction

Enzyme A biological (protein) catalyst

Equilibrium In a reversible reaction, the point where the forward and backward (reverse) reactions are taking place at the same rate

Ester An organic compound containing the –COO– group, formed when a carboxylic acid reacts with an alcohol

Ester linkage The –COO– group found in polyesters

Eutrophication The process caused by excess nitrates and/or phosphates which leads to the death of aquatic life

Evaporation The change from liquid to vapour state below the boiling point

Excess Being greater than something else in amount

Exothermic A transfer of thermal energy to the surroundings from a reaction mixture leading to an increase in the temperature of the surroundings

Experimental yield The amount in moles or grams of a specific product formed from a specific amount of reactants under experiment conditions

F

Fermentation Often refers to the breakdown of glucose by yeast in the absence of oxygen to form ethanol and carbon dioxide. Other fermentations can also occur.

Fertiliser A substance added to the soil to replace essential elements lost when crops are harvested

Filtrate The liquid or solution that has passed through a filter paper

Filtration Separating a solid from a liquid by using a filter paper

Flue gas desulfurisation The removal of sulfur dioxide from the waste gases produced in furnaces by reacting it with bases or carbonates

Forces of attraction Forces which tend to bring atoms, ions or molecules together

Formula (chemical formula) A representation of a compound or molecule using chemical symbols and numbers

Fossil fuel Fuel (coal, oil, natural gas) formed from the remains of tiny, dead, sea creatures or plants over millions of years

Fraction A group of substances with similar boiling points, distilling off at the same place in a fractionation column

Fractional distillation (fractionation) The separation of different substances in a liquid by their different boiling points

Freezing (solidifying) The change of physical state from liquid to solid

Fuel A substance that releases energy (usually when burned)

Fuel cell A cell where hydrogen and oxygen undergo reaction to produce an electric current

Functional group An atom or group of atoms that determines the chemical properties of a homologous series

G

Galvanising Coating a metal (usually iron) with a protective layer of zinc

Gas syringe A piece of glassware for measuring the volume of gases given off in a reaction

General formula A formula that applies to all the compounds in a particular homologous series of organic compounds

Giant covalent structure A structure with a continuous three-dimensional network of covalent bonds, e.g. diamond

Giant ionic structure A structure with a continuous three-dimensional network of ionic bonds

Global warming The warming of the atmosphere due to greenhouse gases trapping infrared radiation radiated from the Earth's surface

Graphite A form of carbon that conducts electricity and is used for inert electrodes

Greenhouse effect The process by which thermal energy is absorbed by the atmosphere after being radiated from the Earth's surface

Greenhouse gas A gas such as methane or carbon dioxide that absorbs infrared radiation radiated from the Earth's surface

Group A vertical column of elements in the Periodic Table

H

Haber process The industrial process for making ammonia

Half-equation See Ionic half-equation

Halide A compound containing an ion formed from a halogen atom

Halogen An element in Group VII of the Periodic Table

Heating curve A graph showing how the temperature changes with time when a substance is heated at a constant rate

Hematite A common ore of iron

Homologous series A family of similar organic compounds with similar chemical properties due to the presence of the same functional group

Hydrated substance A substance chemically combined with water

Hydrocarbon A compound containing only carbon and hydrogen

Hydrogenation The addition of hydrogen to an unsaturated compound

Hydrolysis The breakdown of a compound by reaction with water. Acids or alkalis speed up hydrolysis.

I

Immiscible (liquids) Liquids which do not mix

Impurity A substance which contaminates a pure substance

Incomplete combustion Combustion when the oxygen supply is limited

Independent variable The variable that you decide to change in an investigation

Indicator A substance that has two different colours depending on the solution in which it is placed. It often changes colour according to the pH of the solution.

Inert Does not react

Inert electrode An electrode that does not react during electrolysis

Inert gas See Noble gas

Insoluble Does not dissolve

Insulator A substance that is a very poor conductor of electricity (or thermal energy)

Intermolecular forces Weak forces between molecules

Ion A species formed when an atom or group of atoms has become positively or negatively charged by loss or gain of electrons

Ionic bond The strong electrostatic attraction between the positive and negative ions in the crystal lattice

Ionic equation A chemical equation showing only the ions taking part in the reaction. The unchanged spectator ions are not shown.

Ionic half-equation Equation showing the oxidation or reduction parts of a redox reaction separately. Often used for reactions at the electrodes in electrolysis.

Ionic lattice A giant structure with a regular arrangement of alternating positive and negative ions

Isomers See Structural isomers

Isotopes Different atoms of the same element that have the same number of protons but a different number of neutrons

K

Kinetic energy Energy associated with movement

Kinetic particle theory The idea that the motion of atoms can explain states of matter, diffusion and rates of reaction

L

Lattice A continuous regular arrangement of particles

Leaching The extraction of substances (usually from a solid) into a liquid

Limiting reactant The reactant that is not in excess

Litmus An indicator used to test if a substance is acidic or alkaline

Locating agent A compound that reacts with a colourless substance on chromatography paper to form a coloured spot

Lone pair A pair of outer shell electrons not involved in bonding

M

Malleable Can be hammered into shape

Mass number The number of protons plus neutrons in the nucleus of an atom

Measuring cylinder A piece of glassware to measure out liquids to the nearest 1 or 2 cm^3

Melting The change in physical state from solid to liquid

Metallic bonding The electrostatic attraction between the positive ions in a giant metallic lattice and a 'sea' of delocalised electrons

Microbes Microscopic organisms such as bacteria, viruses and single-celled creatures

Miscible Able to be mixed together (usually refers to liquids)

Mixture Two or more substances which are mixed but are not chemically combined and can be separated by physical means

Molar gas volume The volume occupied by one mole of any gas – 24 dm^3 at room temperature

Molar mass (in g/mol) Mass of substance in g divided by the amount of substance in moles

Mole Unit of amount of substance that contains 6.02×10^{23} specified particles (atoms, molecules, ions or electrons)

Molecular formula A formula showing the number and type of different atoms in a molecule, e.g. C_2H_6

Molecule A particle made up of two or more atoms held together by covalent bonds

Monatomic Consisting of one atom

Monomer A small molecule that combines to form a polymer

N

Neutral Neither acidic nor alkaline. A neutral solution has pH 7.

Neutralisation The reaction of an acid with a base to form a salt and water. For an acid reacting with an alkali this is represented by $H^+ + OH^- \longrightarrow H_2O$

Neutron A particle in the nucleus of the atom which has no charge

Noble gas An element in Group 0 (Group VIII) of the Periodic Table

Noble gas configuration This is obtained when atoms in molecules or ions have a complete outer shell of electrons

Non-biodegradable Not able to be broken down by organisms

NPK fertilisers Fertilisers containing nitrogen, phosphorus and potassium

Nucleon A particle (proton or neutron) in the nucleus of an atom

Nucleon number The total number of protons plus neutrons in the nucleus of an atom. (Also known as the Mass number.)

Nucleus The central part of an atom containing protons and neutrons

O

Octet (of electrons) Having eight electrons in the outer shell

Ores Rocks containing a metal or metal compound from which a metal can be economically extracted

Oxidation (Core definition) The gain of oxygen by a substance

Oxidation (Supplement) The loss of electrons or increase in oxidation number of a substance

Oxidation number A number that describes how oxidised an atom is

Oxidising agent A substance that oxidises another substance by removing electrons, and is itself reduced in the process

P

Percentage abundance The percentage of each isotope in a given sample of an element

Percentage composition by mass The percentage by mass of each element in a compound

Percentage purity The mass of a given substance in a mixture divided by the overall mass of substances present, expressed as a percentage

Percentage yield The actual yield divided by the theoretical yield expressed as a percentage

Period A horizontal row of elements in the Periodic Table

Periodic Table An arrangement of elements in periods and groups in order of increasing proton number so that some elements with similar properties are arranged in the same groups

Petroleum A mixture of hydrocarbons present under the Earth's surface as a black sticky liquid. Sometimes called petroleum.

pH scale A scale of numbers that describes how acidic (or alkaline) a substance is

Photochemical reaction A reaction that depends on the presence of light

Photochemical smog A smoky fog caused by the reaction of oxides of nitrogen, ozone and hydrocarbons in light

Photosynthesis The process by which plants make glucose (and oxygen) from carbon dioxide and water in the presence of sunlight

Physical change A change in which no new substance is formed, e.g. melting, condensation

Physical property A property which does not involve a chemical reaction, e.g. melting point

Pipette See Volumetric pipette

Pollutant A substance that contaminates (makes less pure) the air, water or soil

Polyamide A polymer with –CONH– linkages

Polyester A polymer with –COO– linkages

Polymer A large molecule built up from many smaller molecules called monomers

Polymerisation The chemical reaction where monomers combine to form a polymer

Position of equilibrium How far the reaction goes towards the products or reactants side

Precipitate A solid formed when two solutions are mixed

Products The substances produced as a result of a chemical reaction

Proteins Natural polyamides made from amino acid monomers

Proton A positively charged particle in the nucleus of an atom. Also used as a term for a hydrogen ion.

Proton acceptor The definition of a base

Proton donor The definition of an acid

Proton number The number of protons in the nucleus of an atom

Pure There is only one substance present

Purification methods Methods used to separate the substance you want from mixtures, e.g. distillation, filtration

Q

Qualitative analysis A way of identifying substances by observing the results of specific chemical tests

R

Rate of reaction The amount of product converted to reactants in a given time

Glossary

Raw materials Substances which have not undergone any processing, e.g. air

Reactant A substance present at the start of a chemical reaction which reacts (usually with another substance) to produce a product or products

Reacting mass The amount of one reactant (in grams or moles) needed to react exactly with another reactant

Reaction pathway diagram A diagram showing the enthalpy change from reactants to products for exothermic or endothermic reactions (the activation energy can also be shown)

Reactivity series A list of elements (usually metals) in their order of reactivity

Recycling The processing of used materials to form new products

Redox reaction A reaction in which there is simultaneous reduction and oxidation taking place

Reducing agent A substance that reduces another substance by addition of electrons, and is itself oxidised in the process

Reduction (Core definition) The loss of oxygen from a substance

Reduction (Supplement) The gain of electrons or decrease in oxidation number of a substance

Refining Purifying a substance; used for purifying metals, e.g. copper, or for separating petroleum into its fractions

Refluxing Heating a reaction mixture in a flask with a condenser in a vertical position to reduce the loss of volatile substances

Relative atomic mass, A_r The average mass of the isotopes of an element compared to 1/12 th of the mass of an atom of ^{12}C

Relative formula mass The sum of the relative atomic masses applied to all substances, both ionic and covalent

Relative mass The mass of one particle compared with another; used when the actual mass is very small, e.g. masses of protons, neutrons and electrons

Relative molecular mass, M_r The sum of the relative atomic masses of all the atoms in a molecule

Repeat unit A regularly repeating part of a polymer

Residue The solid that remains after carrying out evaporation, distillation, filtration or any similar process

Respiration The reactions that release energy in all living things. The overall reaction is glucose combining with oxygen to form carbon dioxide and water.

Reversible reaction A reaction in which the products can react together to re-form the original reactants

R_f The distance travelled by a substance during chromatography divided by the distance travelled by the solvent

r.t.p. Room temperature and pressure

Rust Hydrated iron(III) oxide formed when iron reacts with air and water. The term only applies to corrosion of iron and steel.

S

Sacrificial protection A more reactive metal is placed in contact with a less reactive metal. The more reactive metal corrodes and saves the less reactive metal from corrosion.

Salts Compounds formed when hydrogen in an acid is replaced by a metal or an ammonium ion

Saturated compound An organic compound made up of molecules in which all the carbon–carbon bonds are single covalent bonds

Saturated solution A solution which contains the maximum concentration of a solute dissolved in the solvent

Sea of electrons Term used for the delocalised electrons in metallic bonding

Sedimentation The process where insoluble particles suspended in a liquid fall to the bottom

Separating funnel Piece of glassware used to separate two immiscible liquids

Simple distillation Purifying a liquid by first converting it into a vapour by heating and then condensing the vapour back into a liquid in a condenser

Solubility The amount of solute that dissolves in a given quantity of solvent

Solute A substance that dissolves in a solvent

Solution A uniform mixture of a solute in a solvent

Solvent A substance that dissolves a solute

Solvent front The furthest position the solvent reaches when moving up the chromatography paper

Species A general name for atoms, molecules, ions or electrons

Spectator ions Ions that do not take part in a reaction

Stainless steel A mixture of iron with other elements that does not rust

State The three physical states of matter are solid, liquid and gas

State symbols The letters (s), (l), (g) or (aq) placed after each formula in an equation to indicate the state of the reactants and products

Steel An alloy of iron containing controlled amounts of carbon and other metals

Stoichiometry The ratios of the reactants and products shown in a balanced chemical equation

Strong acid/alkali Acids or alkalis that are completely dissociated in aqueous solution

Structural formula A formula that gives an unambiguous description of the way the atoms in a molecule are arranged, for example CH_3CH_2OH

Structural isomers Compounds with the same molecular formula but different structural formulae

Subatomic particles Particles which are inside the atom, e.g. electrons, protons

Substitution reaction A reaction in which one atom or group of atoms is replaced by another atom or group of atoms

Surroundings In energy transfers, anything that is not the reactants and products, e.g. test tube, air, solvent

T

Theoretical yield The amount in moles or grams of a specific product formed from a specific amount of reactants by calculation assuming a 100% yield

Thermal decomposition The breakdown of a compound into two or more substances by heating

Titration A method for finding the amount of substance in a solution using a burette to add one solution to another

Transition element A block of elements between Groups II and III in the Periodic Table with specific properties such as high density and formation of coloured compounds

Triple bond Two atoms share three pairs of electrons between them

U

Unsaturated compound An organic compound made up of molecules in which one or more carbon–carbon bonds are double bonds or triple bonds

V

Volatile Easily vaporised

Viscosity A measure of the resistance of a substance to flow

Volumetric flask A flask with a graduation mark for making up solutions accurately

Volumetric pipette A pipette used to measure out solutions accurately. A pipette for putting drops of liquid into a test tube is called a teat pipette.

W

Water of crystallisation The water molecules present in crystals

Weak acid/alkali Acids or alkalis that are partially dissociated in aqueous solution

Word equation An equation with the reactants and products written as their chemical names

Y

Yeasts Single-celled microorganisms related to fungi

Yield The amount of product obtained in a reaction

Index

A
accuracy of measurement 15, 111
acid–base indicators 134–6, 142–3
acid–base titrations 142–3
acidic oxides 144, 145
acid rain 200–1, 210
acids 134–47
    carbonate reactions 137, 206
    dibasic acids 198
    ethanoic acid 134, 141, 143, 218–20, 237, 240–1
    making fertilisers 196–7
    making salts 136–9, 148–9
    neutralisation reactions 137–40, 144, 150, 196–8
    pH scale 134–5
    properties 136–7
    reaction rates 116–17
    reactions with metals 175
    sulfuric acid 197–9
activation energy (E_a) 98–9, 115
addition polymers 244–7
addition reactions 235, 238, 239, 244
air pollution 103, 200, 204–5
alcohols 23, 42, 218, 219, 235–9
alkali metals 164–5
alkalis 134–8
alkanes 219–21, 230–3
alkenes 218, 219, 223, 233–5
alkyl groups 221
alloys 188–91
aluminium 80, 92–3, 158–9, 176, 180, 190
aluminium oxides 52, 144–5, 176, 178, 232–3
aluminium salts 55, 57, 72, 145
amide linkages 248, 249, 251
amino acids 250–1
ammonia 11, 41, 54, 138
  as a base/alkali 134, 137, 138, 143, 150, 156
  identification tests 154
  production 125, 139, 194–5
ammonium salts 52, 55, 57, 152, 157, 196–7
amphoteric oxides 145
anaerobic respiration 239
anions 44, 158–9
anodes 81, 82
apparatus 14–15, 254–6
aqueous solutions 8, 9
    concentration 8, 76–7, 116–17
    electrical conduction 44
    electrolysis 82, 84–5, 87–9
    reaction equations 58–9
    separation of solids 20–1, 23
    testing for water 9
A_r see relative atomic mass
atmosphere 204–11
atomic mass see mass number; relative atomic mass
atomic number 27
atoms 2, 26–37
average rate of reaction 112
Avogadro constant 64

B
bacteria
    biodegrading plastic 245
    fermentation 237
    in natural water 213, 214
    oxidising ethanol 237
    producing methane 209
balanced equations 56–7, 66–7
bases 137–43
    see also alkalis
basic oxides 144, 145
biodegradability 244–5
boiling points 4, 6, 7, 18, 19
    see also distillation; physical properties
bond energies 100–1
bonds/bonding 32–3, 38–51
branched chain organic compounds 222–3

C
calorimeters 102, 103
carbon
    reducing metal oxides 177–9, 184–5
    structures 32, 46–7
    water purification 213
carbonates
    acid reactions 68–9, 116–17, 137, 201, 206, 240
    ion identification 158
    making salts 149, 150
    naming compounds and ions 53, 55
    thermal decomposition 185, 206
carbon compounds 41–3
    see also organic chemistry
carbon dioxide
    acidic nature 144, 200
    in the atmosphere 200, 204, 206–10
    from fuel combustion 102–3, 204, 206, 225, 230
    identification test 155, 158
    molecular structure 42, 43
carbon monoxide 103, 145, 178–9, 184–5, 205, 211
carboxylic acids 218, 219, 237, 240–1
    see also ethanoic acid
catalysts 56, 114–15, 124, 170, 194–5, 235
    see also enzymes
catalytic converters 211
catalytic cracking of alkanes 232, 233
cathodes 81, 82
cations 44, 156–7
centrifugation 20
changes of state 4–7
charges
    ions 2, 38–9, 44, 45
    subatomic particles 25–6
chemical calculations 62–79
chemical equations 56–9
chemical formulae 52–5
chemical properties of elements 30–1, 35, 162–71
chemical symbols 28, 32, 52
chlorination of water 213
chlorine 29, 32–3, 39–41, 155, 231
chromatography 16–17, 18, 258
climate change 208–9
closed systems 123
coal 102, 103
collision theory, reaction rates 115, 117, 119
colour changes
    electrolysis 89
    halogens/halogen reactions 10, 166, 167, 231, 234
    pH indicators 134–6, 142–3
    redox reactions 130–1
    tests 155–8, 234
    titrations 77, 142–3, 150
coloured compounds 16–17, 170
combining powers of elements 52
combustion 102, 103
    alkanes 230–1
    carbon dioxide formation 102–3, 204, 206, 208, 225, 230
    complete 102–3, 204, 206, 230
    hydrocarbons 225
    incomplete 103, 178–9, 205, 211, 230–1
compounds 32–3
    formulae 52–5
    ionic 38–9, 47
    naming 53
    structure and bonding 38–51
concentration 8–9, 76–7, 116–17, 124
condensation polymers 248–9
condensation reactions 248, 250–1
condensing/condensers 4, 7, 22, 96, 226, 237
conduction of electricity 34–5, 44–5, 47, 49, 80–1, 92–3
constants in experiments 110–11
Contact process 198–9
cooling curves 7
copper
    electrolysis 88–9, 90
    extraction from oxide 126, 176–8
    reactivity 174–7, 180
    uses 80, 190, 191
corrosion 91, 176, 186–7, 191
covalent compounds and bonding 40–5
cracking alkanes 232–3
crystallisation 20–1, 23

D
data interpretation 112–13, 257–8
decanting 20
decomposition
    biological 205, 208–9
    electrolysis 81–3
    thermal 20, 97, 122, 123, 185, 206, 232
delocalised electrons 47–9
diatomic molecules 42, 56–7, 166
dibasic acids 198
dicarboxylic acids 248, 249
diffusion 10–11
diols 249
discharge series of ions 84–5
displacement reactions 167, 176, 180–9
displayed formulae 54, 220
distillation 22–3, 204, 224–7, 239
dot-and-cross diagrams 38–43
double bonds 42–3
ductility 35, 49, 170, 189–91

E
E_a see activation energy
electrical conductivity 34–5, 44–5, 47, 49, 80–1, 92–3
electrical fuel cells 104–5
electrical insulators 80–1
electricity 80–95
electrodes 81, 82
electrolysis 81–95, 104, 179
electrolytes 81
electronic configuration (structure) 30–1,

Index

38–43
electrons 26–7, 30–1
  covalent bonding 40–3
  delocalised 47–9
  ionic bonding 38–9, 46–7
  oxidations and reductions 86–7
electron shells 26, 30–1, 38–9, 162–3, 181
electron transfer, redox reactions 127, 128
electroplating 90–1
elements 26–37, 52
  see also metals; non-metals; periodic table
elimination reactions 248
empirical formulae 54, 74–5
endothermic reactions 96–101, 125
energy
  chemical energetics 96–107
  electron shells 30
  fuels 102–3
  making and breaking bonds 100–1
  transfer in chemical reactions 96–9
enthalpy changes (ΔH) 97–101
environmental chemistry 200–1, 204–17
enzymes 114–15, 238, 250
equations 56–9
equilibrium reactions 122–5
error reduction, using correct apparatus 255
ester linkages 249, 251
esters 236, 240–1
ethanoic acid 134, 141, 143, 218–20, 237, 240–1
ethanol 23, 218–20, 236–9
ethene 43
eutrophication 215
evaluating evidence 258–9
evaporation 4, 22
exothermic reactions 96–103, 125
experimental yield of reactions 72–3
experiments 108–11, 254–61

F
fermentation 236, 238–9
fertilisers 196–7, 214–15
filtration 20, 23, 72, 148–9, 213
flame tests for cations 157
flue gas desulfurisation 210
formulae and equations 52–61
formulae of organic compounds 219–23
fossil fuels 200–1, 204–11, 224–5
fractional distillation 22–3, 224–7, 239
freezing 4, 7
fuel cells 104–5
fuels
  air pollution 103, 200–1, 204–5
  carbon dioxide production 102–3, 204, 206, 208, 225, 230
  energy 102–3
  fossil fuels 200–1, 204–11, 224–5
  hydrocarbons 224–7
  producing oxides of sulfur and nitrogen 200–1, 205, 210–11
functional groups of organic compounds 218–19

G
galvanising iron 187
gases 2–7
  changes of state 4, 7
  collecting 154

compression 3
diffusion 10–11
hydrocarbons 224
identification 154–5
kinetic particle theory 3, 5
noble gases 31
pressure 3, 5
volume 3, 5, 15, 70–1
general formula of homologous series 219
giant lattice structures 44–7
global warming 208–9
graphite 35, 46–7, 80–3, 89, 92, 93
graphs 112–13, 257
greenhouse gases 208–10
Group I (alkali) metals 164–5
groups in periodic table 30–1, 162–8
Group VII elements (halogens) 166–7

H
Haber process 194–5
halide test 158
halogens (Group VII elements) 166–7
heating curves 6–7
homologous series of organic compounds 218–19
hydrocarbons 102, 103, 218–35
  see also fossil fuels
hydrogen 28, 103–5, 154
hydrogenation reactions 235
hydrogen ions 83, 84, 136, 140
hydroxide ions 53, 83, 84
hydroxides 92, 134, 137–9, 140–2, 144

I
immiscible/miscible liquids 21, 22
impurities 9, 18–19, 73, 88, 92, 185, 210, 213
incomplete combustion 103, 178–9, 205, 211, 230–1
insoluble substances 9, 20, 138, 148–9, 213
intermolecular forces 45
ion discharge series 84
ionic bonding 38–9, 44
ionic compounds 32–3, 44–5
ionic equations 58–9
ionic half-equations 86–7, 127, 128
ionic lattices 44, 45, 153
ions 2, 38–9, 44, 58–9
iron 91, 176–7, 181, 184–7, 190–1
isomers 222–3, 236–7
isotopes 28–9, 62

K
kinetic energy 3, 7, 96, 100, 117
kinetic particle theory 3–7, 10–11

L
leaching, toxic metals 214
limiting reactants 68–9
liquids 2–9
  changes of state 4, 6, 7
  diffusion 10
  molten ionic compounds 44, 45, 81–4, 87, 93, 127
  separation 22–3
  volume measurement 14–15
litmus indicator 136, 142, 143

M
malleability 35, 49, 170, 189–91
masses

calculating empirical formulae 75–6
chemical calculations 66–8
following reactions 108
measurement 14
see also relative atomic mass; relative molecular mass
mass number 28–9
measurement 14–15, 111, 256
melting points 4, 6, 7, 18–19, 80
see also physical properties
metallic bonding 48–9
metals 34–5
  alloys 188–91
  corrosion 91, 176, 186–7, 191
  electrical conductors 35, 49, 80, 88
  electrolysis 88–93
  extraction/refinement 88, 92–3, 178–9, 184–5
  Group I/alkali 164–5
  making salts using acids 148
  oxides 137–9, 144, 149, 176–7
  periodic table 34–5, 162, 163
  plating 90–1, 187
  reactions with oxygen 176
  reactions with water/steam 164–5, 171, 174–5
  reactivity 148–9, 164, 171, 174–83
  transition elements 162, 170–1
  uses 190–1
  water pollution 214
methane 41, 205, 209
methanol 42, 218, 219, 236
microbes
  water pollution 214–15
  see also bacteria; yeast
minerals, in natural water 212
miscible/immiscible liquids 21, 22
mixtures 8, 16–23, 33
molecular formulae 54, 75
molecular mass see relative molecular mass
molecules/molecular compounds 2, 32
  see also covalent bonding; organic chemistry
moles 8, 64–79
molten ionic compounds 44, 45, 81–4, 87, 93, 127
monomers 244, 247
motor vehicles 206, 211
M_r see relative molecular mass

N
naming compounds 53, 220–1, 223
natural gas 102, 103
neutralisation reactions 137–40, 144, 150, 196–8
neutral substances 134, 145
neutrons 26, 27–8
nitrates 53, 152, 158–9, 196, 214–15
nitrogen compounds, see also ammonia
nitrogen molecules 42
nitrogen oxides 200–1, 205, 211
noble gas configuration 31, 38–41
noble gases 31, 168, 169, 204
non-metal oxides 144
non-metals 34–5, 162
nucleon number 28

302

nucleus of atoms 26–9
nylon 248–9, 251

O
organic chemistry 218–53
oxidation numbers 128–9
oxidation reactions 126
    alcohols 237
    alkali metals 164
    electrolysis 86–7
    loss of electrons 127
    metals 176
    see also redox reactions
oxidation numbers 127, 128, 171
oxides
    acidic 144, 145
    amphoteric 145
    basic 138, 144, 145
    making salts 149
    metals 137–9, 144, 149, 176–7
    sulfur and nitrogen 200–1, 205, 210–11
oxidising agents 130–1
oxygen 42, 154, 212–13, 215

P
particle theory 2–13
particulates in the atmosphere 205
percentage composition by mass 67
percentage purity 73
percentage yield of reactions 72–3
periodic table 27, 30–1, 52–3, 55, 162–73
petroleum 102, 103, 124–7, 224, 226–7, 233
pH indicators 134–6, 142–3
phosphates 196, 212, 214–15
phosphorus 168, 196
photochemical reactions 231
photochemical smog 205
photosynthesis 206–7
pH scale 134–5
physical changes, energetics 96
physical properties
    giant lattice structures 46–7
    hydrocarbons 224–7
    ionic versus molecular compounds 44–5
    metals 34–5, 49, 164, 170, 171, 188–91
    non-metals 34–5, 166, 168–9
    organic compounds 219, 222, 224–7, 230
    trends in periodic table 164–71
planning investigations 260–1
plastics see polymers
pollution 103–4, 200–1, 204–5, 214–15, 245
polyamides 248–9, 251
polymerisation 244, 248
polymers 244–53
    addition type 244–7
    condensation type 248–51
    plastics as electrical insulators 80
    plastics in the environment 215, 244–5
    proteins 250–1
    structures 246–7, 249
position of equilibrium 123
precipitation reactions 110, 152–3
pressure of gases 3, 5, 116, 117, 125, 195
products
    chemical calculations 68–9
    chemical equations 56–9
    electrolysis 82–3

stoichiometry 66–7
proteins 238, 250–1
proton number 27, 28
protons 26–7
purification
    copper electrolysis 88
    of mixtures 18–23
    water treatment 152, 213
purity 18–19, 73

R
radioactive isotopes 28
rates of reaction 108–21
    collision theory 115, 117, 119
    concentration 116–17
    following progress of reactions 108–10
    surface area 114
    temperature 118–19
reactants 56–9, 66–7
reaction equations 56–9
reaction pathway diagrams 98–9
reaction rates see rates of reaction
reactivity, periodic table 162–7
reactivity series 84–5, 148–9, 174–83
recording results 256–7
recycling 191, 245
redox reactions 126–31, 211
reducing agents 130–1
reductions 126
    electrolysis 86–7
    gain of electrons 127
    metals 177
refluxing 237
relative atomic mass (A_r) 62, 71
relative formula mass 62–3, 66–8
relative molecular mass (M_r) 11, 62–3, 71
renewable energy 210
respiration 206
reversible reactions 122–5
rusting 91, 186

S
safety, using apparatus 254
salts 136–9, 148–61, 166–7
saturated solutions 8, 9
sedimentation, water purification 213
separating substances 14–25
sewage, water pollution 214–15
smog 205
sodium chloride 32–3, 38, 44, 84–5
soil
    acid neutralisation 139, 144, 197
    contamination 201, 214, 245
    fertilisation 196, 214
solids 2–7, 20–1
solubility 9, 23, 152–3
solutes 8–9, 20, 76–7, 117
solutions 8–9
    concentration 8, 76–7, 116–17
    separation 19–23
    see also aqueous solutions
solvents 8, 16–22, 149, 236
species, redox reactions 128
spectator ions 58–9
states of matter 2–13
state symbols 58–9
steel alloys 190–1
stoichiometry 66–9, 71, 73

strong acids 141, 142, 143, 197
strong alkalis/bases 141, 142, 143
structural formulae 222–3
structural isomers 222–3, 236–7
subatomic particles 26–31
substitution reactions 230
sulfates 152, 159
sulfites 159
sulfur dioxide 155, 200–1, 205, 210
sulfuric acid 197–201
surface area, reaction rates 114
symbol equations 56–7

T
temperature
    changes of state 4, 6–7
    gases 5
    heat transfer 96
    measurement 14, 256
    reaction rates 118–19
    shifting equilibrium 125
tests 131, 154–9, 225, 234
theoretical yield of reactions 72–3
thermal decomposition 20, 97, 122, 123, 185, 206, 232
thermal energy 96–7
    see also endothermic reactions; exothermic reactions
thermit reaction 180
time measurement 14
titrations 76–7, 142–3, 150–1
transition elements 170–1
triple bonds 42–3

U
universal indicator 134–5
unsaturated hydrocarbons 234–5

V
variables in experiments 110–11
viscosity of hydrocarbons 224–5
volume of gases 70–1, 109, 112, 116, 125
volume measurement 14–15, 77, 109

W
water
    clean 212–13
    diffusion of particles in 10
    electrolysis 83
    molecular structure 32
    pollution 201, 214–15
    pure/distilled 9, 19
    purification treatment 152, 213
    reaction with alkali metals 164–5
    testing for 9
    see also aqueous solutions
weak acids/bases 141, 143
word equations 56

Y
yeast 238–9
yield of reactions 72–3, 195

Z
zinc 170, 174, 175, 177–81, 187, 190